**SCIENTISTS,
SOCIETY,
AND STATE**

William McGucken

SCIENTISTS, SOCIETY, AND STATE

The Social Relations of
Science Movement in Great Britain
1931–1947

Ohio State University Press: Columbus

Portions of Chapters 2, 3, 5, and 6, and of the Epilogue, appeared originally in "The Social Relations of Science: The British Association for the Advancement of Science, 1931-1946," by William McGucken, in the *Proceedings of the American Philosophical Society*, Vol. 123, No. 4 (August, 1979), pp. 236-64. They are reprinted herein by permission.

Portions of Chapter 7 appeared originally in "The Royal Society and the Genesis of the Scientific Advisory Committee to Britain's War Cabinet, 1939-1940," by William McGucken, in *Notes and Records of the Royal Society of London*, Vol. 33, No. 1 (August, 1978), pp. 87-115. They are reprinted herein by permission.

Portions of Chapter 8 appeared originally in "The Central Organisation of Scientific and Technical Advice in the United Kingdom during the Second World War," by William McGucken, in *Minerva: A Review of Science, Learning and Policy*, Vol. 17, No. 1 (Spring, 1979), pp. 33-69. They are reprinted herein by permission.

Portions of Chapter 9 and the Epilogue appeared originally in "On Freedom and Planning in Science: The Society for Freedom in Science, 1940-46," by William McGucken, in *Minerva: A Review of Science, Learning and Policy*, Vol. 16, No. 1 (Spring, 1978), pp. 42-72. They are reprinted herein by permission.

Extracts from the periodical *Nature*—from an anonymous editorial entitled "Science and the Social Problems," Vol. 132, pp. 653-55; "Science and Warfare," by Lord Rayleigh, Vol. 142, pp. 336-38; "Social and International Relations of Science," by Rainald Brightman, Vol. 142, pp. 310-11; "Science and World Order," by Sir Richard Gregory, Vol. 148, p. 331; "The Prospect before Us," by L. J. F. Brimble, Vol. 148, p. 497; "British Association Conference on Mineral Resources," by C. H. Desch, Vol. 150, pp. 171-73; and an anonymous editorial entitled "Organization of Science for Warfare," Vol. 150, pp. 302-5—are copyright © 1933, 1938, 1941, and 1942 by Macmillan Journals Limited, and are reprinted herein by permission.

(Continued on page vi)

In gratitude to my parents
William and Ethel McGucken

(Continued from page iv)

The extracts from "Science News," by R. D. Potter, in *Science*, Vol. 87 (January 7, 1938), p. 10, are copyright © 1938 by the American Association for the Advancement of Science, and are reprinted herein by permission of the AAAS.

The two quotations from "A Narrative Written by P. G. H. Boswell for His Wife," on deposit with the Papers of the late Professor P. G. H. Boswell, O.B.E., F.R.S. (1886-1960 . . .). . . in the Archives of the University of Liverpool (ref: D.4/1), are reprinted herein by permission of the Archives.

Quotations from the Joseph Needham papers on deposit in the Cambridge University Library are used by permission of Dr. Joseph Needham, F.R.S., F.B.A., Director of the Needham Research Institute in Cambridge, and the Syndics of Cambridge University Library.

Quotations from the J. D. Bernal Papers on deposit in the Cambridge University Library are used by permission of Mr. Francis Aprahamian and the Syndics of Cambridge University Library.

Quotations from the papers of Lord Hankey are used by permission of the Master, Fellows, and Scholars of Churchill College in the University of Cambridge.

Quotations from the papers of Sir Henry Tizard in the Imperial War Museum in London are used by permission of the Trustees of the Museum.

Copyright © 1984 by the Ohio State University Press
All rights reserved

Library of Congress Cataloging in Publication Data

McGucken, William
  Scientists, society, and state.

  Includes bibliographical references and index.
  1. Science—Social aspects—Great Britain—History.
I. Title
Q175.52.G7M38   1984   306′.45′0941   83-8320
ISBN 0-8142-0351-5

They are writing a new chapter; and it is a chapter of novel importance, because as they extend our record of the facts of Nature they find themselves compelled at the same time to consider a new problem—the relation of those facts to society and to the government of nations.

Sir William Bragg
President, The Royal
Society of London
30 November 1940

CONTENTS

	Introduction	3
Chapter 1	The Public and Private Organization of Science in Britain to 1930: A Sketch	9
Chapter 2	The British Association for the Advancement of Science during the Depression	27
Chapter 3	The Parliamentary Science Committee	51
Chapter 4	Left-wing Scientists	71
Chapter 5	The British Association Again, and Further International Developments	95
Chapter 6	The British Association's Division for the Social and International Relations of Science	119
Chapter 7	Scientists and War	155
Chapter 8	Scientists and the Central Direction of the War Effort	215
Chapter 9	Freedom and Planning in Science	265
Chapter 10	The Central Organization of Science in Peacetime	307
	Epilogue	343

Acknowledgements

In preparing this book I have enjoyed much assistance from numerous individuals and institutions, and it is with gratitude that I acknowledge my indebtedness to all of them, especially the following. For their valuable help in various scholarly ways, I thank John R. Baker, Peter Collins, Robert Filner, June Fullmer, Margaret Gowing, Kay MacLeod, Joseph Needham, and Gary Werskey. For her constant encouragement and frequent help, I thank my wife, Emilia. I also appreciate the assistance of my daughter, Natalie, and son, Elliot. For cheerfully and skillfully typing the various drafts of the manuscript, I thank Garnette Dorsey, and especially Dorothy Richards. For permission to use the collections of papers listed in the notes on primary sources, I am thankful to Lady Blackett, John R. Baker, Joseph Needham, Lady Todd (Sir Henry Dale Papers), the British Association for the Advancement of Science, the University Library and Churchill College in Cambridge, the Royal Society of London, the University of Chicago, University of Liverpool, University of Sussex, and University of Warwick. For research grants that allowed me to travel to distant archives and for two leaves that were used for research and writing, I am much obliged to the University of Akron. Finally, I thank the editors and publishers of *Minerva*, *Notes and Records of the Royal Society of London*, and *Proceedings of the American Philosophical Society* for permission to include in revised form materials that I have published as articles. Chapter 9 and the Epilogue incorporate "On Freedom and Planning in Science: The Society for Freedom in Science, 1940–46," *Minerva* 16, no.

1, Spring 1978, pp. 42–72; chapter 7 incorporates "The Royal Society and the Genesis of the Scientific Advisory Committee to Britain's War Cabinet, 1939–1940," *Notes and Records of the Royal Society of London* 33, no. 1, August 1978, pp. 87–115; chapter 8, "The Central Organisation of Scientific and Technical Advice in the United Kingdom during the Second World War," *Minerva* 17, no. 1, Spring 1979, pp. 33–69; and chapters 2, 3, 5, and 6 and the Epilogue incorporate "The Social Relations of Science: The British Association for the Advancement of Science, 1931–1946," *Proceedings of the American Philosophical Society* 123, no. 4, August 1979, pp. 236–64.

William McGucken

Note on Primary Sources

The following is a list of the collections of papers used and repositories in which they are found.

Papers of the Association of Scientific Workers, Modern Records Centre, University of Warwick

J. R. Baker Personal Papers

J. D. Bernal Papers, Cambridge University Library

Lord Blackett Papers, Royal Society of London

P. G. H. Boswell Papers, University of Liverpool

Papers of the British Association for the Advancement of Science, Bodleian Library, Oxford

British Government Documents, Public Record Office, Kew

Sir Henry Dale Papers, Royal Society of London

A. C. G. Egerton Papers, Royal Society of London

Allan Ferguson Papers, Department for the History and Social Studies of Science, University of Sussex

A. V. Hill Papers, Churchill College, Cambridge, and Royal Society of London

Lord Hankey Papers, Churchill College, Cambridge

Joseph Needham Papers, Cambridge University Library

Michael Polanyi Papers, Joseph Regenstein Library, University of Chicago

Papers of the Royal Society of London, Royal Society of London

SCIENTISTS, SOCIETY, AND STATE

Introduction

In Britain, interest in the social relations of science goes back to the seventeenth century—to Francis Bacon and the Royal Society of London. However, a major phase of this concern began in 1931 and continued into the immediate postwar years.[1] During this period, unlike any earlier one, a social relations of science movement was widely recognized. This movement constitutes my subject. It involved British scientists, and its principal characteristic was that, in addition to their traditional pursuit of investigating the natural world, these scientists became concerned with a series of issues involving the relationships between science and society.

It should be stated at the outset that during the course of the movement the phrase "social relations of science" was sometimes used when "social relations of technology" was actually intended, that is, technology was confused with science. At other times, the intended meaning was the social relations of science *and* technology, that is, science was taken to stand for itself *and* technology. In what follows, especially chapter two, there are instances where the word *science* and its derivatives are not used in a strict sense. In such instances, I am usually following what I have found in the sources. The loose use of the words has existed, I suppose, since the nineteenth century when the sciences of electricity and chemistry began to be applied to yield new technologies, and today it shows no sign of abating. The practice is understandable; for example, some might speak of a radar set as the product of science, or of applied science, or as a piece of technology; and of the builder of the set as a scientist, or applied scientist, or technologist. I trust that in following the usages found in the sources I have introduced no major confusions into my narrative.

Initiated by a catastrophic economic depression, the social relations of science movement was shaped by other international developments of which the most important were the enthusiastic cultivation and application of science in the new Soviet state, the rise of fascism with its persecution of scientists and "misuse" of science, and the Second World War. The movement involved, and in several cases altered, Britain's major scientific organizations—the Royal Society, the British Association for the Advancement of Science, the British Science Guild, and the Association of Scientific Workers. It also saw new organizations created, both outside and inside government—the Parliamentary Science Committee (later the Parliamentary and Scientific Committee), the Society for Freedom in Science, the Scientific Advisory Committee to Britain's War Cabinet, and its peacetime successor, the Advisory Council on Scientific Policy. Unlike others who have written on the social relations of science movement, I have proceeded by examining its connections with all of these and other bodies.[2]

The relationships that existed between science and the state when the movement hesitantly started in 1931 had been considerably developed and strengthened by the time it began to lose momentum in the early postwar years, the developing and strengthening of the relationships and the loss of momentum being directly related. The first chapter sketches the general evolution of these relationships, and also the emergence of the principal scientific organizations, to 1930. The succeeding chapters deal in detail with the various phases of the movement whose chief aims came to be increased governmental support for, and the central organization of, science in Britain. Both goals were realized by 1947.

There were many currents and cross-currents in the social

relations of science movement. For example, one individual might at a given time be associated with the quite different activities of two or more organizations; or a program undertaken in one organization at a given time and at a certain stage of thought might continue for a considerable time after a new stage of thought had prompted new activities within other organizations. However, the movement may be regarded as having had five major phases, each involving the activities of one or more organizations.

The initial phase was shaped by the Depression and is considered in chapters two through four. Chapter two deals with the activities of the British Association during the years 1931-34. Prior to 1931, the Association had been little concerned with the social relations of science, but by 1934 it had developed a social consciousness and was encouraging its members to contribute papers on the social impact of science at that year's annual meeting. Although the Association avoided involvement in national politics concerning science, the British Science Guild and Association of Scientific Workers believed that scientists should contribute to the consideration of issues relating to science arising in Parliament. Their successful creation in 1933 of an independent Parliamentary Science Committee—the first organization to be created in connection with the social relations of science movement—and the work of its early years form the subjects of chapter three. The next chapter, and the final one related to the first phase of the movement, deals with the spread of a Marxist view of science among British scientists, taking particular note of the views of the leading left-wing scientist, J. D. Bernal; the entry into, and the domination of, the Association of Scientific Workers by left-wing scientists; and, finally, Bernal's work for the Parliamentary Science Committee regarding the financing of British scientific research.

The second phase of the movement was shaped, as was the first, by international developments affecting Britain, namely, the rise of fascism in Europe with its persecution of scientists and use of gas warfare. This phase, the subject of chapters five and six, is most clearly seen in the increasing concern with the social relations of science within the British Association. Chapter five deals with the Association's activities during the years 1935-37. In this period, international interest in the social relations of science was growing, and the chapter also considers the parallel activities of the American Association for the Advancement of Science, stimulated by those of the British Association, and those of the International Council of Scientific Unions. In addition, the chapter shows that the question of the scientist's responsibility for the uses of science, especially in the creation of weapons, became a troubling and insistent one. By 1938, as is seen in chapter six, scientists had concluded that all members of society, not only scientists, bore responsibility for the uses of science. Also in 1938, the Association took an unprecedented step in creating its Division for the Social and International Relations of Science in the belief that the study of those relations had become as important as the scientist's traditional study of the natural world. The creation of the Division and its work down through the war years are the chief subjects of chapter six. The Division's principal objective came to be to impress upon politicians the importance of cultivating science in the national interest.

As with the first two phases, the third was also shaped by a further major international development, namely, the Second World War. This phase, dealt with in chapters seven and eight, concerns principally the Royal Society, now drawn into the movement; the Association of Scientific Workers; and to a lesser degree, the Parliamentary and Scientific Committee, which evolved from the Parliamentary Science Committee in 1939. The theme of chapter seven is the desire expressed by the Royal Society and Association of Scientific Workers that the country should make the fullest use of its scientific resources in the war effort. It relates how scientists exerted pressure on the government to create in 1940 a central body, the Scientific Advisory Committee to the War Cabinet, to coordinate science in the war effort. Through its creation scientists made their most important wartime advance within the hierarchy of government. Nevertheless, scientists were still only advisers to the political and military leaders who determined the strategy of the war effort. In the belief that scientists should also be involved in the making of strategy on an equal footing with them, scientists lobbied in Parliament and the press in mid-1942, in what was the most political period of the social relations of science movement, for the creation of a scientific general staff. This is the subject of chapter eight. Although the government refused to form such a staff, it did appoint a team of three scientific advisers to the recently created Ministry of Production. During the last two years of the war, the Royal Society and Association of Scientific Workers, now generally satisfied with the organization and use of science in the war, began to turn their thoughts to the postwar organization of British science.

Meanwhile, a fourth phase of the movement had been initiated. It involved a major polemic among scientists themselves concerning freedom and planning in science and the consequent formation of a new organization, the Society for Freedom in Science. The history of the Society from its founding in the winter of 1940-41 to the end of the war is the subject of chapter nine. The Society's principal opponents were found in the left-wing Association of Scientific Workers.

The fifth and final phase of the social relations of science movement occurred during the period 1943-47 and concerned the central organization of science in peacetime. Chapter ten tells how the Royal Society and the Association of Scientific Workers independently lobbied a willing Labour government to create in 1947 the Advisory Council on Scientific Policy, the peacetime successor to the Scientific Advisory Committee.

The epilogue deals with the fading of the movement in the postwar years. In addition to creating the Advisory Council on Scientific Policy, the Labour government also demonstrated in other concrete ways that it fully appreciated the importance of developing and using science in the national interest in peacetime. As the British Association's Division had striven to inculcate such an appreciation in government, it found itself without a purpose and its activities quickly declined. Finally, within government and among scientists, including the Association of Scientific Workers, there was widespread support for the idea of freedom in science upheld by the Society for Freedom in Science during the war.

---

1. Neal Wood says that the movement "seemed almost to dominate the British scientific world between 1932 and 1945" (*Communism and British Intellectuals* [New York: Columbia University Press, 1959], p. 121).

2. In addition to Wood, these include: J. G. Crowther, *The Social Relations of Science* (London: Macmillan & Co., 1941); Hilary Rose and Stephen Rose, *Science and Society* (Harmondsworth, England: Penguin Books, 1969); Gary Werskey, "British Scientists and 'Outsider' Politics, 1931-1945," *Science Studies* 1 (1971): 67-83; Gary Werskey, "The Visible College: A Study of Left-Wing Scientists in Britain, 1919-1939," (Ph.D. diss., Harvard University, 1973); Robert Earl Filner, "Science and Politics in England, 1930-45: The Social Relations of Science Movement," (Ph.D. diss., Cornell University, 1973); Elizabeth Kay MacLeod, "Politics, Professionalisation, and the Organisation of Scientists: The Association of Scientific Workers, 1917-1942," (Ph.D. diss., University of Sussex, 1975); Robert Earl Filner, "The Roots of Political Activism in British Science," *Bulletin of the Atomic Scientists* 32 (1976): 25-29; Roy and Kay MacLeod, "The Social Relations of Science and Technology 1914-1939," in Carlo M. Cipolla, ed., *The Fontana Economic History of Europe, The Twentieth Century, Part-One* (Glasgow: Collins/Fontana Books, 1976), 5:301-363; Robert Earl Filner, "The Social Relations of Science Movement and J. B. S. Haldane," *Science and Society* 41 (1977): 303-16; Gary Werskey, *The Visible College: The Collective Biography of British Scientific Socialists of the 1930s* (New York: Holt, Rinehart, & Winston, 1978); Peter M. D. Collins, "The British Association for the Advancement of Science and Public Attitudes to Science, 1919-1945," (Ph.D. diss., University of Leeds, 1978); Maurice Goldsmith, *Sage: A Life of J. D. Bernal* (London: Hutchinson, 1980); and Peter M. D. Collins, "The British Association as Public Apologist for Science, 1919-1946," in Roy M. MacLeod and Peter M. D. Collins, eds., *The Parliament of Science. Essays in Honour of the British Association for the Advancement of Science, 1831-1981* (Northwood, England: Science Reviews, 1981), pp. 211-36.

1

The Public and Private Organization of Science in Britain to 1930: A Sketch

When the social relations of science movement began in 1931, science was a firmly established and highly respected, if not universally loved, institution in British society. Science had evolved its own organizations, had become a major force in general culture, particularly in higher education, had contributed much to industry, and was supported by the government. As the social relations of science movement expressed itself largely through scientific organizations, and as its principal concern came to be to further the development of the bonds between science and government, this first chapter sets the stage by dealing in a general way with the origins and purposes of the relevant organizations—the Royal Society, British Association for the Advancement of Science, British Science Guild, and Association of Scientific Workers—that existed prior to 1931 as well as with the ties that had developed between science and the state down to that year.

The oldest existing scientific organization in Britain and the world is the Royal Society, chartered in 1662. Although its advice on scientific matters affecting the British nation has been frequently sought throughout its history by successive governments, the Society has always been independent of state control. Today its fellows are elected for their high scientific achievement and are the quintessence of British scientists; but this was not always so. During the late eighteenth and early nineteenth centuries, science in Britain progressed largely independently of the Society's leadership.[1]

Partly in reaction to the decline of the Royal Society, the British Association for the Advancement of Science was founded 1831.[2] One of the founders, the Reverend William Vernon Harcourt, canon of York, declared at the time, "The Royal Society no longer performs the part of promoting natural knowledge by any such exertions as those which we now propose to revive. As a body, it scarcely labours itself, and does not attempt to guide the labours of others."[3] The aims of the new Association were "to give a stronger impulse and a more systematic direction to scientific inquiry; to promote the intercourse of those who cultivate Science in different parts of the British Empire with one another and with foreign philosophers; to obtain more general attention for the objects of Science and the removal of any disadvantages of a public kind which impede its progress."[4]

In its early years, the Association had its critics too, including the *Times*, which followed the Association's activities "with the most uncompromising hostility, refusing at last to print the lucubrations of the new philosophers unless inserted as advertisements, but continuing its sneering paragraphs or contemptuous articles."[5] Though the criticisms were often deserved, the members of the Association "could not realize that they had really a ludicrous side; that their feasting and holiday-making, their frequent mutual laudation, and above all, the opening which their meetings afforded to any hobby-rider to air his crotchets, were features which could not but strike the nonscientific outsider, who, if he could not appreciate the science, might not unnaturally form but a poor estimate of the usefulness of the Association."[6] By mid-century, however, the Association had become a more respected organization, as had the Royal Society, whose fellowship was in the process of being transformed from a largely nonscientific into an almost exclusively scientific one.[7]

The waxing of the Royal Society's scientific reputation coincided with an important innovation in the state's support of science. In 1849, the government decided to make an annual grant of £1,000 for the support of scientific investigations to be administered by the Royal Society. Although the Society had frequently received state funds, these had always been for scientific projects specified by the government.[8] Now the council of the Society could allocate government money to support researches that it deemed meritorious.[9]

In 1849, the British Association, which had at various times approached the government with specific proposals, requests, and recommendations, formed a Parliamentary Committee composed of selected members who sat in Parliament.[10] The committee was soon considering "whether any measures could be adopted by the Government or Parliament that would improve the position of Science or its Cultivators in this Country."[11] Nine of the ten suggestions advanced in its report of July 1855 concerned means of improving science education and the establishment of privileges and rewards for scientists. The most ambitious suggestion was the tenth one, namely, "That a Board of Science shall be constituted, composed partly of persons holding offices under the Crown, and partly of men of the highest eminence in science, which shall have the control and expenditure of the greater part at least of the public funds given for its advancement and encouragement, shall originate applications for pecuniary or other aid to science, and generally perform such functions as are above described, together with such others as Government or Parliament may think fit to impose upon it."[12] Although copies of the Parliamentary Committee's report were sent to the Prime Minister, Lord Palmerston, and other Cabinet members, the council of the British Association was nevertheless aware that the suggestions "touching the questions of rewards to be given in various shapes to the cultivators of science, and more especially that of the creation of a Board of Science ... have yet to receive ... sanction from public opinion; and more especially from the opinion of men of science themselves!"[13]

On this, as on other occasions, the Association had the cooperation of the Royal Society.[14] The general subject of the Association's report was taken up by the Society whose president, the Earl of Wrottesley, was also chairman of the Association's Parliamentary Committee. The Society drew up a series of resolutions that incorporated the committee's principal suggestions and had the approval of the council of the Association. These resolutions were communicated to the government by Wrottesley in his capacity as president of the Royal Society, but nothing came of them. J. B. Morell has observed: "Invariably the mid-Victorian Parliament did not move until a compelling case had been irrefutably substantiated by some private or official body: the norm was self-help; any proposed deviation from that norm had to be buttressed with irresistable evidence and argument."[15]

A movement initiated within the British Association in 1866 by Colonel Alexander Strange did, however, lead to the appointment in 1872 of a Royal Commission chaired by the Duke of Devonshire and charged with inquiring into the advancement of science and science instruction in Britain.[16] In its eighth and last report, published in 1875, the Commission recommended, *inter alia*, the creation of state laboratories, the increase of research grants for scientists, and the establishment of a Ministry of Science and Education with a Council of Science to assist it.[17] Although few of the recommendations were implemented, the government grant administered by the Royal Society was increased to £4,000.

The development of Treasury policy toward science at this time has been examined recently. Employing a broad definition of science, R. M. MacLeod calculates that government expenditure on civil scientific activities increased from £70,115 in 1859–60, to £261,184 in 1869–70, to £346,528 in 1879–80.[18] He observes that energetic appeals for funding "when launched and sustained by men of science well placed in society, who could marshall effective utilitarian, cultural, or 'prestige' arguments to the support of a united case, usually succeeded." But he finds that on balance the Treasury's approach "probably delayed innovation and discouraged many good men." MacLeod concludes that the two major stumbling blocks to the endowment of science by the State were "the widespread inability of civil servants to appreciate the difficulty, purposes, and ultimate value of fundamental research, coupled with the inability of men of science to recognize the difficulty and necessity of accountability."[19]

Following the sitting of the Devonshire Committee, state-related activities within the British Association declined during an extended period of comparative quiescence. No longer were the claims of science vigorously pressed upon the government. Wrottesley's death in 1867 marked the beginning of the decline of the Association's Parliamentary Committee.[20] A quarter of a century was to pass before an attempt was made by Norman Lockyer to rekindle the earlier concerns of the Association.

Lockyer, the founder and editor of *Nature*, had been secretary of the Devonshire Commission. He felt strongly about the Commission's recommendation to create a national science council, and for many years wasted no opportunity of stressing the need for such an organization.[21] In 1903, at a time of intense dissatisfation with technological development in Britain, Lockyer spoke on "The Influence of Brain-Power on History" in his presidential address to the British Association.[22] He argued that science had had a much greater influence than politics on history and expressed the hope that the realization of this would lead to an extension of the public activities of the Association.[23] The meeting was sufficiently persuaded to adopt the resolution:

> That, as urged by the President in his Address, it is desirable that scientific workers, and persons interested in Science, be so organised that they may exert permanent influence on public opinion, in order more effectively to carry out the third object of this Association originally laid down by the founders, viz.: "to obtain a more general attention to the objects of Science, and a removal of any disadvantage of a public kind which impedes its progress," and that the Council be recommended to take steps to promote such organisation.[24]

To Lockyer's disappointment, however, the council took no action, so he sought to achieve his goal through some new organization. This would be the British Science Guild, whose first president was the eminent Lord Haldane. Lockyer himself occupied the influential position of chairman of committees.[25]

The Guild was not a scientific society, but rather a nonparty and nonpolitical imperial organization composed of representatives of science, industry, and education that attempted to coordinate the activities of these three fields for the imperial good. In regard to general policy, the Guild initially had four objectives: to bring together in the Guild all persons within the Empire interested in science and in scientific methods in order that they might convince others of the necessity of applying such methods to all areas of human endeavor and thereby further the progress and increase the welfare of the Empire, to bring before the executive officers of British national and departmental organizations the scientific aspects of all matters affecting the national welfare, to promote and extend the application of scientific methods to general administrative and other purposes, and to promote scientific education by encouraging the support of universities and other institutions that advance, or devise new applications of, science.[26] These objectives were pursued through inquiry, lectures, and publications. Inquiry, the principal activity of the Guild, was carried on by means of specially appointed committees composed of distinguished representatives of science, industry, and education. Many of the reports prepared by these committees were forwarded to appropriate government ministers whose areas of responsibility coincided with, or overlapped, those dealt with in the reports.[27] To single out one notable example of its work, the Guild had an important influence on the government's decision in 1915 to create an Advisory Council for industrial and scientific research, the forerunner of the Department of Scientific and Industrial Research.[28]

Following the outbreak of World War I, Britain quickly became aware of its dependence on other nations for supplies of vital manufactured items. In particular, Britain discovered to its alarm that it was dependent on Germany for magnetos, certain types of optical glass, pharmaceutical preparations, and tungsten for use in steel production.[29] The Board of Trade appointed a committee chaired by Lord Haldane to consider and advise how British industry could be adequately supplied in the new international situation. At the same time, scientists were urging the government to take action.[30] On 23 July 1915, a "white paper" on the subject was presented to Parliament, and five days later the creation of the Department of Scientific and Industrial Research as a new state organization was authorized. In its first annual report, the Committee of the Privy Council for Scientific and Industrial Research acknowledged that the British Science

Guild had been active in trying to impress upon the nation the necessity for government organization of science, adding that the shock of war had been required to make the need manifest to Parliament and the nation.[31]

In the postwar years, the British Science Guild considered the standing of science in the nation to have been enhanced to such a degree that it sought a new role for itself. In 1922, J. J. Robinson, honorary secretary of the Guild's Parliamentary Committee, explained that the work the Guild had set out to do at its founding "when it was a voice crying in the wilderness, has in considerable measure been begun."[32] Referring to the "new and separate organisations which have come into existence since 1905 with the avowed objects of getting the results of scientific research and knowledge made accessible to a special industry, interest, or propagandist group," he continued: "How many new Committees inside and outside State Departments have been set on foot for scientific research alone; and for practical application of its fruits. Something had been done up to 1914 in this way; since 1914—a beneficent Deluge, as compared with the timid and tentative tricklings of the previous years.[33] . . . Science is honoured where previously it was scouted. Our daily newspapers have their scientific correspondents, and a month rarely passes now without some new special advance in the practical applications of science. This is all to the good."[34]

The first quarter of the twentieth century saw a great increase in the government's involvement with science, many innovations being made in addition to the creation of the Department of Scientific and Industrial Research. At the turn of the century, a few museums promoted the spread of scientific knowledge. These were the British Museum, including the Natural History Museum, founded in 1753; the Museum of Practical Geology, created in 1837; and the Science Museum, established in 1853.[35] The Imperial Institute, dating from 1887, provided scientific and technical advice regarding the economic development of the natural resources of the Empire. A few government departments—the Admiralty, the Post Office, the Royal Observatory (founded in 1675, the first of the state-aided institutions), the Geological Survey (founded in 1835), and the Department of the Government Chemist (originally founded as a branch of Customs and Excise in 1842)—used scientific knowledge in their areas of responsibility. Beyond this there was little organized effort to apply the discoveries of fundamental research.

In 1900, however, the national need for physical and engineering research, and for investigation into methods of precise measurement, led to the creation of the National Physical Laboratory under the control of the Royal Society. The provision of funds and facilities to enable government scientists to undertake investigations that industry was considered incapable of conducting, yet would benefit from, was a new departure.[36] Four years later, the Post Office began organized research in the application of discoveries in physical science in the development of the telegraph and telephone systems.[37]

The Development Commission appointed in 1909 was given responsibilities in the areas of agriculture, fisheries, and rural economy. In 1911, it framed a comprehensive scheme of agricultural research, advisory services, and postgraduate scholarships that led to the formation of most of the agricultural research institutions that would serve the country over the next half century. The Commission also appointed a Scientific Advisory Committee on Fishery Research that examined the whole field of fishery investigation and made recommendations that led to a substantial strengthening of the facilities for research.

In 1913, in connection with the national health insurance scheme, and recognizing the need for a central organization to be responsible for promoting medical research, the government established the Medical Research Committee. Seven years later, this became the Medical Research Council, the medical counterpart of the Department of Scientific and Industrial Research.[38]

In the same year, 1920, the Building Research Board was created. Thus, in the early 1920s, the British Science Guild could rightly claim that in the area of government science much had been initiated since the turn of the century. More was to come. Guided by Lord Balfour, Lord President of the Council, the government of Stanley Baldwin created in 1925 the Committee of Civil Research, whose function was to give "connected forethought from a central point of view to the development of economic, scientific and statistical research in relation to civil policy and administration."[39] The committee has been described as a constitutional innovation, being the first attempt to create "a 'research department' at the highest level, able to recruit economic and scientific specialists to tackle problems outside the jurisdiction of particular government departments."[40] Before the committee was transformed into the Economic Advisory Committee in 1930, its eighteen subcommittees had considered a broad range of imperial and domestic matters, including tropical disease and the shortage of radium supplies. The Subcommittee on the Coordination of Research reported on the history of government involvement in science but did not make any recommendations for future policy. However, one of its reports was to lead to the creation in 1931 of the third major research organization, the Agricultural Research Council.

Well satisfied with the burgeoning of state science, the British Science Guild had decided by 1924 that its chief aim should be to promote the application of scientific methods in public affairs:[41] "It is among public administrators, those who control, in public affairs and businesses, in private enterprises, in organisations of all sorts, that the help of the Guild is most urgently needed to get efficient service in the place of inefficient, and science and correct method in place of ignorant and slovenly method."[42] The Guild was convinced that no other science organization shared its new goal. The most likely one to do so was the British Association; however, the Guild's *Journal* explained that the Association "operates through one reservoir of scientific knowledge and interest, annually filled and released, and treats solely of scientific issues, or other issues regarded scientifically, in the most valuable and helpful way. But it has no direct connection with the administrative pastimes or pursuits of the unscientific world. Its functions, however admirable and fertilizing, cover no part of the Guild's objective; and indeed its form of organisation prevents anything of the kind being even contemplated."[43] Despite its newly charted course, the Guild never recovered its earlier vitality. Ironically, significant future developments would occur along its original course, but they required other organizations and altered times, as will be seen in later chapters.

From 1905, the Guild and the British Association by and large had pursued separate courses; but in December 1926, the council of the Association appointed a committee to consider the advisability of establishing closer relations with the Guild.[44] On the committee's recommendation, the Guild was asked to consider "whether, having regard to the close community of scientific interests between the Association and the Guild, their objects would as the [British Association] Council believe, be more fully attained by means of a working union between the two societies."[45] A joint committee of the Association and the Guild was formed. It produced a report recommending union of the parent bodies that the British Association council urged the general committee to adopt at the annual meeting of 1927.[46] However, within the general committee the recommendation met with organized opposition.[47] The proposals had included the stipulation that the Guild come "into the British Association on, so to speak, 'equal terms,' the name being included in the title of the Association."[48] These proposals "led to stormy scenes, and Sir Richard Gregory [chairman of the Guild's executive committee and editor of *Nature*], the protagonist, was unable to carry them through." The recommendation was also opposed on the grounds that the Association should concern itself only with science and not with its social consequences, which might be left to other organizations.[49] The general committee resolved that the council should continue its negotiations with the Guild and submit definite proposals to the general committee at the following year's meeting.[50] But the matter did not come before the general committee until it was revived again some nine years later.

Meanwhile, in 1929, the Association and the Guild cooperated with the Association of Scientific Workers in a new attempt to promote the interests of science in Parliament. The Association of Scientific Workers had evolved from the National Union of Scientific Workers, founded at the end of 1917.[51] A recruiting memorandum issued by the Union at its founding explained: "one of the main reasons why science does not occupy its proper place in national life is that scientific workers do not exercise in the political and industrial world an influence commensurate with their importance. It is also widely held that the reason why they do not exercise such influence is that they have not hitherto adopted the form of organisation which, in a democratic community, is necessary to obtain it."[52]

The Union's first branch was formed in Cambridge, holding its initial meeting on 17 February 1918. By October 27, when the

Union held its first general meeting in London, eleven branches, with a membership of some five hundred, had been formally constituted. During the early 1920s, the Union organized public meetings on various aspects of science and strove to impress upon Parliament the importance of science for the nation; but its principal work was the advancement of the economic interests, and thereby, it believed, the standing and influence, of scientists. In 1927, dissatisfied that its members numbered only about eight hundred and desirous of adding new ones, the Union abandoned its trade union status and became the Association of Scientific Workers.

In the general election of 1929, which returned the second Labour government to office, party leaders expressed general agreement that the nation and Empire would benefit if science and scientific methods were applied to national and imperial problems. As this view, which the British Science Guild would naturally have supported, was also one that the Association of Scientific Workers had espoused from its inception, the time seemed opportune to the Association to take further action.[53] In the hope of promoting "scientific research in all its aspects" within the country, it decided to form a parliamentary science committee representative of all political parties. The committee's function would be to gather "information concerning science in industry, research institutions, and higher educational institutions" that related to imperial affairs and bills before the House of Commons.[54]

Through a fortunate coincidence, the general secretary of the Association, Major Archibald G. Church, was again elected as a Labour member of Parliament at this time. When previously in the Commons (1923–24), he had served as a member of the Medical Research Council. In the summer of 1929, following the general election, he and Professor B. W. Holman, of the Royal School of Mines, wrote on behalf of the Association to those members of the new Parliament whom they thought might be interested in joining the proposed Parliamentary Science Committee. The response was most encouraging, and in August 1929 Church was able to publish a list of twenty-eight Conservative, twenty-three Labour, and six Liberal acceptances.[55] The new committee was modeled after others already in existence, namely, the Agricultural, the Medical, the Commercial, and the

League of Nations Parliamentary committees. The membership of these committees was confined entirely to members of Parliament. The eminent physicist Lord Rayleigh was the first chairman of the Parliamentary Science Committee.

It was soon realized that a body representing science was needed to work conjointly with the committee, and so a Scientific Advisory Council of about twelve members was formed.[56] These included representatives of the British Association and British Science Guild as well as the Association of Scientific Workers. One of the council's purposes was to draw up an advisory panel of scientists, each of whom would be willing to address the Parliamentary Science Committee as and when required.

That members of Parliament had need of such scientific services was as obvious to Church from his parliamentary experience as the political inexperience, and even naivety, of most scientists.

> Both appreciation and destructive criticism of research or of the applications of science must be based upon knowledge, and hitherto, with a few notable exceptions, members of Parliament have lacked this knowledge. Again, few constituents demand that their members should take an intelligent interest in science or scientific research. Scientific workers, if they were properly organised, could create this interest in the constituencies, but at present, unlike other professional workers, they display little concern with politics and little appreciation of the potentialities of the political machine. They hold themselves aloof from every subject of current controversy, fearful lest they should lay themselves open to the charge of taking sides. That they are citizens, that they are often more the servants of the public than are other professional workers, that it is their duty to make known the truth that is in them without fear that this truth will be interpreted by party politicians as evidence in support of one or other set of political theories does not occur to them. Either that, or they assume, quite wrongly, that governments and politicians are well acquainted with their work and will call upon them for their advice and assistance whenever necessary.[57]

Such considerations led Church to suggest a second equally useful purpose for the Parliamentary Science Committee. Not only would it provide a forum for the discussion of appropriate matters already before Parliament, but it would also afford scientists an opportunity of initiating parliamentary action. Scientists would show members of Parliament how science could be appropriately applied to industrial and social problems "such as smoke pollution in cities, the contamination of . . . rivers by sewage and noxious effluents from industrial works, food adulteration, dangerous trades, the sterilization of the unfit [!], the control of venereal diseases, and the control of plant and animal diseases."[58]

The creation of the Parliamentary Science Committee signaled an awareness among some scientists of a new state of affairs, namely, that scientific matters of importance both for the profession and the nation were increasingly coming before Parliament. For selfish as well as unselfish reasons, these individuals recognized the desirability of bringing scientists and politicians together. As was to become clear rather quickly, however, their efforts met with little success.

To create a parliamentary science committee had seemed sound, and the initial response of members of Parliament had been encouraging, but attendance at meetings of the Parliamentary Science Committee proved to be "distinctly disappointing." Henry W. J. Stone, an experienced Parliamentary observer, explained that members had merely joined the Committee because to have refused might have alienated some votes.[59] But having no real interest in scientific matters, they readily invented excuses to miss Committee meetings, the more so as no division lists were published to reveal their attendance or absence. In the unlikely event of the work of the Committee being referred to in a member's constituency, the member could affect a sympathetic interest and mention his membership on the Committee. In contrast, participation in the meetings of the Agricultural and Commercial committees was enforced because, since many constituents were interested in the work of these committees, many votes were consequently involved. With science it was quite otherwise. One vote in a thousand—Stone's "liberal" estimate of the professional scientific component of the electorate—was hardly likely to make the difference between the reelection or defeat of a member. Thus, scientists were bluntly told that politicians were not yet convinced that science controlled enough votes, "the politician's oxygen," for serious attention to be paid to it.

Although politicians could survive without caring about science, science and scientists, if divorced from government, would not fare as well as they otherwise might. Church's parliamentary experience had convinced him of this, and he persuaded the Association of Scientific Workers to his view. So in 1929, the Association also formed a Parliamentary Committee with Holman as chairman and Church as secretary.[60] Its composition was significantly different from that of the Parliamentary Science Committee. The nine members of Parliament who were considered to be eligible for membership, because they were thought to have an interest in science, were this time offered honorary membership in the Association of Scientific Workers. Of this handful, seven accepted—fifty fewer than those who responded initially to the invitation to join the Parliamentary Science Committee. The Association's Parliamentary Committee must not have been as unsatisfactory as the latter, for when Church became organizing secretary of the British Science Guild in 1930 he revived its Parliamentary Committee.

In 1930, there was general satisfaction with both the organization and condition of science in Britain. The Royal Society had evolved as the most prestigious scientific organization and the British Association as the most popular and populous one. Both bodies were devoted to the advancement of science but little concerned with its cultivation and use by the state. Both were flourishing, as was science itself. Indeed, the rapid advance of science alarmed some laymen including the Bishop of Ripon, who, in his sermon during the official church service at the British Association meeting in 1927, suggested that every physical and chemical laboratory be closed for ten years to enable society to assimilate the staggering amounts of new scientific knowledge.[61] Scientists paid little heed to the Bishop's suggestion. At ensuing meetings of the Association, mere mention of a "scientific holiday" was certain to elicit derisive laughter.

One organization, the British Science Guild, had been formed early in the twentieth century to encourage the national and imperial cultivation and use of science because the British Association had refused to do so. However, by the mid-twenties, in the Guild's view, science was held in such esteem nationally and the government had done so much since the beginning of the century by way of putting science to work in national life, through the creation of such state institutions as the Department of Scientific and Industrial Research and the Medical Research Council, that the Guild adopted the new goal of promoting the application of scientific methods in public affairs. By contrast, the Association of Scientific Workers, which had been formed in 1917 as a trade union, turned its attention toward government. It recognized that issues relating to science were increasingly coming before Parliament, and from 1929 strove to contribute to both the discussion and the raising of such issues through the creation of an effective Parliamentary science committee.

1. Dorothy Stimson, *Scientists and Amateurs: A History of the Royal Society* (New York: Henry Schuman, 1948), p. 179.

2. For discussion of both the founding of the Association and of recent historical scholarship on it, see Susan Faye Cannon, *Science in Culture: The Early Victorian Period* (New York: Dawson and Science History Publications, 1978), chapters 6 and 7. See also Jack Morrell & Arnold Thackray, *Gentlemen of Science: Early Years of the British Association for the Advancement of Science* (Oxford: Clarendon Press, 1981).

3. O. J. R. Howarth, *The British Association for the Advancement of Science: A Retrospect, 1831-1931* (London: The British Association, 1931), p. 19.

4. *Advancement of Science* 17 (1960): 339.

5. Howarth, *The British Association*, p. 35.

6. Ibid.

7. Sir Henry Lyons, *The Royal Society, 1660-1940* (Cambridge: Cambridge University Press, 1944), p. 233.

8. Ibid., pp. 266-67.

9. For a history of the grant, see Roy M. MacLeod, "The Royal Society and the Government Grant: Notes on the Administration of Scientific Research, 1849-1914," *Historical Journal* 15 (1971): 323-58.

10. Howarth, *The British Association*, pp. 212, 220; David Layton, "Lord Wrottesley, F. R. S., Pioneer Statesman of Science," *Notes and Records of the Royal Society of London* 23 (1968): 234.

11. *Report to be presented by the Parliamentary Committee to the British Association for the Advancement of Science at Glasgow, on the question, Whether any measures could be adopted by the Government or Parliament that would improve the position of Science or its Cultivators in this Country* (London: British Association, 14 July 1855).

12. Ibid., p. 18.

13. Howarth, *The British Association*, p. 223.

14. Ibid., p. 224.

15. J. B. Morell, "The Patronage of Mid-Victorian Science in the University of Edinburgh," in G. L'E. Turner, ed., *The Patronage of Science in the Nineteenth Century* (Leyden: Noordhoff International Publishing, 1976), p. 57.

16. D. S. L. Cardwell, *The Organisation of Science in England; A Retrospect* (London: William Heinemann, 1957), pp. 97-98; Roy M. MacLeod, "The Support of Victorian Science: The Endowment of Research Movement in Britain, 1868-1900," *Minerva* 9 (1971): 197-230.

17. Cardwell, *The Organisation of Science in England*, p. 97.

18. R. M. MacLeod, "Science and the Treasury: Principles, Personalities and Policies, 1870-85," in Turner, ed., *The Patronage of Science in the Nineteenth Century*, p. 116.

19. Ibid., p. 160.

20. Howarth, *The British Association*, pp. 231, 227.

21. T. Mary Lockyer and Winifred L. Lockyer, *Life and Work of Sir Norman Lockyer* (London: Macmillan & Co., 1928), p. 85.

22. For details of this dissatisfaction, see Cardwell, *The Organisation of Science in England*, pp. 147-51.

23. *Nature* 68 (1903): 439-46.

24. British Association Council minutes, 6 November 1903.

25. For an account of the formation of the Guild and a sketch of its activities during its first decade, see A. J. Meadows, *Science and Controversy: A Biography of Sir Norman Lockyer* (Cambridge, Mass.: MIT Press, 1972), pp. 272-78, 303.

26. "The British Science Guild," (4 pp.) appended to British Science Guild, *Engineers' Study Group on Economics: First Interim Report on Schemes and Proposals for Economic and Social Reforms* (London, 1935).

27. Subjects investigated included agricultural research in the United Kingdom, awards for medical discovery, British chemical industries, the British dye industry, bureaucratic control of education, conservation of natural sources of energy, coordination of charitable effort, design and manufacture of microscopes, explosives, fisheries development, industrial research and trained scientific workers, introduction of the metric system, medical research, national education, patent law reform, position of scientific and professional staffs in public services and industry, prevention of the pollution of rivers, provision of glass and other laboratory ware, science and labour in the modern state, synchronization of clocks, technical optics and the manufacture of optical instruments, utilization of science in public departments, veterinary research (ibid.).

28. Meadows, *Science and Controversy*, p. 303.

29. Sir Harry Melville, *The Department of Scientific and Industrial Research* (London: Allen & Unwin, 1962), p. 24. See also Ian Varcoe, "Scientists, Government, and Organised Research in Great Britain, 1914-16: The Early History of the DSIR," *Minerva* 8 (1970): 192-216.

30. See, for example, Sir William Ramsay's address to the annual meeting of the British Science Guild, 1 July 1915. *Journal of the British Science Guild*, no. 1 (September 1915).

31. Sir Richard Gregory, "Education and National Development," in Lockyer and Lockyer, *Life and Work of Sir Norman Lockyer*, p. 449.

32. J. J. Robinson, "1922 and the Future of the Guild," *Journal of the British Science Guild*, no. 15 (May 1922), p. 69.

33. For a brief discussion of science and technology during World War I, see Arthur Marwick, *The Deluge: British Society and the First World War* (Boston: Little, Brown, & Company, 1965), pp. 226-39.

34. Robinson, "1922 and the Future of the Guild," pp. 66–67. See also Robinson's "Scientific Method in Public Affairs," *Journal of the British Science Guild*, no. 16 (February 1923), p. 73.

35. Advisory Council on Scientific Policy, *Government Scientific Organisation in the Civilian Field* (London: H.M.S.O., 1951), p. 3.

36. Russell Moseley, "The Origins and Early Years of the National Physical Laboratory: A Chapter in the Pre-History of British Science Policy," *Minerva* 16 (1978): 222.

37. Advisory Council on Scientific Policy, *Government Scientific Organisation in the Civilian Field*, p. 3.

38. The origins of the Committee and Council are described in A. Landsborough Thomson, *Half a Century of Medical Research, Volume 1: Origins and Policy of the Medical Research Council (UK)* (London: H.M.S.O., 1973).

39. R. M. MacLeod and E. K. Andrews, "The Committee of Civil Research: Scientific Advice for Economic Development 1925–30," *Minerva* 7 (1968–69): 688–89.

40. Ibid., p. 704.

41. *Journal of the British Science Guild*, no. 17 (February 1924), pp. 14, 28.

42. Ibid., no. 15 (May 1922): 68. See also Gregory, "Education and National Development," p. 451.

43. "The Distinctive Purpose of the Guild," *Journal of the British Science Guild*, no. 17 (February 1924), p. 28.

44. British Association Council minutes, December 1926.

45. Ibid., 4 February 1927.

46. Ibid., 4 March 1927; ibid., 31 August 1927.

47. *Scientific Worker* 3 (1927): 95.

48. "A Narrative written by P. G. H. Boswell for his wife," (typescript, 1942–ca. 1948), p. 241. Boswell was general secretary (1931–35) and treasurer (1935–42) of the British Association. The narrative is in the archives of Liverpool University.

49. *Nature* 132 (1933): 654.

50. British Association General Committee minutes, third meeting, Leeds, 6 September 1927.

51. Roy and Kay MacLeod, "The Contradictions of Professionalism: Scientists, Trade Unionism, and the First World War," *Social Studies of Science* 9 (1979): 1–32.

52. "The A.S.W.—Twenty Years History," *Scientific Worker* 11 (1939): 68.

53. *Scientific Worker* 5 (1929): 77–78.

54. Ibid., pp. 49–50.

55. Ibid., pp. 78–79.

56. British Science Guild Council of Management, *Annual Report* (1929–30), p. 9.

57. *Scientific Worker* 5 (1929): 92.

58. Ibid., p. 93.

59. *Progress and the Scientific Worker* 2 (1933–34):214–16.

60. *Scientific Worker* 10 (1938): 20.

61. *Times*, 5 September 1927, p. 15. For the reception of the Bishop's idea in the United States, see Carroll Pursell, "'A Savage Struck by Lightning': The Idea of a Research Moratorium, 1927–37," *Lex et Scientia* 10 (1974): 146–58.

2

The British Association for the Advancement of Science during the Depression

    The stirrings of the social relations of science movement, initiated by the economic depression of the 1930s, were first seen within the British Association for the Advancement of Science, the British Science Guild, and the Association of Scientific Workers. The form that the initial phase of the movement took within the British Association was markedly different, however, from that seen in the joint actions of the other two. Maintaining the practice it had followed since the last quarter of the nineteenth century, the British Association carefully avoided becoming involved in governmental affairs relating to science. In contrast, the other two organizations sought such involvement, in keeping with the earlier practice of the Guild and the recent efforts of the Association of Scientific Workers in connection with the Parliamentary Science Committee. This chapter deals with the British Association's activities, and chapter 3 with those of the Guild and Association of Scientific Workers, during the initial phase of the social relations of science movement.

  Jolted by the harsh social consequences of the depression, caused according to some by an uncontrolled science and technology's overproduction of goods, the members of the British Association were driven to reexamine the Association's function. The issue was first raised in 1931 and dominated the annual meetings of the next three years. After contending with doubt and confusion of purpose in 1932, and after engaging in extended internal debate during the next two years, the Association implemented, at its widely noted 1934 annual meeting, its new resolve to add to its traditional study of the natural world that of the study of the social impact of science.

The British Association celebrated its centennial at the annual meeting of September 1931. As was customary, the meeting was opened by the president, on this occasion the South African statesman General Jan Christiaan Smuts, giving his presidential address. Smuts spoke on "The Scientific World-Picture of Today" to what some considered to have been the most distinguished gathering in the Association's history.[1]

Toward the end of his remarks, Smuts turned briefly to a troubling subject that would be given much attention at Association meetings throughout the decade and beyond, namely, the threat of science to society. Smuts thought that one of the greatest future tasks for mankind would be to link science with ethical values.[2] There were many who would have agreed with him that already a serious lag existed between man's rapid scientific advance and his "stationary ethical development," as evidenced by World War I—"the greatest tragedy of history." Scientists, Smuts argued, had to help in closing this dangerous gap that threatened "the disruption of our civilization and the decay of our species."[3]

In 1931, civilization was not being disrupted by world war, but it was suffering from the effects of a worldwide economic depression; and Smuts could not avoid mentioning the fact when later at the Association's meeting he opened a discussion on aspects of imperial agriculture. Here he spoke to industrial Britain of the farmer's view of the universal calamity. It was, he said, as though another war had ravaged the world with science as the aggressor, "as though science had achieved such devastating successes that the world agricultural applecart had been overturned."[4] A solution was not to be hoped for through an attempt to arrest the advance of science, for that could not be stopped. Rather, an organization was needed that would coordinate the results of science and keep the entire system of science, production, and consumption in harmony. But what the nature of this organization might be, and who should create it, Smuts did not say. There were some who held that only socialism could provide the desired solution, but that view found little support within the British Association.

Smuts' remarks were given emphasis by contemporary events in British national life. At the beginning of 1931, the United Kingdom, with two and one-half million persons unemployed, was already gravely affected by the economic depression. This situation was made worse by the financial crisis of August and September of that year, caused by American and French banks calling in large sums of money previously advanced to British banks. To make repayment, the British were forced to deplete their gold reserves. The ensuing threat to the pound sterling precipitated a political crisis that led to the collapse of the second Labour government and the formation under Ramsay MacDonald of the National government. In an unsuccessful attempt to stimulate foreign trade and the home economy, the National government abandoned the gold standard in September 1931.[5]

According to J. G. Crowther, who at the time was science correspondent of the *Manchester Guardian*, the British Association's centennial meeting was almost swept out of the public consciousness by the national economic crisis.[6] Much later, Crowther, a socialist, would write: "Against this background, the Victorian form of the British Association seemed rather irrelevant. In the situation the institution had little to offer. In fact, Brewster's British Association of 1831 appeared more modern. The government of the Association, and many individual scientists, felt uncomfortable, and began to discuss what science could do to help the nation in the crisis, and what action the Association and scientists should take to make themselves more effective in this direction."[7] As Crowther also clearly saw, the economic crisis of 1931 placed British science and scientists in a new perspective.[8] Scientists and laymen alike were again being forced by events to concern themselves with the relationships that existed between science and society. A new era was beginning in the history of science, in which social and political dimensions would be added to scientists' traditional concerns.[9]

Its development, however, was initially slow. Smuts' coordinating body was never defined by anyone, let alone created, and the British Association's centennial meeting ended without the Association offering the country any solution to its economic difficulties. The scientific community was initially divided as to whether it should do anything other than advance and disseminate science, its traditional roles. Those who were convinced that scientists should do more were confused about what this should be and through what means it should be attempted. But the uneasiness present at the British Association meeting persisted beyond its closing, as is illustrated by a letter published in *Nature* by Henry E. Armstrong, Emeritus Professor of Chemistry at the City and Guilds College, South Kensington. In his view, the British Association's proceedings had been, from begining to end, "Neronic: little effort was made by anyone to pay serious attention to our present earthly situation or to consider how we are to get out of the muddle 'science' has created."[10]

It was hardly surprising that an editorial written by H. E. A[rmstrong]. appeared the following week in *Nature*.[11] Under the editorship of Sir Richard Gregory, the journal was to continually support and frequently lead those who held that the responsibilities of scientists extended beyond laboratory walls. With the objective of increasing public interest in science, Gregory and others had from 1916 made popular lectures a permanent feature of British Association meetings.[12] Gregory and Armstrong had been acquainted since the turn of the century when, with others, they were successful in having the education section created within the British Association.

Armstrong's leader appealed to scientists to turn their attention to affairs of state. It told them that they would have to recognize that they had been selfish in their too exclusive devotion to experimental study and the immediate development of their discoveries. "In future," it continued, "the scientific worker, to be worthy of the name, must justify himself through social service in the first instance. Our situation is so grave that he must be militant without delay in every quarter."[13] About the same time, Cecil Henry Desch, Professor of Metallurgy at Sheffield University, privately circulated a pamphlet entitled "The Social Function of Science" in which he stated that scientists could not disclaim responsibility for the purposes to which the knowledge they created was applied.[14]

There was, nevertheless, uncertainty as to the appropriate actions scientists might take, as was indicated by a letter written in reply to Armstrong's editorial. It was most stimulating, said the writer, to read that scientists should become militant without delay and in every quarter, but what precisely were they to do? There was not a single major issue of statesmanship on which the scientific profession was in unanimous agreement. All political opinions were represented among scientists. What then, he pointedly asked, could they as a profession achieve in the world of statesmanship in view of "such radical disagreement" among themselves?[15]

Within a month, Armstrong was urging, in a second editorial in *Nature*, that the Royal Society should take the lead in organizing the country's scientific forces. "It should forthwith constitute its fellows a parliament of scientific opinion ... also co-opting a limited number of specially competent persons from outside the Society. Our first act would be to learn to take counsel together for the common good—a hitherto unknown art unpracticed by scientific workers."[16] However, the Royal Society was to be unmoved by the considerations that stirred the British Association at this time.

Concern about the social consequences of the applications of science was expressed again at the Association's 1932 annual meeting. It could hardly have been otherwise in that depression year, which was perhaps the one most disturbed by hunger marches and other demonstrations. These were accompanied by frequent baton charges on the part of the police and by numerous arrests.[17] In this grim social context, the eminent and elderly engineer, Sir Alfred Ewing, gave his presidential address to the Association. In it he displayed doubt, bewilderment, and hesitation, with great impact both within and beyond the Association:

> In the present-day thinker's attitude towards what is called mechanical progress we are conscious of a changed spirit. Admiration is tempered by criticism; complacency has given way to doubt; doubt is passing into alarm. There is a sense of perplexity and frustration, as in one who has gone a long way and finds that he has taken the wrong turning. To go back is impossible: how shall he

proceed? Where will he find himself if he follows this path or that? An old experimenter of applied mechanics may be forgiven if he expresses something of the disillusion with which, now standing aside, he watches the sweeping pagent of discovery and invention in which he used to take unbounded delight. It is impossible not to ask, Whither does this tremendous procession tend? What, after all, is its goal? What its probable influence on the future of the human race?[18]

In the immediate situation, Ewing was particularly troubled by unemployment, the potential for highly destructive war, and the danger of the public's having leisure time without the knowledge to use it well.[19] He shared the view of Smuts and others that man was ethically unprepared for the tremendous responsibilities that the engineers' many gifts entailed. The command of nature had been placed in man's hands before he knew how to command himself. The only certain solution to the problem seemed to be to arrest the development of science, but Ewing, like almost everyone else, regarded this as impossible even if desirable. The continual growth of science and technology was now a permanent feature of civilization.

Unsettled by his remarks, laymen, including the Sheriff of York, where the meeting was held, pressed Ewing to make a public statement that would serve as "an 'apologia' for the life scientific, and a guide for general public action."[20] To assist Ewing in his task, a meeting was called of leading scientists and others including Sir William Bragg, Director of the Royal Institution of Great Britain, Sir Richard Glazebrook, the first director of the National Physical Laboratory, Sir Walter Fletcher, Secretary of the Medical Research Council, Lt. Col. Sir Arnold Wilson, managing director of the D'Arcy Exploration Company, and the secretaries of the Association, P. G. H. Boswell, Professor of Geology at the Imperial College of Science and Technology, and O. J. R. Howarth. However, following much desultory discussion, Ewing was unable to frame a useful statement. Later during the meeting he could do no better than feebly suggest that the "only way we should find our salvation was that industry should be carried on in the spirit and teaching of Christ."[21]

In an editorial on Ewing's address, the *Times* posed the rhetorical question whether science was as confident of itself now as it had been a century before when the Association was founded. It might be argued, said the paper, that pure science had done little more than unsettle beliefs that were at least of pragmatical value, and that applied science had loaded the ordered civilizations with "benefits" under which they were being crushed. However, it was at least to the good, allowed the *Times*, that a scientific leader should forewarn society that the "brave new world" (a phrase that was the title of Aldous Huxley's timely and disturbing novel published earlier that year) into which mechanical progress might lead was not necessarily a happy or a good one.[22]

The views of Ewing and the *Times* contributed to the marked sense of the insufficiency of science that pervaded the British Association's meeting that year.[23] Not only the physical sciences and engineering, but also the social sciences, came under attack. Those present at a discussion on economics were startled when one august lay member rose and scathingly exclaimed that he was tired of economics and of politics. "There are millions," he cried, "of unemployed and of people suffering privation. If this is what economics and politics can do, I say damn them."[24] He wanted, he said, less economics and more humanity.

Yet it was the engineers who suffered the severest criticisms; and doubtless it was for this reason that they were first to react in a positive way. That year's president of the engineering section, Miles Walker, Professor of Electrical Engineering at the University of Manchester, delivered an address to his fellow engineers entitled "The Call to the Engineer to Manage the World."[25] The engineer and the scientist, he said, had shown the way to material plenty. All of the necessities and comforts of life could potentially be produced and distributed in sufficient quantities to make the inhabitants of the world ten times more wealthy than they now on the average were. The vast difference between what was possible and what had been achieved, charged Walker, was in great measure due to the incompetence of national and local government officials. Contrasting the successful management of large engineering businesses with "the way in which things were muddled in the world at large," he confidently argued that if engineers and scientists were to take a greater part in the management of the world's affairs, they would make a greater success of it.

Walker naively proposed that the government should found

an experimental self-supporting colony under the auspices of engineers, scientists, and economists. The objective would be to ascertain to what extent it was possible, with contemporary knowledge and the best manufacturing methods, for a group of about one hundred thousand persons to maintain themselves and to increase their wealth continually when freed from the restraints and social errors of modern civilization. To the credit of his colleagues, this unrealistic proposal found little support.

His provocative remarks nevertheless sparked vigorous discussion both of the proposal and of means by which scientists might participate in government. Several scientific leaders declared that scientists should abandon their traditional noninvolvement in affairs of state. Sir Richard Gregory, for example, explained that scientists had created "the machine" but had left others to manage it, and that these latter had failed. However, if the government were to call upon scientists, as it had during World War I, and set them the task of "managing the machine," he was certain that they could successfully do so.[26] The government, however, was currently under no pressure to take such a step, and so did not.

The outcome of the ferment at the British Association meeting was the adoption of a more realistic resolution by the Engineering Section, which declared

> That this meeting of the British Association feels that the present position of Great Britain calls for far closer cooperation between the scientific community and the Government.
>
> Further, it suggests, as a possible means to this end, that the Government should invite the leading scientific institutions and societies to appoint representatives to co-operate with it to formulate plans for dealing with the present pressing problems facing the country.[27]

The resolution, of course, had yet to receive the support of the entire Association, which meant that it had to be adopted first by the committee of recommendations and then by the general committee. Some advocates set immediately to work canvassing the support of the members of both.[28]

The general committee included, as its name implies, a wide representation of Association members running into several hundreds.[29] One of its functions at each annual meeting was to formally appoint the committee of recommendations and subsequently to consider its report. This committee included the president and vice-president of the Association, the general secretaries, the general treasurer, past presidents of the Association, and current presidents of the various sections.[30]

As a member of the committee of recommendations considering the engineers' resolution, Walker insisted that the Association face its responsibilities. Ewing, however, was influential in arguing that were the Association to forward the resolution to the government it would be exceeding its rights. And so, some said, the engineers' resolution died in the committee of recommendations largely through Ewing's opposition.[31] In addition, although this may also have been the basis of Ewing's position, there was apparently a widespread feeling, both on the committee and within the Association, that passing such a resolution would involve the Association in national politics, which was considered most undesirable.[32] Thus, the engineers' reasonable resolution lost by a large majority, with only the economics and education sections supporting the engineers.[33]

In this action, the British Association maintained its practice, which dated from the last quarter of the nineteenth century and which had led some of its members to form the British Science Guild, of not becoming involved in national affairs relating to science. Throughout the period considered here, this unwritten policy would be adhered to closely. Other British scientific organizations, including the British Science Guild, the Association of Scientific Workers, and, at a later time, the Royal Society, placed no such restriction upon themselves.

Among those for whom political involvement on the part of scientists and the British Association was not undesirable was Major Archibald Church. In addition to being a Labour member of Parliament and secretary of the Association of Scientific Workers, Church was a member of the British Association's general committee. Sir Richard Gregory's biographer regarded Church and Gregory as "the *avant-garde* of the most formidable revolutionaries of our days, for they were trying to establish the right of the scientist to make his voice heard in political affairs."[34] In late 1930, they had both given evidence on behalf of the Association of Scientific Workers before a Royal Commission examining the Civil Service, and had proposed that the government appoint a Minister of Science.

In Church's view, the British Association had missed a great opportunity to initiate an attack on what he considered was surely the greatest scientific problem of all. It could have given "expression to the feeling of the vast body of science—that science has a duty to humanity, and a responsibility for seeing that the instruments it creates are not abused or used to the detriment of society."[35]

That the Association had not done so was to Gregory's *Nature* not so surprising for a reason unconnected with the avoidance of national political involvement. This was pointed out by Gregory's chief leader writer, Rainald Brightman, an employee of Imperial Chemical Industries.[36] Science was now so specialized, he said, that academic scientists were intent upon their specialties and quite unconcerned with the activities and problems of science in general. Yet Brightman thought it at least encouraging that, with the passage of the resolution in the engineering section, the responsibility of scientists had been publicly admitted before such a representative gathering.[37]

However, not so the *Times*, which maintained that the proper concern of the scientist was science. It alleged that the British Association was always subject to the temptation to prefer width to depth, that is, as it explained, topics that attracted many to topics that few could grasp, and it hoped that the Association would maintain and extend its influence by attracting the makers rather than the camp followers of science. The paper suggested that Walker's remarks had "gone to the heads of some enthusiasts," and it misrepresented the situation in suggesting that there was a desire for a scientific fascism to run the state.[38] The resolution that had come from the entire engineering section was much more moderate than its president's original proposal, in which some might have seen scientific fascism. But the teachings of technocracy, which were shortly to have a brief and limited appeal in the United States, found little support in Britain.[39]

To the left of the political spectrum, the *Daily Herald*—through its energetic, outspoken, and influential science correspondent, Ritchie Calder—complained that not a single great social issue had been faced by the British Association.[40] However, from his knowledge of unofficial activities at the Association's meeting, Calder was able to report that a strong movement had been organized to ensure that, at the following year's meeting, each section, instead of giving itself over to "academic considerations and to frivolous papers," would organize discussions on "the things that matter."[41]

Fully aware of this movement, Brightman pointed out in a *Nature* editorial a week later that the British Association's council, which between annual meetings replaced the general committee as the executive body of the Association, could itself take independent action on the engineers' resolution if it wished.[42] The council did not choose to take this particular course, but it did resolve in early November "that the President be authorized and requested to call a conference comprising the available Past Presidents of the Association, the President-elect, the General Officers, and other persons in his discretion" to consider the general questions raised by the recent meeting and to take such action as they might consider desirable.[43] No public announcement, however, was made of this decision.

That changes were under way nevertheless became apparent in January 1933 when Ewing introduced his successor as president of the British Association, Sir Frederick Gowland Hopkins, the Cambridge biochemist who had shared the 1929 Nobel prize in physiology and medicine. It now seemed to Ewing not unlikely, and probably desirable, that at future meetings of the Association members should make a more conscious effort to relate their studies to social problems.[44] Science was now playing so large a part in human life, both for good and for evil, he said, that scientists could not logically stand aloof. It had created new powers and grave new dangers of which the general public was scarcely yet aware. Clearly, concluded Ewing, it was the duty of scientists to point these out.

These changed views foreshadowed the "extraordinary evolution of opinion" among scientists of all ranks that occurred between the British Association's meeting of September 1932 and that of September 1933, as noted by Ritchie Calder.[45] During the period, informal meetings of scientific leaders were held to discuss ways in which science might assist mankind; as a result the leading members of the Association now, according to Calder, fully realized the gravity of their responsibilities. To

Calder this change was nothing short of miraculous, and he attributed it in large part to the influence of the new president, Hopkins, who was also president of the Royal Society and, as such, chief scientific adviser to the government. Hopkins doubtless had the support of Gregory, also a member of the council, which went to the 1933 annual meeting with a plan. It knew the general direction it wanted the Association to take—toward social involvement of individual members through their work, without, however, national political involvement for the Association as a whole—but it wanted first to gauge, and perhaps mold, the Association's mood.

The meeting began in a positive fashion with Hopkins' presidential address showing none of the pessimism of his predecessor's. Hopkins wanted to see machines continue to replace human labor; and although it was probable that this would lead to a restructuring of society, there need be, he said, no revolution if there was planning.[46] In this planning the scientist would have to play his part. His vision of the future might be very limited, but in regard to material progress and its possible consequences the scientist had at least better data for prophecy than others. These data could form a reservoir of synthesized and clarified knowledge on which statesmen and politicians could draw. In particular, suggested Hopkins, when civilization was in danger and society in transition might not a "House," with functions similar to Solomon's House envisioned in Francis Bacon's *New Atlantis*, be recruited from the best scientific and nonscientific intellects in the country. Such a body would avoid politics and devote itself to appraising the progress of knowledge and to synthesizing existing knowledge. It would also concern itself with the bearing of knowledge on social change.

Clearly, Hopkins' outlook was quite different from Ewing's of the previous year, for these suggestions were in the spirit of the resolution that had been adopted by the engineers in 1932 only to be opposed by Ewing. Yet in one respect Hopkins did not go as far as the engineers. He did not call upon the Association to appeal to the government to set up his proposed house, nor did he ask the Association to create this house by itself. Perhaps this was because he anticipated that his general ideas would strike many as unrealistic.[47]

In this he was correct. In by now predictable fashion, the

Times observed that distinguished scientists who had sat in Parliament had seldom played a prominent part in its affairs. The paper recalled Bertrand Russell's "pleasant" gibe that, apart from their specialized knowledge, such men suffer from the common load of prejudices. It claimed, moreover, that anyone with experience of scientific committees could testify that the more distinguished the individual members, the more difficult it was to find agreement![48]

The *Manchester Guardian*, to the contrary, was most sympathetic, and it urged that Hopkins' suggestion not be dropped—although, however, it would be. There was a need, said the *Guardian*, for an organization of the type Hopkins had proposed, because its potential place in British life was not adequately filled by any existing body, including the Royal Society and British Association.[49] Too often, the paper explained, the "scientist" had casually unloaded his "inventions" on a world in which, because it was nobody's business to attend to them, they had taken long to adapt, often painfully and expensively, to the uses of the community.

Meanwhile, in an invited evening discourse at the British Association's meeting, another member of the council, Sir Josiah Stamp—chairman of the London, Midland, and Scottish Railway, Director of the Bank of England, and a future president of the Association—also tried to spur scientists to action in social matters. Evening discourses had a deliberately wide appeal, and on this occasion, having been asked to address himself to social problems, Stamp spoke on the pertinent subject, "Must Science ruin Economic Progress?" Economic life, he believed, would have to pay a heavy price for the gains of science unless all classes became economically and socially minded, and unless there was much planning of social affairs and international cooperation.[50] He took care to note that this did not mean government by scientific technique, technocracy, or any other transferred technique. He thought that scientists might contribute much by cooperating with social scientists in attacking the intrinsic problems of social science, and by giving a greater proportion of brilliant minds to its study. However, for a long time to come, sociology would receive little support from the British Association. Only in 1959 was a section created for it.

A third member of the council, Gregory, saw in the addresses of Hopkins and Stamp, and doubtless they shared his view, a "fundamental challenge to the whole system of British politics, economics, and education."[51] If it was a challenge it was a weak one, and British politics, economics, and education continued unchanged by it. But the supposed challenge having been given, it was now time for the council to attempt to accomplish what it could in its own bailiwick. Arrangements were made to place a resolution before the committee of recommendations—and, it was hoped, the general committee—requesting "the Council to consider by what means the Association, within the framework of its constitution, [might] assist towards a better adjustment between the advances of science and social progress, with a view to further discussion at the [1934] meeting."[52] This reflected the view that scientific advance was responsible for the social disruptions of the depression.

In the eyes of Ritchie Calder, who enthusiastically assessed the situation prior to the committees' consideration of the resolution, the British Association was approaching one of the most important decisions in the entire history of science. It would decide whether, and in what way, science would fulfill its responsibility to society, and how science could help master the chaos that discovery and invention had created.[53] Although this was a grossly exaggerated view of the situation, it nevertheless served to emphasize the new concerns of scientists.

The council's tactics proved to be sound, and the resolution was adopted by both committees. To the exuberant Calder, the Association had taken the most important step in its history. For the first time, scientists would give a lead to the world in attacking social chaos.[54] Hopkins was encouraged: "The feeling of this year's Association and the reaction of the General Committee have left us with no doubt that science intends to do something to help in the problems of society."[55] Precisely what that something was to be would be dictated by what the council might learn informally at the meeting about the desires of the membership. Furthermore, the council intended to translate its conclusions into action during the 1934 meeting rather than merely discussing them then as the resolution required.

The adoption of the resolution was followed by considerable discussion at the meeting about what might be attempted, and *inter alia* it was suggested that each of the Association's thirteen sections should concentrate on the immediate problems of society insofar as they concerned its branch of science.[56] This appeared to have the council's support, and as the meeting progresed it gained general approval with the modification that, in addition, any section might cooperate with another to study problems of mutual concern.[57] But as Stamp explained to Calder, the Association did not contemplate invading the domain of any other institution, and in particular would not develop a "violently political" program.[58] The Association's leaders were still determined to keep it out of governmental affairs and possible controversy.

Calder was nevertheless encouraged by the new developments, and from the editorial page of the *Daily Herald* he appealed to the general membership to support its leaders.[59] He explained how in the past, that although humanity had been greatly concerned about science, scientists had nevertheless told him that they had no concern with humanity. Now, however, the leaders of science whom he had met at the British Association meeting no longer believed that scientists should isolate themselves from general affairs. They now admitted that science had a "social conscience." They accepted a moral responsibility for science and agreed that something must be done to help humanity. Calder believed that in a vague, uneasy way most members of the Association agreed with them. He had learned that many of the members' misgivings were due to the fear that they would trespass into the domains of finance and politics. In regard to the former, the specific fear was that the results of their investigations might not fit into the existing economic system. To Calder, however, science, which was devoted to the pursuit of truth, should not be afraid of that. As for politics, Calder did not remind them of what Stamp had said, for he was opposed to it. Instead, he asked if they were to stand by and see politicians exploit or retard science either through ignorance or for the benefit of a particular group rather than for humanity. In the British Association, he stressed, they had a democratic movement which by its collective knowledge might help statesmen guide democracy.

The general membership had, of course, to await the further action of the council, which met on November 3 to consider the recommendations approved by the general committee. The council appointed a distinguished committee consisting of the president (Hopkins), the general officers (Stamp, F. J. M. Stratton, Professor of Astrophysics at Cambridge University, and P. G. H. Boswell), Sir Henry Dale, Director of the National Institute for Medical Research, Allan Ferguson, a university lecturer in physics, Sir Henry Fowler, Assistant to the Vice-President (Research and Development) L. M. S. Railway, J. L. Myres, Wykeham Professor of Ancient History at Oxford University, Sir John Russell, Director of the Rothamsted Experimental Station and of the Imperial Bureau of Soil Science, and Henry Tizard, Rector of Imperial College of Science and Technology, to consider and report what help could be invited from the sections in regard to the recommendation.[60] This committee reported back a month later, and as a result a memorandum was issued by the council to the organizing sectional committees responsible for arranging the sectional programs at the following year's annual meeting.[61]

The memorandum explained that the council had limited its interpretation of the phrase "adjustment between the advances of science and social progress" in the resolution, to that aspect which appeared to them to be within the purview of the Association, namely, "an understanding of the relations between the advance of Science and the life of the community." In other words, the Association might study these relations; but, in conformity with the attitude indicated by Stamp, it would have to leave to other institutions the making of any adjustments. Also, in omitting the phrase "social progress" the council avoided involving the Association in internal arguments over what might or might not be regarded as social progress. The implications of the amended resolution nevertheless appeared many, and so the collaboration of all sections of the Association was invited. The organizing committee of each section was asked to select appropriate social problems and "possibly to devote to such problems a portion of their programmes for the [1934] meeting by means of special papers, discussions, or symposia."

It should be noted that the council's amendment could, of course, be interpreted to suggest also that the life of the community, that is, the nature of the society, affected the advance of science, but that view, although pressed by certain left-wing scientists and by the Association of Scientific Workers, was suppressed within the British Association. The Association's conservative response to the challenge raised by the depression resulted in little internal change. Members were encouraged to examine the social impact of science but without the Association itself becoming politically involved.

Sir Richard Gregory's changing attitude to these developments is of interest. One week prior to the meeting of the council on November 3, he wrote in an anonymous editorial in *Nature*:[62]

> It may greatly be doubted ... whether the British Association either through its Council or Sections, is competent to undertake the proposed inquiry into the ethical and social consequences of scientific progress. The Association is only in session for a week annually, and though the Council meets several times during the intervening period, and may act independently, the Sections themselves are in being only during the annual meeting. Moreover, while the members of the Council, or of sectional committees, are familiar with their own particular fields of scientific work, few of them take an active interest in other branches of science, and fewer care to concern themselves with ethical or social standards, except as citizens in a civilized community.[63]

Gregory concluded that "how to adjust social and economic conditions to progressive scientific knowledge is not the function of the British Association or any other body concerned mainly with the promotion of such knowledge. What could undertake such work more appropriately is a council or committee having upon it leading representatives of the social, as well as the natural, sciences."[64]

Gregory was a strong supporter of the British Science Guild and chairman of its council. A month after his editorial appeared, and three weeks after the council of the British Association had deliberated, he wrote complainingly to Allan Ferguson, a member of the Association's council, that "the Association has been urged over and over again to deal with [the subject of science in relation to social progress], and that the British Science Guild was founded purely because the Association would not take it up."[65] Now, he continued, the Association apparently proposed "to do what it formerly declined to do, and without any reference whatever to the British Science Guild which grew out of its loins." The Guild had made no recent attempt to deal with the subject because there had been a suggestion that to do so would be to trespass in the field of the British Association. However, soon after the Association had decided upon its course of action, Gregory eagerly explained in *Nature* that the proposed studies of the factors affecting human welfare and economic life would be illuminating and would afford profitable guidance for the future.[66]

Although some small but vocal opposition to the Association's taking any action was expressed at the meeting of the organizing sectional committees, a majority of the committees responded favorably and gave careful consideration to the council's memorandum.[67] As a result, the 1934 meeting in Aberdeen was a somewhat different one and was widely considered to be most successful. There was certainly a larger attendance than usual.[68] Members, however, did not find a radically different program, for social aspects of science had occasionally been considered at previous meetings. They had, however, a program more heavily weighted with social concerns.

In a *Nature* editorial, Brightman observed that the program reflected a welcome assumption by scientists of their civic responsibilities, and indicated that an influential section of the scientific world was beginning to think with intelligence and imaginative insight about social affairs.[69] The attention given to relations between the advance of science and the life of the community might well, added Brightman, make the meeting one of the most notable in the British Association's history. In other quarters, however, the reduction in the number of technical papers and the attention given to the effects of science on everyday life caused concern.[70] That year's president of the British Association, Sir James Jeans, in assessing the meeting warned that: "While the Association had taken social welfare so largely into its field of vision, it was not to be forgotten that the Association had another trust, that of purely scientific research, and they would be betraying their trust if they ever lost sight of that."[71] In spite of such remarks, the *Times* could say that the program "differed in no salient respect from that of other meetings." But it indirectly indicated otherwise in sternly adding that "the scientific investigator does not know, nor is it his business to care, whether his work may be turned to good or evil purposes, whether it will create labour or create new industries, whether it will be used for war or for peace. As an investigator he is neutral and impartial, but as a citizen he may himself take a share and encourage others to do their part in the application of knowledge for the benefit of mankind."[72]

Many scientists were clearly of a different mind, for as Ritchie Calder wrote: "The success of the BA's experiment of applying science to social problems this year has aroused great enthusiasm among the scientists and has hastened the movement to bring them actively into public affairs."[73] The monthly journal *Discovery*, which aimed to interest its readers in both the sciences and the humanities, commented that the general public interest of the program could not be too strongly stressed, and that the Association had more than maintained its reputation as a public benefactor.[74] Contrary to the *Times*, *Discovery* argued that social welfare and economic recovery had to be principal concerns of scientists in the contemporary world, and the Association's program had shown that British scientists were not unmindful of this.[75] Among Association officials there was general satisfaction that the consideration of relations between scientific advance and social life had been "widely recognized and esteemed."[76]

To illustrate the program, the engineering section considered the development of inventions as a stimulus to economic recovery, and the mathematics and physics section discussed photoelectric cells, their applications, and the repercussions of such scientific and technological developments in the economic and general social life of the community. Relations between advances in geography and the life of the community were considered in the geography section; in the agriculture section, its president examined scientific progress and economic planning in relation to agriculture and rural life.

Ever since the last quarter of the nineteenth century, the British Association had shown little interest in the social relations of science, but from 1934 that changed. The economic depression of the 1930s cast its shadow over the Association's centennial meeting in 1931. By the following year's annual meeting, some Association members believed the charge that science and technology were responsible for the overproduction of goods and consequent unemployment associated with the depression. In addition, they felt a sense of helplessness because they also believed that the progress of science and technology could not be halted. Others, however, exonerated scientists and engineers and placed the blame on politicians, alleging that the latter did not understand how to properly incorporate the products of science into the life of the nation. Some members wanted the Association to suggest to the government that it invite scientific organizations to cooperate with it in combatting the effects of the depression, but the Association was determined not to involve itself in national political affairs. However, it did decide in 1933 to add to its traditional study of the natural world that of the study of the social impact of science.

Thus, within the British Association, the depression sparked a reexamination of the Association's function, which led in 1934 to a modest change in its annual program. To the world at large with which members of the Association were now concerned, this change, if noticed, would have appeared insignificant. Nevertheless, it signaled the beginning of a new era both for scientists and in the history of the Association, characterized by an awakened social consciousness.

1. *Manchester Guardian*, 24 September 1931, p. 13.

2. These views had earlier been expressed by others. See, for example, Raymond B. Fosdick, *The Old Savage in the New Civilization* (New York: Doubleday, Doran, & Co., 1929).

3. *Nature* 128 (1931): 527.

4. *Manchester Guardian*, 26 September 1931, p. 15. Within two years, an American professor of agriculture would write to Sir Daniel Hall: "Ten million acres of cotton and some thousands of tobacco have been ploughed under. The latest move is the killing of some 5 million pigs weighing under 100 lbs. and the slaughter of some 200,000 prospective mother sows. If this will bring national prosperity I have wasted my life." Quoted in Sir Daniel Hall, *The Organisation of Agriculture* (London: British Science Guild, [1933]), p. 1. Five years later, another British scientist, noting "the belief, held in many quarters, that, but for the activities of scientific researchers, there would be no world unrest in industry and politics," went on to describe problems in both agriculture and industry: "Planters have been heard to complain that new varieties have doubled their output, while new methods have halved the costs, with the result that their product is redundant, their expenditure on land laid out in its production, wasted; and the capital embarked in their enterprise, lost. The like complaint is uttered also by corporations engaged in the production of minerals. Scientific methods, replacing haphazard prospecting, have brought vast new deposits to light; while innovations in metallurgy have reduced the expenses of treatment and rendered obsolete the fixed capital sunk in excavation, buildings and plant. That there is foundation for such beliefs is idle to deny.

Again, the application of science to mass-production has enormously reduced the number of labourers employed per unit of output for almost every article of commerce, and the belief is widespread that the progress of science tends to create unemployment by substituting the machine for the man, and to discourage skill by replacing the highly-trained operative by the unskilled worker" (R. W. Western, "How Can Science Help?", *Scientific Worker* 3 [1936]: 58).

5. F. C. Dietz, *An Economic History of England* (New York: Henry Holt & Co., 1942), pp. 564, 571.

6. J. G. Crowther, *Fifty Years with Science* (London: Barrie & Jenkins, 1970), p. 83.

7. Ibid.

8. Ibid., p. 84.

9. This was not solely a British phenomenon. In the United States, F. B. Jewett told of his and his generation's awakening to the social effects of science in "Social Effects of Modern Science," *Science* 76 (1932): 23-26. Also in the United States *Living Age* initiated, with its issue of August 1933, a section on the sciences and society.

10. *Nature* 128 (1931): 761.

11. Ibid., pp. 809-11.

12. W. H. G. Armytage, *Sir Richard Gregory* (London: Macmillan & Co., 1957), pp. 32, 77.

13. *Nature* 128 (1931): 811.

14. Ibid., p. 824. Later Desch wrote: "The pamphlet was entirely without effect; I sent out nearly 500 copies, and received five replies" (J. G. Crowther, *Fifty Years with Science*, p. 236).

15. *Nature* 128 (1931): 909.

16. Ibid., pp. 1053-55.

17. C. L. Mowat, *Britain between the Wars, 1918-1940* (Chicago: University of Chicago Press, 1955), p. 472.

18. *Nature* 130 (1932): 349.

19. Ewing spoke again on these themes in a Hibbert Lecture given at the University of Cambridge in February 1933 and published as "Science and Some Social Problems," *Hibbert Journal* 31 (1933): 321-36. In regard to warfare, Ewing had been influenced by the unsettling views of Winston Churchill, which he quoted (p. 331): "It is established that henceforward whole populations will take part in war, all doing their utmost, all subjected to the fury of the enemy. . . . It is probable, very certain, that among the means which will next time be at their disposal will be agencies and processes of destruction wholesale, unlimited and perhaps, once launched, uncontrollable. Mankind has never been in this position before. Without having improved appreciably in virtue or enjoying wiser guidance, it has got into its hands for the first time the tools by which it can unfailingly accomplish its own extermination" (Churchill, *Thoughts and Adventures* [London: Butterworth, 1932], p. 247).

20. "A Narrative written by P. G. H. Boswell for his wife," (typescript, 1942-ca. 1948), p. 210, in archives of Liverpool University.

21. *Manchester Guardian*, 7 September 1932, p. 4.

22. *Times*, 1 September 1932, p. 11.

23. *Manchester Guardian*, 5 September 1932, p. 12.

24. *Daily Herald* (London), 2 September 1932, p. 1.

25. *Nature* 130 (1932): 355-56.

26. *Daily Herald*, 5 September 1932, p. 11.
27. British Association, General Committee minutes, 2 September 1932.
28. *Daily Herald*, 6 September 1932, p. 9.
29. In 1922 there were some 700 members, and the figure for 1932 would not have been all that different. The 1922 figure is given in Howarth, *The British Association for the Advancement of Science*, p. 27.
30. British Association, General Committee minutes, 2 September 1932.
31. *Daily Herald*, 7 September 1932, p. 6; 8 September 1932, p. 1.
32. *Manchester Guardian*, 7 September 1932, p. 10; J. G. Crowther, *The Social Relations of Science*, rev. ed. (London: Dufour, 1967), p. 434.
33. *Daily Herald*, 7 September 1932, p. 9.
34. Armytage, *Sir Richard Gregory*, p. 97. See also pp. 93–96, 106, 110.
35. *Daily Herald*, 7 September 1932, p. 9.
36. Gary Werskey, *The Visible College: The Collective Biography of British Scientific Socialists of the 1930s* (New York: Holt, Rinehart, and Winston, 1978), p. 30.
37. *Nature* 130 (1932): 413–15.
38. *Times*, 8 September 1932, p. 11.
39. For a short account of technocracy in the United States, see Edwin T. Layton, Jr., *The Revolt of the Engineers: Social Responsibility and the American Engineering Profession* (Cleveland: The Press of Case Western Reserve University, 1971), pp. 226 ff. In Britain, *Nature* (131 [1933]: 87) borrowed from an article in the *Times* (5 January 1933, pp. 11–12) in explaining and criticizing the "new economic doctrine" technocracy.
40. *Daily Herald*, 8 September 1932, p. 1.
41. Ibid.
42. *Nature* 130 (1932): 414.
43. British Association, Council minutes, 4 November 1932.
44. *Nature* 131 (1933): 50.
45. *Daily Herald*, 11 September 1933, p. 2.
46. *Nature* 132 (1933): 394.
47. Ibid.
48. *Times*, 7 September 1933, p. 11.
49. *Manchester Guardian*, 11 September 1933, p. 8.
50. *Nature* 132 (1933): 432.
51. Ibid., pp. 797–99.
52. *Report of the British Association for the Advancement of Science* (1933), p. xliv; *Nature* 132 (1933): 653–55.
53. *Daily Herald*, 11 September 1933, p. 2.
54. Ibid., 13 September 1933, pp. 1, 2.
55. Ibid.
56. Ibid., 11 September 1933, p. 2.
57. Ibid., 12 September 1933, p. 2; 13 September 1933, pp. 1, 2.
58. Ibid., 13 September 1933, pp. 1, 2.
59. Ibid., 15 September 1933, p. 10.
60. British Association, Council minutes, 3 November 1933.

61. Ibid., 1 December 1933.

62. Peter M. D. Collins, "The British Association for the Advancement of Science and Public Attitudes to Science, 1919-1945" (Ph.D. diss., University of Leeds, September 1978), p. 89.

63. *Nature* 132 (1933): 654.

64. Ibid., p. 655.

65. R. A. Gregory to Allan Ferguson, 27 November 1933. Ferguson Papers, Department for the History and Social Studies of Science, Sussex University.

66. *Nature* 132 (1933): 981-83.

67. "A Narrative written by P. G. H. Boswell for his wife," p. 224.

68. *Daily Herald*, 13 September 1934, p. 6. The attendance figures for the years 1931 to 1935, with the locations in parentheses, were respectively: 5,702 (London—the Centennial meeting); 2,204 (York); 2,268 (Leicester); 2,938 (Aberdeen); and 2,321 (Norwich).

69. *Nature* 134 (1934): 301.

70. *Daily Herald*, 13 September 1934, p. 6; *Manchester Guardian*, 13 September 1934, p. 12.

71. *Manchester Guardian*, 13 September 1934, p. 12.

72. *Times*, 13 September 1934, p. 11.

73. *Daily Herald*, 13 September 1934, p. 6.

74. *Discovery* 15 (1934): 298-301.

75. Ibid., p. 241.

76. British Association, Council minutes, 7 December 1934.

3

The Parliamentary Science Committee

In Britain, a scientific organization's past practices largely determined its responses to the economic depression. For over half a century, the British Association had eschewed involvement in governmental affairs and continued to do so, even though from 1934 it encouraged study of the social impact of science. The British Science Guild, on the other hand, had been concerned primarily with the national and imperial cultivation and use of science, and during the two decades following its formation in 1905 had made frequent approaches to successive governments. In 1929, the Association of Scientific Workers had taken a further step in endeavoring to establish an effective mechanism linking scientists and Parliament. The joint response of the Guild and Association of Scientific Workers to the depression was to renew the attempt to create such machinery. Realizing that some scientists were no longer content to leave decisions on matters involving science to nonscientific members of Parliament and civil servants, one parliamentary wit would remark that scientists should be on tap, but not on top. However, scientists desired neither; what they wanted was meaningful participation in parliamentary affairs involving science. How, as a second part of the first phase of the social relations of science movement, they sought to achieve it, is the subject of this chapter. It deals with the formation in 1933 and history to 1939 of the second and much more successful Parliamentary Science Committee.

The years 1929 to 1932 were lean ones for the Association of Scientific Workers. To economize, it abandoned publication of its journal, the *Scientific Worker*, in February 1930. But in spite of this and other measures, the Association went into debt, and some eighteen months of strenuous struggle were required to put it in credit again by October 1932.[1] During these difficult times, the Association developed, and eventually published, a draft policy and related program of action. Professor B. W. Holman, of the Royal School of Mines, was a principal figure in the revitalization of the Association, which now declared two main objectives. The first was to secure a fuller recognition of the value of scientists to society and thereby a wider application of science and scientific method to industry, education, and government. The second was to develop the Association, "a professional society of qualified men and women," into a central unifying body sufficiently powerful to advance the interests of science and scientists as essential elements in the life and progress of the nation.[2] In regard particularly to the first objective, the Association declared that a new outlook was essential: science, "the handmaiden of progress," could not be kept separate from industry, administration, or social problems. So the Association joined forces with the publishers of the journal *Progress*, the British Institute of Social Service, in publishing *Progress and the Scientific Worker*, designed to give voice to the "new citizenship" that would combine the scientific and humanistic outlooks.[3]

The rejuvenated Association of Scientific Workers began a fruitful collaboration with the British Science Guild. In October 1931, following the national financial and political crisis, the Guild's council of management, chaired by Sir Richard Gregory, invited the nation's chief scientific organizations to cooperate with it in establishing "an authoritative body able to present to Parliament and to the country the collective opinion of the scientific community on matters affecting the country's interest."[4] Unwilling to have the British Association become involved in national affairs, the council of the Association decided to take no action on the request for cooperation and financial support. In contrast, the Association of Scientific Workers, attracted by the prospect of involvement in national affairs, readily cooperated with the Guild through the agency of the Scientific Advisory Council previously formed in 1929 in conjunction with the Parliamentary Science Committee. The Guild and Association decided to form what initially was referred to as a "national," and sometimes "national parliamentary," science council, representative of the country's scientific and engineering organizations.[5]

About this time H. E. Armstrong, the elderly chemist and a fellow of the Royal Society, appealed in a *Nature* editorial to the Society to take the lead in organizing the country's scientific forces.[6] Two months later, on 9 January 1932, Major Archibald Church opposed this suggestion, also in a *Nature* editorial. Church argued that the Royal Society was not constituted even to engage in the campaign required, let alone act as a focus for it. "Having been consulted by the central government for many years past, and being in receipt of Government Grants, the Society [was] not in the position of independence essential for the consideration of criticism of public offices or affairs."[7] Furthermore, the fellowship of the Royal Society excluded many whose cooperation Church deemed essential for the production of a practical scheme for national and imperial reorganization and reconstruction. Church believed that the British Science Guild, because of its catholic membership, could produce a plan that would command the country's attention. With the Guild's proposed national science council in mind, he added that much more weight would be carried by a scheme produced by and published with the authority of representatives of the country's various scientific and technological societies.

The Association of Scientific Workers and the Guild worked through the spring and summer of 1932 designing a national science council. Their goal of providing an adequate channel for bringing the views of scientists before Parliament embodied the dual purposes that Church had envisioned for the earlier Parliamentary Science Committee. However, the organization was this time modeled on the Federation of British Industries in that it would be the liaison body between the country's scientific organizations and Parliament in regard to scientific and technical matters of concern.[8] By September 1932, a specific set of objectives and program had been drafted for the national science council, intended to be an independent, self-supporting body composed of representatives of all organizations subscribing to it.[9] This draft was approved by the British Science Guild's council of management and the executive committee of the Association of Scientific Workers.[10]

The approval and cooperation of Parliament would also have been necessary for the principal objectives to have been realized. These were, first: "To promote the spirit and methods of science as essential elements in public and political life, and to participate actively in deliberate planning of the affairs of the nation and of industry,"[11] and second: "to formulate and assist in carrying out a definite programme of legislation dealing with those objects which the (National Science) Council regards as important." The remaining three objectives were less ambitious and more realistic. The council, as the Parliamentary Science Committee and the Parliamentary Committees before it, would supply members of Parliament with information on scientific matters of current legislative interest. It would also urge all political parties to increase the number of parliamentary and local government candidates having training in, and knowledge of, science and its applications. Finally, the council would supply relevant scientific information to its constituent associations.

Before further action was taken, however, the council of management of the Guild resolved in October 1932 to cease its cooperation with the Association of Scientific Workers and to proceed independently with the formation of the national science council.[12] This decision was amicably accepted by the Association which offered its cooperation to the Guild in any future action in which it might be invited to participate.[13]

Although the reason for the Guild's action is unclear, it may have been related to recent developments within the British Association. It was at that year's meeting of the Association that the engineering section had resolved "that the Government should invite the leading scientific institutions and societies to appoint representatives to co-operate with it to formulate plans for dealing with the present pressing problems facing the country."[14] Disappointed that the resolution did not receive the support of the Association, Gregory had declared: "If the BA is prepared to divorce itself from public affairs and social problems some other body must intervene, and the British Science Guild is going to do it."[15]

At the meeting of the Association of Scientific Workers that was informed of the Guild's decision, it was explained that Gregory's main concern in attempting to form a national science council was the application of science to everyday life. "With the Council pursuing this policy," said the chairman of the Association's executive committee, Captain Hume, "the outlook would be safe, it would leave the [Association] with the parliamentary and professional interests of scientists to look after."[16] It would appear then that the science council would, as the Guild, but with more authority because of its wider representation, attempt to promote science for the benefit of the nation—while on the other hand, the Association, as a professional body, would concern itself primarily with the welfare of scientists. Both organizations would be involved with Parliament, but for different reasons. It was therefore suggested at this same meeting that the Association re-form its Parliamentary Committee that had been allowed to lapse during the joint planning of the national science council.[17]

In November 1932, the Parliamentary Committee recommended that for the purpose of maintaining day-to-day scrutiny of parliamentary proceedings directly or indirectly affecting scientists, the practice of employing a parliamentary secretary to the Association be resumed. The appointment of a secretary, the Committee argued, would further the Association's objective of establishing contact between scientists and Parliament and make possible the collection of material on which desired legislation could be based. One week later Henry W. J. Stone became Parliamentary Secretary to the Association.[18] Although not a scientist, Stone had been associated with parliamentary affairs for twenty-six years. For most of that time, he had been political private secretary to a member of Parliament who was for many years deputy-chief government whip. Stone's years in Downing Street had made him familiar with the various government departments. In addition, he had had considerable experience as an organizer, public speaker, and journalist. It was hoped that the Association would benefit from his extensive experience.[19]

The British Science Guild, it will be recalled, had also had a Parliamentary Committee, but it, too, had lapsed while the national science council was being planned. It was, however, resuscitated in June 1933 by which time further action on the council was in abeyance.[20] That this latter was so, and permanently so, may again be explained by developments within the British Association, in particular by the remarkable change that had occurred in the outlook of the Association's leadership over the winter of 1932–33.[21] The Association's council, of which Gregory was a member, had become convinced that the Association should become concerned with the social relations of science. This was also to have been the general area of concern of the national science council. Yet, as the Association would avoid becoming involved in parliamentary affairs, there was still a role for parliamentary committees.

In re-forming its Parliamentary Committee, the Guild held to the idea of a united effort on the part of scientific organizations and appealed to other bodies for support. Within the Association of Scientific Workers, the appeal was considered by the Parliamentary Committee which formulated a resolution later accepted by the executive committee. This was to the effect that the Association should inform the Guild that it was prepared to contribute ten guineas for the purposes of its Parliamentary Committee. If, however, the Guild was willing "to re-constitute the joint parliamentary committee" (presumably the joint committee on the national science council) and agree to the employment of a part-time secretary solely for the work of the committee, then the Association would give greater financial support.[22] This resolution was favorably received by the Guild, and the matter was referred to the honorary secretaries of both bodies to draft a constitution for the proposed joint committee.[23]

Many meetings followed. These discussed not only the formation of a parliamentary science committee—either jointly between the Guild and Association or on behalf of a number of subscribing organizations—but also the natures of the Guild and Association.[24] In the belief that there were already too many parliamentary committees representing isolated bodies, the

Association's executive committee argued that it would be all to the good if these independent bodies were consolidated to form a parliamentary science committee representative of science and technology as a whole. In the event of that occurring, they were prepared to disband their own Parliamentary Committee and to give general and financial support to the new committee.[25] These suggestions were accepted by the Guild and a second, quite different Parliamentary Science Committee was launched in October 1933.

Shortly before the formation of the new Committee, Sir Frederick Gowland Hopkins observed in his presidential address to the British Association that though statesmen might have wisdom adequate for the immediate and urgent problems with which they dealt, there should nevertheless be a reservoir of synthesized and clarified knowledge on which they could draw.[26] Soon after the creation of the Committee, Hopkins spoke again on the same theme. He believed that scientific opinion in Britain should be so organized that when it was expressed on matters within its competence it should be with such authority that no government or legislative house could ignore it. A cabinet minister had recently wittily restated a common view in asserting that scientific opinion should be "on tap" but not "on top." In Hopkins' view, the trouble was that statesmen had never known how to turn on the tap or even to find it. So he stressed the desirability of having a widely representative and full-time chamber of science. This body would be consulted by the government, yet would be independent of it. It would criticize governmental policy affecting science and on appropriate occasions make its own suggestions. However, as to the actual formation of the body Hopkins said no more than that it could be formed by a judicious combination among existing organizations.[27]

This was generally what the Guild and Association of Scientific Workers were actively attempting to achieve with the new Parliamentary Science Committee, which when formed had the support of ten organizations in addition to the Guild and Association.[28] The Committee's constitution provided for a general committee representative of all the affiliated organizations and a small executive committee.[29] The Guild and Association each elected two members to the general committee; the other organizations each elected one member only. The dominant position of the Guild and Association is further seen in the stipulation that the chairman and vice-chairman were to be appointed by the Guild and Association respectively.[30] In addition, the Guild and the Association each supplied one of the two joint secretaries and each two of the four remaining members of the executive committee. Initially, the chairman of the Parliamentary Science Committee was Commander L. C. Bernacchi, physicist, explorer, and member of the council of management of the Guild; the vice-chairman was Holman, honorary general secretary of the Association; and the joint parliamentary secretaries were Stone of the Association and Albert Howard, honorary secretary of the Guild. On the executive committee, the Guild was further represented by Miles Walker, Professor of Electrical Engineering at Manchester, and Sir James B. Henderson, also an engineer; and the Association by V. H. Blackman, professor of plant physiology at Imperial College, and R. W. Western.[31] In November 1934, Bernacchi resigned the chairmanship and was replaced by Lt. Col. Sir Arnold Wilson, M. P., a member of the council of management of the Guild.[32] An energetic and influential chairman, Wilson remained in office until 1938. Through his association with members of both Houses of Parliament, and following the earlier practice of the Association's and Guild's Parliamentary Committees of approaching only members with a genuine interest in scientific matters, contact was made with Parliament. The number of members of Parliament associated with the Parliamentary Science Committee was always small, with the maximum number of three peers and fourteen members of the Commons being reached in 1938.[33] The Earl of Dudley became president of the Committee in 1936. Ramsay MacDonald, the former Prime Minister, accepted an invitation to join the executive committee in 1937 just a month before his death.[34]

As was to be expected, *Nature* gave its support to the Parliamentary Science Committee. Within two months of its formation, Henry W. J. Stone wrote in an editorial that, a start having been made in creating an effective link between science and Parliament, it remained only to emphasize the paramount importance of science presenting a consolidated front. A united Parliamentary Science Committee would perforce be treated with respect by politicians. It was imperative, therefore, that as many scientific organizations as possible join the Committee in order "to give it the necessary strength to make it the spearhead of science as a whole, and a worthy co-partner with our rulers at Westminster."[35] But support came slowly, and by April 1934 only two additional organizations had joined the Committee.[36] During this period, and continuing through April 1935, the Committee was, as already indicated, dominated by the Guild and Association of Scientific Workers and also housed at the Guild's headquarters. In these circumstances, certain institutions felt that it was not "consonant with their dignity" to belong to the Committee. Consequently, "with great good sense" the two parent bodies, having been assured that through their action the Committee would gain adequate additional support, estabished it as an independent federal body.[37] This occurred in April 1935, following which there was a notable increase in the number of supporting organizations.[38] By the end of the year, twenty-three "leading institutions and societies" with a combined membership of over 100,000 had joined the Committee.[39] This was one less organization than the maximum number of twenty-four reported in 1938.[40]

Each affiliated organization elected two representatives to the new Committee. Officers were elected at the annual meeting held in November each year—the only occasion when the entire Committee convened. An executive committee composed of the officers and six other elected members conducted the Committee's affairs at other times.[41] As Wilson, Holman, and Stone continued together in office through 1937, the influence of the Guild and Association of Scientific Workers remained strong within the Committee.[42] A Parliamentary Committee was made up of those members of Parliament who were engineers or scientists by training or profession and of those who, like Dudley and Wilson, were interested in scientific matters.[43]

The Parliamentary Science Committee confined itself initially in 1933 to exploring only those subjects already under preparation by the constituent bodies for presentation to Parliament. At the same time, it began to issue to its member organizations copies of "Science in Parliament," an indexed summary of all parliamentary proceedings affecting science collated from Hansard from the opening of the 1932–33 session.[44] This work had been begun by Stone when he became parliamentary secretary to the Association of Scientific Workers in late 1932 and was continued by him in his capacity as parliamentary secretary to the Parliamentary Science Committee.[45]

Within a few months of its formation, the specific aims and aspirations of the Committee had crystallized.[46] The Committee would promote discussions in both Houses of Parliament on scientific matters related to economic policy and national well-being, arrange periodic addresses by scientists to the chief parliamentary committees and groups, consider bills before Parliament that involved the application of scientific method, and urge the proper representation of science on government departmental and other public committees.

The asking of questions in Parliament became a frequent activity of the Committee. Many of these questions were framed by Stone and posed by members of Parliament who were actively connected with the Committee in an elective or honorary capacity.[47] For example, during the 1934-35 parliamentary session questions were asked concerning agricultural and horticultural research, the interchangeability of water supplies, the International Locust Conference, milk pasteurization, the production of aeroplane engines, the gas grid system, the possibility of constructing earthquake-proof buildings in India, the research capabilities of the agricultural marketing boards, technical education, and grants for industrial and agricultural research.[48] Questions continued to be asked throughout the life of the Committee.[49]

On several occasions the Committee became involved, through its members, in parliamentary business initiated by others—for example, in debates on the Herring Industry Bill and the Metropolitan Water Board General Powers Bill in 1934-35.[50] In the debate on the Agriculture Bill in 1937, amendments were tabled by parliamentary members of the Committee's executive committee to ensure that the chief veterinary officer should not be subordinate to an administrative civil servant and that he should have direct access to the Minister. On the latter point, the Committee was successful, and on the former a lack of friction was guaranteed.[51]

Outside Parliament, but within government, the Committee undertook various investigations such as how the Patent Office library could be brought up to date so as to serve the interests and convenience of those members of the public who used the library. In this case a report was completed, and a deputation from the Committee, led by Wilson and including other members of Parliament who were also members of the Committee, saw the parliamentary secretary of the Board of Trade on the matter.[52]

The promotion of science and scientists in the affairs of the nation was urged by the Committee in letters written to influential government officials. In May 1934, it wrote to the Prime Minister, Ramsay MacDonald, asking him "in the national interests, to use [his] powerful influence to secure that all Royal Commissions, Committees, and Departmental Committees appointed to deal with matters in which scientists and technicians would obviously have to be consulted, should include a representation of scientists and technicians on the personnel of the Commissions or Committees, rather than to be content that a non-scientific personnel should hear evidence from experts."[53] The letter went on to draw attention to two recent instances where this recommended practice had not been followed and one where it might yet be implemented. The first of the former instances was the appointment early in 1933 of the Post Office Advisory Committee. No scientist or technician had been appointed, and yet the Committee would probably have to advise on many scientific and technical questions, including telephony and telegraphy. The Parliamentary Science Committee urged that a current vacancy on the Advisory Committee be filled by someone with suitable scientific qualifications. The second case in which the practice had not been followed was the appointment of the Sea-Fish Commission in late 1933. Although the commission would have to investigate the catching, landing, storage, and treatment of fish, none of the four commissioners appointed had any scientific or technical knowledge. This telling observation was not, however, followed by a call for immediate change but by the timid suggestion that any future vacancy on the commission be filled by a pisciculturist. The Committee urged that the practice it was recommending could conveniently be implemented in connection with the government's decision to create a committee to consider the development of television and to advise on the conditions under which any public television service should be provided. "The public advantage of the proposed Committee including amongst its personnel an adequate representation of scientists and technicians is so obvious as to require no argument."[54]

In the House of Commons, Alan E. L. Chorlton, M. P., an industrialist who was later to be Deputy Chairman of the Committee, asked about the general practice recommended in the letter. The Lord President of the Council, Stanley Baldwin, replied on behalf of the Prime Minister that the objective in selecting personnel "is always to obtain the best possible guidance, and, while in view of the variety of circumstances it would be inadmissible to lay down any general rule—the importance of including scientists and technicians in appropriate cases is and will be fully borne in mind."[55] A reply was subsequently received from the Prime Minister's office to the specific items raised in the letter. It said that the Post Office Advisory Committee was not intended to be a technical body, that it was not possible to have technical representation of the different aspects of the industry and trade on the Sea-Fish Commission, and that the Television Development Commission, now appointed, did include expert representation.[56]

Other letters also brought some satisfaction. It had long been a grievance with scientific and technical officers in the civil service that their opportunities of rising to the higher administrative posts were negligible. In the Committee's view, this situation severely hindered the application of scientific methods to public administration.[57] Accordingly, at an opportune moment early in 1935, it wrote to the Treasury suggesting that, in the process of selecting assistant principals within the civil service, the claims of scientific and technical officers "*possessing the necessary administrative ability*" should not be overlooked. The Committee had in mind such departments as the Patent Office, the Ministry of Agriculture, and others employing such officers.[58] This correspondence resulted in the Treasury recognizing the principle urged by the Committee.[59]

In these various ways then, the Committee pursued its goal of forging a link between science and Parliament for the benefit of the nation and the advancement of scientists in national affairs. Yet the Committee could not claim any great achievements. The attention of scientists had been turned in a new direction, but with little or no effect upon the nation, for during this time neither Parliament not government had been moved to make any significant innovations in introducing science or scientists into the nation's affairs. Nevertheless, scientists were convinced that they were proceeding in the right direction, and they did not expect too much too soon. The Committee's chairman, Wilson, told the general committee at the House of Commons in 1935: "It will take time to evolve a suitable mechanism and a live organisation, but, if sufficient support is forthcoming and the membership widened to cover science as a whole, there is no reason why we should not be of real use and value to the nation; for it is in Parliament, and nowhere else, that the balance between science and ethics has to be settled, day by day, in terms of statutes and regulations."[60]

By this time, the Committee was already engaged in what was to be its most ambitious project, namely, a report embodying novel recommendations on the finance of British scientific research. This had its origins in the exhaustion in 1932 of what was known as the Million Fund, administered by Britain's Department of Scientific and Industrial Research (DSIR).[61] The opening passage of the white paper presented to Parliament in July 1915 and recommending the creation of the DSIR read: "It is impossible to contemplate without considerable apprehension the situation which will arise at the end of the war unless our scientific resources have previously been enlarged and organized to meet it. It appears incontrovertible that if we are to advance or even maintain our industrial position we must as a nation aim at such a development of scientific and industrial research as will place us in a position to expand and strengthen our industries and to compete successfully with the most highly organized of our rivals."[62] These views were still current in 1917 when, under the sponsorship of the DSIR, the formation of research associations was facilitated by a grant-in-aid of one million pounds sterling voted by Parliament (the Million Fund). The associations were industrial organizations formed by firms in a given industry, or by firms in a group of related industries, for the purpose of conducting cooperative research. They maintained laboratories in which this research was conducted mainly on problems of immediate interest to the industries they represented, but also on more fundamental matters with a view to the future development of the industries.[63]

By 1933, at which time there were twenty-four research associations, their valuable work and its contribution to the national welfare were widely recognized.[64] This work had been financed in part by the Million Fund and in part by contributions from the industries themselves, largely on a pound for pound basis. Down to March 1932, about one million pounds had come from the Fund and about £1,350,000 from industry. Throughout this period the policy of the DSIR had been to encourage the associations to become self-supporting; but the associations were still dependent on government support when the Million Fund became exhausted. This occurred at the depth of the depression when industry certainly could not double its contribution to make up the loss of government support. As one consequence, many scientists in the associations found themselves in 1933 actually under, or likely to receive, notice of dismissal.[65]

In that year the DSIR could only afford to make special grants out of its own funds in urgent cases. But its funds had been increased by £70,000 for the following year in order that it might assist the research associations. Welcome as this was, many firms making contributions to their research associations nevertheless feared that the DSIR would become entirely responsible for the associations. The firms were afraid that they would then be unable to appoint the staffs of the associations and organize their own researches. Worse still, they feared that the research associations would quickly lose touch with industry.[66] Thus the important question—for the nation, industry, and science alike—of the satisfactory financing of industrial research in Britain was once again raised.

The situation of the research associations was reviewed in *Progress and the Scientific Worker* in June 1933 by B. W. Holman, honorary secretary of the Association of Scientific Workers. During the following month, the Association's executive committee, on Holman's proposal, resolved to invite the British Science Guild to cooperate with the Association in establishing a joint committee to consider the best means of stabilizing the finance of the associations.[67] The invitation was accepted, and subsequently Holman and Howard composed an extensive memorandum on the finance of industrial research.[68] Copies were circulated to the research associations, and a conference of their directors was called in London. Fifty-one persons attended but gave the memorandum only poor support. It was therefore decided that the Association and Guild should concentrate their attention on the finance of the DSIR.[69]

A new joint committee of the Association and Guild had been created for this purpose by October 1933, the month in which the Parliamentary Science Committee was formed. Sir Richard Gregory and Sir Richard Redmayne, an authority on mining and president of the Association in 1930, shared the chairmanship, Howard was secretary, and Holman was one of the members.[70] By December, the committee was actively engaged in collating information on the finance of industrial research with the purpose of formulating a satisfactory future policy. It was intended that a memorandum outlining the policy would be forwarded to the newly-formed Parliamentary Science Committee which would, it was hoped, persuade Parliament and the electorate of the wisdom of the policy.[71]

The joint committee, however, held only one formal meeting—in January 1934. This was of "such a discouraging character" that the matter lay dormant until February of the following year when at the annual council meeting of the Association of Scientific Workers it was decided that the undertaking be referred to the Parliamentary Science Committee together with all relevant documents. This was agreed to by the British Science Guild, and the material that had been gathered reached the Committee in June 1935.

The task, however, had effectively to be restarted, as during the intervening time the situation had changed in such a way as to make much of the material useless.[72] The principal change involved a new system of financing industrial research recently implemented. This was the datum line system under which the DSIR paid a research association pound for pound on all expenditures above a certain fixed datum considered necessary for the upkeep of the association on a minimum basis and up to a limit originally fixed at twice, but subsequently raised to three times, the datum figure. This scheme was to prove effective in checking the decline of the associations and enabling them from

1935 to register a distinct increase in their incomes.[73] Other developments, discussed in the next chapter, also had an important bearing on the Parliamentary Science Committee's study of the finance of industrial research, and consideration of it is best left to the end of that chapter.

The Parliamentary Science Committee was the first new organization to be spawned by the social relations of science movement. As an independent body, it was a unique creation. The British Association had had an active parliamentary committee in the mid-nineteenth century, and the British Science Guild had had a similar committee in its first two decades, although it too had fallen into inactivity, following the Guild's change of purpose in the early 1920s, only to be revived in 1930. The Association of Scientific Workers had from 1929 tried to fashion an effective parliamentary science committee and now in cooperation with the Guild had finally succeeded. The Parliamentary Science Committee became an independent body with some twenty affiliated scientific and technical organizations. The appearance of a new body devoted exclusively to parliamentary and governmental affairs concerning science was a principal development in the early phase of the social relations of science movement. However, as with the new social consciousness within the British Association, its appearance was more to be remarked than its early accomplishments. But its modest successes encouraged its leaders, who anticipated future achievements in an area deliberately avoided by the British Association, which did not, incidentally, become affiliated with the Parliamentary Science Committee until 1936.

1. *Progress and the Scientific Worker* 1 (1932): 95.
2. Ibid., pp. 54–55.
3. *Nature* 130 (1932): 339.
4. British Association, Council minutes, 6 November 1931; British Science Guild Council of Management, *Annual Report* (1931–32), p. 4.
5. *Progress and the Scientific Worker* 1 (1932): 56.
6. *Nature* 128 (1931): 1053–55.
7. *Nature* 129 (1932): 38.

8. Ibid., p. 824. On the Federation, see S. E. Finer, "The Federation of British Industries," *Political Studies* 4 (1956): 61–84.
9. *Progress and the Scientific Worker* 1 (1932): 96.
10. Ibid., 2 (1933–34): 151.
11. Ibid., 1 (1932): 55.
12. Ibid., p. 95.
13. Association of Scientific Workers, Executive Committee minutes, 23 November 1932.
14. British Association, General Committee minutes, 2 September 1932.
15. *Daily Herald*, 9 September 1932, p. 3.
16. *Progress and the Scientific Worker* 1 (1932): 96.
17. Ibid.
18. Association of Scientific Workers, Executive Committee minutes, 23 November 1932.
19. *Progress and the Scientific Worker* 1 (1933): 20.
20. Ibid., p. 151.
21. See above, p. 37.
22. Association of Scientific Workers, Executive Committee minutes, 4 July 1933.
23. *Progress and the Scientific Worker* 2 (1933–34): 217–18.
24. Ibid., p. 240.
25. Ibid., pp. 267–68.
26. *Nature* 132 (1933): 394.
27. Ritchie Calder, *The Birth of the Future* (London: Arthur Barker, 1934), pp. xiii–xiv.
28. The ten were: Joint Council of Qualified Opticians, Pharmaceutical Society of Great Britain, Institute of Physics, Royal Institute of British Architects, Society of Engineers, Institution of Professional Civil Servants, South Eastern Union of Scientific Societies, Institution of Naval Architects, Oil and Color Chemists' Association, Institute of Metals (*Progress and the Scientific Worker* 2 [1933–34]: 240).
29. British Science Guild Council of Management, *Annual Report* (1933–34), p. 3.
30. *Progress and the Scientific Worker* 2 (1933–34): 240.
31. Ibid., p. 307.
32. British Science Guild Council of Management, *Annual Report* (1934–35), p. 6.
33. S. A. Walkland, "Science and Parliament: The Origins and Influence of the Parliamentary and Scientific Committee," *Parliamentary Affairs* 17 (1964): 308–20, 389–402, 311.
34. *Nature* 140 (1937): 721. After MacDonald resigned as Prime Minister and First Lord of the Treasury in 1935, he became Lord President of the Council and as such was responsible to Parliament for the reports of the Department of Scientific and Industrial Research, in the work of which he took particular interest. He had been elected a Fellow of the Royal Society in 1930 (*Nature* 140 [1937]: 839).
35. *Nature* 132 (1933): 982.
36. *Progress and the Scientific Worker* 2 (1933–34): 307; British Science Guild Council of Management, *Annual Report* (1933–34), p. 3. The two bodies were the Institution of Mechanical Engineers and the Institution of Structural Engineers.
37. *Progress and the Scientific Worker* 3 (1935): 78.

38. *Scientific Worker* 9 (1935-37): 172.
39. Ibid., p. 20; *Progress and the Scientific Worker* 3 (1935): 78.
40. *Scientific Worker* 10 (1938): 21.
41. Ibid.
42. Ibid., 9 (1935-37): 130. The Guild amalgamated with the British Association in 1936, so after that Wilson could represent only the ideals of the Guild. In late 1936, following the amalgamation, the British Association finally became a constituent member of the Parliamentary Science Committee (*Nature* 138 [1936]: 1005).
43. *Scientific Worker* 9 (1935-37): 65.
44. *Nature* 132 (1933): 668.
45. *Progress and the Scientific Worker* 2 (1933-34): 308.
46. *Nature* 132 (1933): 981-82.
47. *Scientific Worker* 9 (1935-37): 20.
48. *Nature* 136 (1935): 945.
49. *Scientific Worker* 9 (1935-37): 94-95, 173.
50. *Nature* 136 (1935): 945.
51. *Scientific Worker* 9 (1935-37): 173.
52. *Nature* 139 (1937): 750.
53. *Progress and the Scientific Worker* 2 (1933-34): 332.
54. Ibid., p. 333.
55. Ibid.
56. Ibid.
57. *Progress and the Scientific Worker* 3 (1935): 80.
58. Ibid., p. 45.
59. Ibid., p. 80.
60. *Nature* 135 (1935): 837.
61. Sir Harry Melville, *The Department of Scientific and Industrial Research* (London: Allen & Unwin, 1962), pp. 23-25.
62. Ibid., p. 23.
63. B. W. Holman, "Finance of Research Associations," *Progress and the Scientific Worker* 2 (1933-34): 171.
64. These research associations, with their dates of formation in parentheses, were: British Scientific Instrument Research Association (1918); Wool Industries Research Association (1918); British Boot, Shoe and Allied Trades Research Association (1919); British Cotton Industry Research Association (1919); British Iron Manufacturers' Research Association (1919); Linen Industry Research Association (1919); Research Association of British Rubber Manufacturers (1920); British Association for Research of Cocoa, Chocolate, Sugar Confectionery and Jam Trades (1919); British Non-Ferrous Metals Research Association (1920); British Refractories Research Association (1920); Scottish Shale Oil Scientific and Industrial Research Association (1920); British Launderers' Research Association (1920); British Leather Manufacturers' Research Association (1920); British Cutlery Research Association (1921); British Electrical and Allied Industries Research Association (1920); British Silk Research Association (1921); British Cast Iron Research Association (1921); Research Association of British Flour Millers (1923); British Colliery Owners' Research Association (1924); British Food Manufacturers' Research Association (1925); Research Association of British Paint,

Colour and Varnish Manufacturers (1926); Industrial Research Council of the National Federation of Iron and Steel Manufacturers (1929); Printing Industry Research Association (1930); Research and Standardisation Committee of the Institution of Automobile Engineers (1931) (ibid., p. 172).

65. Ibid., pp. 171, 174.
66. Ibid., p. 174.
67. Association of Scientific Workers, Executive Committee minutes, 4 July 1933.
68. Ibid., 20 October 1933; *Scientific Worker* 10 (1938): 110.
69. Association of Scientific Workers, Executive Committee minutes, 20 October 1933; *Progress and the Scientific Worker* 2 (1933–34): 240.
70. *Progress and the Scientific Worker* 2 (1933–34): 240.
71. *Nature* 132 (1933): 982.
72. *Scientific Worker* 9 (1935–37): p. 173; *Nature* 138 (1936): 714.
73. The Parliamentary Science Committee, "Memorandum on the Finance of Research" (1937), *Scientific Worker* 10 (1938): 117.

4

Left-wing Scientists

Although marking new departures for the organizations involved, the developments described in the preceding two chapters pleased but did not satisfy certain radical younger scientists who espoused a Marxist view of science and who from 1932 became increasingly visible within the social relations of science movement already set in motion by others.[1] Some of the most influential of these left-wing scientists would become fellows of the Royal Society, most would participate in the new socially-oriented activities of the British Association, but the principal organization through which they would express themselves was the Association of Scientific Workers. The present chapter, concerning the third and final aspect of the first phase of the social relations of science movement, deals with the spread of a Marxist view of science among British scientists, taking particular note of the views of the leading left-wing scientist, J. D. Bernal; the entry of the vocal left-wing scientists into the Association of Scientific Workers and, simultaneously, the social relations of science movement; and, finally, Bernal's work for the Parliamentary Science Committee regarding the financing of British scientific research.

With the stirrings within the British Association, British Science Guild, and Association of Scientific Workers, 1931 marked a watershed in British scientific life. An additional reason for regarding it as a turning point is that from that time a novel philosophy and historiography of science began to influence certain British scientists. From 29 June to 4 July 1931, the Second International Congress of the History of Science and Technology was held in London. Being attended as it was by a Soviet delegation of "executives and scientists," it afforded British scientists an opportunity of furthering their acquaintance with the theory and practice of Soviet science.[2] In writing about the Congress, J. D. Bernal, the thirty-year-old Cambridge crystallographer and Communist, explained how he had been impressed during the first session by the wholly different attitude displayed by the Soviets toward the history of science: "The history of science was plainly vitally important to them; it was not only an academic study but a guide to action. They proceeded integrally with the social aspect dominant, in the past as in the present."[3]

In subsequent sessions, the Soviets argued that the growth of pure science was dependent on the development of "economics and technics both for the problems they present it with and for the means provided for their experimental study."[4] This argument was supported by, among others, Boris Hessen's famous and influential study of the social and economic roots of Newton's *Principia*.[5] The Soviets also stressed the close relationship that existed between science and industry in the planned economy of the Soviet Union, and Bernal was persuaded that both science and industry profited "not only industry from science through the rapid solutions of problems and the suggestions of new processes, but science gains from industry by the vastly greater funds at its disposal, by its more coherent organisation, by the possibility of experiments on large-scale factory lines, but most of all by the inspiration from the problems of actual practice and by the intellectual co-operation of the workers."[6] To Bernal, the most convincing illustration of this was the contemporary work of the Soviet botanist N. I. Vavilov, who through field expeditions and experiments had established, with the help of two thousand assistants, the seven chief distribution centers of the world's cultivated plants. Vavilov's work would also be of great assistance to a rational agriculture.

Although Bernal himself had been impressed by the Soviet papers, he admitted that the Soviets' influence at the Congress was limited. This was because their appeal to the dialectic, to the writings of Marx and Engels, instead of impressing the audience, had disposed it not to listen, its attitude appearing to be that anything so ungentlemanly and doctrinaire had best be politely ignored. Bernal was one of the few who had reacted differently, and he informed his readers that to continue to ignore the Soviet view would be "for our own sakes" a great mistake. He believed that the more intelligent of "bourgeois scientists" realized the apalling contemporary inefficiency of non-Soviet science, tied as it was to academic and impoverished universities, secretive and competitive industries, and national governments. They could look to the Soviet Union where, in contrast, 850 linked research institutes and 40,000 research workers were said to be contributing to a rapidly growing and relatively efficient "mechanized" science.[7] Bernal concluded that he and other British scientists had to face two insistent questions. First, were their individualistic methods in science not obsolete and effectively doomed, and thus not worth saving? Next, was "it better to be intellectually free but socially totally ineffective or to become a component part of a system where knowledge and action are joined for one social purpose?"[8]

Eight years later, Bernal explained to Beatrice Webb, co-author with her husband, Sidney, of *Soviet Communism—A New Civilization?* (1936), the profound effect which the Congress had had on him and two other leading left-wing scientists:

> I can say that the inspiration for my own work and that of many others in science, notably [J. B. S.] Haldane, [Professor of Biometry, University College, London], and [Lancelot] Hogben, [Regius Professor of Natural History, University of Aberdeen], can be traced definitely to the visit of the Marxist scientists to the History of Science Congress in 1931.[9] We did not understand all that they said, in fact I now suspect that they did not understand it entirely themselves, but we did recognize that here was something new and with immense possibilities in thought, and that, as it were, the whole range of our understanding could be multiplied by working out the suggestions they offered.[10]

This is what Bernal proceeded to do, in the conviction that "the approach of Marxism seems to be as important a reformation as that of the birth of science in the sixteenth century, and that the older views need fighting as much as did those of the old Aristotelians and Churchmen even though they have like their predecessors much good to be said for them. But the new must drive out the old and the process is bound to be unsettling."[11]

Later in July 1931, Bernal was one of a party of British scientists and physicians who visited the Soviet Union for three weeks and returned to find themselves in the midst of Britain's financial crisis. Also in the party was Julian Huxley, the biologist and writer, who afterwards described his experiences and impressions in a book. Huxley observed that the Soviet Union's First Five Year Plan was "a symptom of a new spirit, the spirit of science introduced into politics and industry."[12] The Plan, he said, heralded the birth of a new kind of society, one coherently planned, adding that "proper planning is itself the application of scientific method to human affairs; and also it demands for pure science a very large and special position in society."[13]

Huxley noted, as Bernal had earlier, that in the Soviet view science had a dual aspect. On the one hand, it was seen as influencing human life and the destiny of society, and on the other it was regarded as "a social phenomenon, its tendencies and its achievements not blossoming out of nowhere in the brains of isolated men of genius, but determined by the social and economic environment of the time."[14] This conception of science had led the Soviets to hold recent conferences on the planning of science, and also on the planning of planning.[15] Thus, in contrast with Britain, the Soviet Union had already progressed beyond the actuality of large-scale planning to the realization that planning itself had principles to be studied and a technique to be improved. Also, in spite of the average economic level being so much lower than that in Britain, a far larger proportion of the national income was being assigned to science than was the case in Britain even during times of prosperity.[16]

Huxley had visited the Scientific Planning Department of the Supreme Economic Council, which he judged to be roughly equivalent to Britain's Department of Scientific and Industrial Research except that the Soviet Department dealt with all of industry and with the great bulk of pure scientific research in physics, chemistry, and engineering. The head of the department, Nikolai Bukharin, had explained how Marxists believed that in the long run all science is of practical value and that it should be made to serve practical needs as thoroughly and as quickly as possible. The Soviets' general goal was to apply science on a larger scale than any other country—their agriculture, industry, mining, the health of their people, their whole national life, all would be based upon science. Achieving this would require more pure science, and therefore they were preparing, as fast as resources would permit, to increase expenditures on pure research to a degree far beyond that attempted in any capitalist country. Already, in Huxley's view, the Soviets were in the first rank in most branches of pure science, were rapidly making fundamental discoveries, and were forcing the rest of the world to think.[17]

> The technique and the very idea of large-scale planning: the socialization of agriculture: the reduction of private profits and class distinctions: the provision of peace-time incentives which shall on the one hand not be merely individualist, and on the other not be centered mainly on the crude worship of national power: the elevation of science and scientific method to its [sic] proper place in affairs—in these and many other ways the new Russia, even at its present embryo stage of development, is in advance of other countries: and if the rest of the world refuses to learn from the object lesson provided by Russia, as well as profiting by her mistakes, so much the worse for the rest of the world.[18]

With Britain's continuing depression on the one hand and the rise of fascism in Germany and Italy with its threat of war on the other, the Soviet example continued to find admirers among left-wing scientists. Of these, Bernal was the most ardent. By late 1933, he had become concerned with "the immediate and compelling dangers of war and fascism" and considered possible courses of action.[19] The only solution he could see to the problems created by fascism was the radical one of the USSR.

> The logic of orderly production for welfare has there shown its value in practice even when weighted against a series of apparently insuperable disadvantages: the ruin of war, the interventions, the lack of capital, the lack of administrators and technicians, the

> backwardness of industry, the inert and conservative peasantry, the suspicious hostility of other powers, the permanent threat of invasion. The U.S.S.R. is alone among the nations moving on. [It] provides the antithesis to the whole catalogue of cultural reaction represented by fascism. In science, in education, in religion, in the family, in the prisons, the U.S.S.R. gives practical embodiment to the progressive ideas of the nineteenth and twentieth centuries. The communists are the heirs and the only defenders of the liberal tradition.[20]

Bernal believed that the achievement of the USSR was the best practical embodiment of the ideals for which scientists work and that British scientists ought to strive for a similar achievement. The task would necessarily be a long and difficult one, but the first step would be the acquisition of effective knowledge. Without an understanding of the working of economic or political factors in the contemporary world, goodwill, idealism, and ability could easily be diverted into fantastic currency schemes or even fascism itself. Scientists would also have to organize themselves effectively. As individuals they were weak and easily intimidated, but once organized their key positions in all modern states would make them a powerful force.[21] Furthermore, participation by the scientist in working-class movements would remove his isolation and provide a firm base of political support. Meanwhile, through implacable opposition to war and fascism in all forms, particularly in his own country, the scientist could show that he really believed in his tradition and was prepared to fight and suffer for it. Here was a call to political action that far exceeded anything heard from the leadership of the British Association, British Science Guild, or Association of Scientific Workers.

By 1934, Bernal was convinced that dialectical materialism was the most powerful factor in the thought and action of his day. He said so in an essay chiefly concerned with showing, first, how dialectical method could be derived from that "co-operative struggle with the material world" which is called science, and second, what dialectical method had done and could do to aid the development of science and its application to human welfare.[22] Hessen's "brilliant" study of Newton's *Principia* is here described as a model of the dialectical method. The Marxian analysis, by explaining simultaneously both the flourishing and the decay of science, is said to show in general what must be done in order to preserve and extend the scope of science.

> It shows that modern science has itself, by its own development, destroyed the possibility of the old methods of scientific discovery by individual, inspired, haphazard work. Science now depends on large-scale, international organisations of laboratories, these almost on the scale of factories, and it implies with that a corresponding economic and political organisation to support these laboratories, to utilize their results, and to suggest the type of problems for them to attack; for science does not necessarily or even usually produce most of its own problems. These conditions are not to be found in a rapidly disintegrating capitalistic society harassed by economics and preoccupied by preparations for war. The future of science lies in the communist state.[23]

Bernal was one of several contributors to a volume of essays published early in 1935 under the title *The Frustration of Science*—a familiar phrase in contemporary scientific circles. The general theme of the essays, including Bernal's entitled "Science and Industry," concerned the misuse and nonuse of science.[24] Bernal gave a critical assessment of the organization in Britain of physical research and its relations with industry. His overall conclusion was that "If science is to help humanity, it must find a new master."[25] The new master, of course, would be socialism—successor to the capitalism that he indicted in general and criticized in particular in the British context. So long as science was used merely to increase profits, its full potentialities could not be glimpsed, he said, let alone developed.[26] "The best application of science is conceived of as producing such a fatuous and stultifying paradise as Huxley's *Brave New World*; at worst, a superefficient machine for mutual destruction with men living underground and only coming up in gas masks."[27] Bernal's source of hope was his Marxist view that science had not been developed in the past for the purposes of human welfare, but rather to increase profits and to secure military superiority. This could all be changed, and the first requirement was the adequate financing of science. He suggested that a tenfold increase in the grant made in Britain to industrial research would still leave the cost of science an insignificant part of general production costs. In government and industry, which together furnished most of the research money, Bernal saw interests that both limited the amount of money and hindered and distorted the application of science.[28] Consequently, he regarded the entire national system as disastrous for science.

Bernal had become so extreme in his views that his prescriptions for science in Britain were beginning to appear unrealistic even to some who shared his concerns about science and society. In the foreword to *The Frustration of Science*, Frederick Soddy, Lee's Professor of Inorganic and Physical Chemistry at Oxford, described Bernal's suggestion that science should find a new master as naive.[29] About the same time, Hyman Levy, professor of mathematics at Imperial College of Science and Technology and like Bernal a Communist, in replying in *Nature* to H. G. Wells, an anti-Marxist, admitted that revolution in Britain was out of the question.[30]

Perhaps it was as a consequence of his Soviet visit, as well as of his moderate political views and scientific eminence, that Julian Huxley was invited by the British Broadcasting Corporation to make a survey of the organization of science in Britain in its relations to social needs. Various circumstances of the day would naturally have suggested this as a topic of general interest both within and without scientific circles. Huxley accepted the invitation and arranged a series of lectures and discussions given first on the air, then published in the *Listener*, and finally issued as a book.[31]

The broadcasts opened and closed with discussions between Huxley and Levy. The first broached the questions that Huxley would explore during the course of visits made throughout Britain. One of the most important involved the nature of science itself. Levy began by espousing the Marxist view; Huxley, his Soviet experiences notwithstanding, responded by defending the traditional view. According to this, science is a body of knowledge based on scientific methods. Part of that knowledge, pure science, has a momentum of its own and goes on growing irrespective of its applications. It is done merely to satisfy the curiosity of the scientists who carry it out; that is, it is done for its own sake, as an end in itself. Further discussion led Huxley and Levy to adopt a compromise definition. Science is a body of knowledge that has been tested experimentally. Historically, it has grown as a result of several factors that affect man both as an individual and as a member of society. These are man's need to exercise some control over the forces of nature; his impulses to manipulate, to be curious, and to understand his place in the universe; and the pleasure he derives from using his faculties in the processes of observing, understanding, and changing nature.[32]

When the two met again after Huxley's various visits, Huxley declared that his investigations had confirmed what he now incorrectly recalled as having been the compromise definition, namely, "that science is a particular method for getting knowledge of and control over nature, and that the form and direction it takes are largely determined by the social and economic needs of the place and period."[33] Indeed, in keeping with this changed perception the chief moral of his investigations was "that science is not the disembodied sort of activity that some people would make out, engaged on the abstract task of pursuing universal truths, but a social function intimately linked up with human history and human destiny. And the sooner scientists as a body realize this and organize their activities on that basis, the better both for science and society."[34] Huxley's new view of science had also been shaped by a second discussant, the physicist and Fabian socialist P. M. S. Blackett.[35] The subject of their broadcast was primarily pure science, and Huxley was more than ever convinced that any line between pure and applied science, even if it could sometimes be drawn, was merely arbitrary. Huxley had found that the bulk of research in Britain was organized from the production end, that is, planned with a view to improving efficiency in technical processes and to reducing costs to the producer or state. He believed that there ought to be much more research organized from the consumption end, that is, research directed towards the needs of the individual. A second major finding was that British research was entirely lop-sided—displaying a great bulge on the side of industry and the physical and chemical sciences that assist industry, being distinctly undeveloped on the biological and health side, and being quite embryonic in the psychological and human sciences. For example, there were actually more research chemists in one of the several laboratories of Imperial Chemical Industries than there were research workers in psychology in the entire country.[36]

The view of British science presented by Huxley's book,

showing as it did the organization, finance, and the relative strengths and weaknesses of the various branches of science, was novel and to many, stunning. A review in *Nature* described the book as being of outstanding interest, fascination, and usefulness.[37] More important, its ideas were echoed and supported by Rainald Brightman in an editorial in *Nature* the following week: "If science is to fulfil its function in the modern State, we must, in fact, regard it as a social activity and not as something apart from the rest of human life and interest. Not only is a sharp distinction between pure and applied science no longer possible, but also the scientific movement as a whole requires scientific study, and its activities must be planned as much as any other social or industrial activity, if the maximum results are to be obtained and its resources wisely exploited."[38] Brightman thought that it should be possible to survey Britain's scientific resources and then make authoritative recommendations for their reorientation and for an attack on neglected problems of outstanding importance. He apparently thought that all of this could be accomplished in a satisfactory manner without a radical transformation of the political system and suggested the recently formed Parliamentary Science Committee as an appropriate body to undertake the task.

Throughout the thirties, the Marxist view of science was widely disseminated in a minor flood of speeches, articles and books by left-wing authors. J. G. Crowther, science correspondent of the *Manchester Guardian*, was the most prolific writer of books. Two large, popular works by Lancelot Hogben, *Mathematics for the Million* (1936) and *Science for the Citizen* (1938), enjoyed an enormous circulation. The former was a best-seller for one week; two impressions of it were printed in the month following its publication.[39] For more rigorous minds, Hyman Levy's *A Philosophy of a Modern Man* (1938) and J. B. S. Haldane's *The Marxist Philosophy and the Sciences* (1939) elucidated the complexities of Marxist theory.

Among scientific organizations, the Association of Scientific Workers began to espouse the Marxist view of science from 1932, as radical scientists, Bernal in particular, joined its ranks. During 1931–32, as described in chapter two, the Association had become a viable organization once again, in large part due to the efforts of its general secretary, B. W. Holman. In his search for new members, Holman went to Cambridge in 1932 where he recruited Bernal, W. A. Wooster, a physicist, and N. W. Pirie, a demonstrator in the Biochemical Laboratory, who became members of the steering committee of the Cambridge branch, soon to be the most lively branch in the Association.[40] Bernal's attendance at its meetings was rare, "but when he took part he did inspire activity."[41] His election in 1934 to the Association's national executive committee further increased his influence.

Bernal had numerous other involvements. He was one of a group of British academics who, disturbed by instances in various countries of interference with academic freedom, formed the Provisional Committee on Academic Freedom in October 1934. The other six founders included Blackett, Levy, and C. H. Waddington, Lecturer and Embryologist, Strangeways Research Laboratory, Cambridge. The committee's successor, the Academic Freedom Committee, whose members included Sir Frederick Gowland Hopkins and Joseph Needham in addition to Bernal, Blackett, and Levy, arranged a conference on academic freedom and allied subjects in Oxford in August 1935. The Committee considered the time opportune to review "the position of school and university teachers and of scientific research workers, not in their professional capacities, but as citizens upon whom devolved special responsibilities." The subject of greatest concern, the utilization of science, was considered in the third and final session of the conference chaired by Julian Huxley, a former president and now a vice-president of the Association of Scientific Workers. Holman and Levy read papers. The conference's organizers were determined to see that something practical was done to further the utilization of science for the general good.[42] In summing up the proceedings, Bernal spoke on this theme:

> It might be thought that the Royal Society should take on this function. That was certainly the intention of the original founders of the Society.... but it has shown, in the last 230 years, no inclination to take a positive leading part in the direction of scientific research or its application to human needs. Nor has the British Association for the Advancement of Science, founded a hundred years ago with a similar purpose, been much more effective in this direction.... Recently it has taken up the question of science

and human welfare and provided some interesting discussions on the subject, but it shows no inclination as yet to take positive action.[43]

A number of smaller organizations—including the Association of Scientific Workers, the Engineers' Study Group on Economics, Political and Economic Planning, the New Fabian Research Bureau, and the Committee against Malnutrition—had, however, independently studied the organization of scientific research and its applications to human welfare. In the view of the conference organizers, the immediate need was to coordinate and extend these efforts. So with the intention of drawing up "an authoritative report and bringing the result of this work to public notice," they formed a committee of representatives of the organizations.

The committee, however, produced no report. Instead, the energies displayed at the conference were soon channeled into the Association of Scientific Workers. Under Bernal's leadership, the Cambridge branch drafted a revised policy for the Association, which, after being approved by the national executive committee was unanimously accepted by the council in October 1935.[44] (At this time, W. A. Wooster succeeded Holman as general secretary.) The main objectives of the new policy, "to promote the interests of the Scientific Worker and to secure the wider application of science and scientific method for the welfare of society," were essentially those of the 1932 policy.[45] In regard to the second objective, the *Scientific Worker* explained:

> The specific contribution that organised scientific workers can make to social welfare lies, in the first place, in what they can do to put their own house in order. We are forced to recognize that at the present time scientific research and teaching are inadequately financed and insufficiently organised and applied. They suffer, in addition, from an initial disadvantage in that the recruiting for science is gravely restricted by deficiencies in the educational system. It is clear that these aspects are closely linked. Unless science is able to show concretely the services it can give to society, it will not be able to demand, or expect to get adequate financial support: unless it receives adequate financial support it will not be effectively organised. But one way out of this impasse is to make quite clear what the present defects are, and what steps would be necessary to deal with them, and this is a function that can be carried out by organised scientific workers and by them alone. . . . Unless scientists

themselves are able to indicate what is the nature of the most effective mode of organisation, that organisation will never come into being.[46]

On 22 May 1936, the Association of Scientific Workers sponsored a public meeting at University College, London, on the utilization of science.[47] Bernal, Blackett, Needham, and representatives of various research and technical organizations participated in the discussion. In connection with the question of whether science was beneficial or destructive, the 1934–35 expenditures on civil and military research were contrasted. The Medical Research Council had received £165,000 and the Department of Scientific and Industrial Research £440,000; the research departments of the Army, Navy, and Air Force obtained £2,750,000. The main conclusions of Sir John Boyd Orr's *Food, Health and Income* (1936), that in one way or another half the British population was badly nourished and that from twenty to twenty-five percent of British children lived on a wholly inadequate diet, were presented as strong incentives for scientists to take action. The *Scientific Worker* claimed that only the organized "effort of all scientific workers" could correct these inequities.[48]

How the new Association of Scientific Workers viewed itself in relation to the "senior" scientific societies was revealed by R. W. Western in a recruiting article. In contrast to the Association, the senior societies were said to follow a tradition by which "(a) men of science are left to look after their own economic interests individually; and (b) no advice is offered to the powers that be unless it is asked for."[49] "I know," wrote Western,

> the atmosphere of the Council of the British Association as well as if I had sat on the Board. It is composed of men who have made their reputations and are secure of their emoluments, who are there to receive homage, deference from the younger men, admiration from the masses, and respectful recognition from the powers that be. They receive municipal functionaries with a ritual of condescension modelled on feudal precedent; but expect to be side-tracked in any political endeavour, because they always have been. The greater part have learnt not to meddle in matters they do not understand. Moreover, the most active are apt to be covetous of titular distinctions.[50]

In contrast, Western saw the political path, in spite of its being strewn with "crafty impediments" with which most scientists were unsuited by temperament to cope, as the only practicable route for the Association of Scientific Workers to take. In doing so, it would be entering "a new and slippery field controlled by people who treat scientists—much as scientists treat the curate—outwardly with every mark of respect, inwardly, with contemptuous indulgence."[51] However, by steadily using what voice they had for impartial exposure of "political humbug," Western believed that scientists might one day come to be treated not only with outward, but also with inward respect.

In the vanguard as always, Bernal had earlier taken the political path in assuming the study of the finance of the Department of Scientific and Industrial Research initiated by the British Science Guild and Association of Scientific Workers and handed over by them to the Parliamentary Science Committee in June 1935.[52] He "most kindly undertook, at an immense output of labour to draft a memorandum to provide the basis for the consideration of the matter anew."[53] His first draft was circulated to members of the executive committee of the Parliamentary Science Committee in March 1936. Before considering it in detail, the executive committee referred it to the councils of the constituent organizations and to the general committee requesting their views. A considerable volume of constructive and polemical criticism was received, and Bernal prepared a second draft incorporating many of the suggestions submitted. Only then, in late 1936 and early 1937, did the executive committee formally commence consideration of the memorandum.[54] After further modifications, a majority of the executive committee agreed to a final draft.[55]

In the words of the secretary of the Parliamentary Science Committee, Henry W. J. Stone, the task had been "brilliantly carried out" by Bernal and had resulted in "a masterly piece of work."[56] But that was not to say that the content reflected only, or even largely, Bernal's views. He had occupied an influential position where he could attempt to press his views; however, the first and second drafts of the memorandum were subject to the review of many who did not share his political outlook. Whatever the nature of these preliminary drafts, the third and final one displayed no traces of Marxist philosophy. The "Memorandum on the Finance of Research" or, as it came to be called, the "Bernal Memorandum," having been completed had now to be sent to the government for its consideration and action. On 29 April 1937, almost four years after the initial efforts of the British Science Guild and Association of Scientific Workers, the memorandum was forwarded by the Parliamentary Science Committee's chairman, Sir Arnold Wilson, together with a covering letter to the Lord President of the Council, Ramsay MacDonald.

The Lord President was the minister responsible to Parliament for expenditures on industrial, medical, and agricultural research. He presided over the Committee of the Privy Council whose function was "to direct, subject to such conditions as the Treasury may from time to time prescribe, the application of any sums of money provided by Parliament for the organisation and development of scientific and industrial research."[57] The other seven members of the Committee were also holders of high ministerial office. Under the Lord President, and directly responsible to him, were the Department of Scientific and Industrial Research and the Agricultural and Medical Research Councils. The Lord President and his Committee received scientific advice from an Advisory Council. The Council had no executive authority, and the Committee of the Privy Council did not have to act on its advice, yet the Committee was required to seek the Council's advice before taking any major action. "The effect of this over the years was that it was rare indeed for anything to be done without the Council's recommendation or agreement."[58] Thus, it was to be expected that the Lord President would refer the Bernal memorandum to the Advisory Council, which he did. The Council, chaired by Lord Rutherford, was composed of academic scientists and of representatives of business and industry, most of whom had an engineering background.[59]

Arguing that the amounts of money given to scientific research in Britain were both inadequate and too liable to fluctuation, the memorandum presented proposals for improving this unsatisfactory state of affairs. It also suggested how government-aided research might be extended to include all and not only a large part of industry.[60] Addressing itself to the current datum line system of financing research, the memorandum admitted that this system had in fact arrested the decline of the research associations and enabled them to register a distinct increase in their incomes. Nevertheless, the system was considered to have unsatisfactory features. The associations' incomes were subject to considerable variations because the government's contributions were directly determined by those of the industries, which would vary with "the more or less regular recurrence of slumps and booms" in the economy.[61] In the current boom there was no immediate danger, but in the depression of the early thirties the Empire Marketing Board had been abolished, the Silk and Cutlery Research Boards and Associations closed down, the activities of the Rubber Research Association suspended, and those of the other associations diminished. Thus, the potentially rapidly fluctuating amounts of money available for research in industry and the related lack of provision for the steady development of such research were the chief reasons for dissatisfaction with the datum line system.

The memorandum argued that in financing research the first requirement was to ensure that work begun would be completed; the second was to make sufficient provision for steady and continuous expansion of research. This led to the conclusion that research should be permanently endowed. The memorandum, therefore, proposed that a fund or trust derived both from government and industry be set up to provide a steady income sufficiently large for the basic needs of scientific research and its expansion at a "definite" rate. It further proposed that the contributions from the government should take the form of block grants for a period of from five to ten years. The amount of money that the government alone would have to provide over the ten-year period 1937–46 to adequately support industrial, medical, and agricultural research was estimated to be between thirty and fifty million pounds. This was a great increase over the earlier one million pounds spent on industrial research over the sixteen-year period 1917–32. Under the proposed scheme, the percentage of the national income spent on research would rise over the ten-year period from the current one-tenth to one-fifth of a percent. This would still be lower than the current estimated percentages of national income spent on research in the United States and the USSR, given as one-half and one percent, respectively.[62]

To control the endowment fund, the memorandum proposed the creation of an independent central authority that would be to scientific research what the British Broadcasting Corporation was to broadcasting—that is, a semi-public body independent of the government. This authority, it was suggested, might have as its nucleus "the present Development Commissioners and be strengthened by representation of the Learned Societies, the Universities, Medical, and Agricultural Associations, Scientific Workers, Industry, Labour, and the Consuming Public."[63] Such a body was needed particularly to ensure a proper distribution of money between industrial, agricultural, and medical research: "It is generally recognized at the present that the relative amount of research work done in the biological and sociological sciences is on far too small a scale to meet the serious problems presented by modern civilization."[64]

In concluding, the memorandum stated that it would have served its purpose if attention were drawn to four primary needs that had to be met if British science was to play the part it should for the benefit of the nation. These were that science should be well financed, that its finance be steady and not influenced by economic fluctuations, that science be better coordinated, and that it be introduced into all of industry, agriculture, and health. The Parliamentary Science Committee hoped that the memorandum's proposals for meeting these needs would form a useful basis for discussion and action by the government.[65]

To judge from a *Nature* editorial by J. W. Williamson on Bernal's first draft of the memorandum, the final draft was, with respect at least to the major proposals, very similar to the first—and this in spite of some sound criticism made by Williamson. In confining his attention to one cardinal point, he had asked: "On what grounds is it possible to justify the removal from all Parliamentary control and criticism of the administration of essentially national funds raised by authority of Parliament?"[66] He suggested that the Parliamentary Science Committee's proposals were not "'practical politics'." On the other hand, the executive committee of the Association of Scientific Workers considered that if the memorandum were acted upon it would make a "profound difference" to scientific research in Britain and would "constitute one of the greatest contributions which our Association has yet made to the progress of science."[67]

When the members of the DSIR Advisory Council received copies of the memorandum, which they were scheduled to consider for the first time on 26 May 1937, they found them *appended* to a severely critical six-page commentary on the memorandum prepared by the Council's secretariat. The commentary began by saying of Bernal and his colleagues: "Unfortunately they are imperfectly informed about the work of the [DSIR] and they have not based their arguments on unquestionable premises."[68] The ensuing specific criticisms were largely to shape the Advisory Council's reply to the Parliamentary Science Committee. The only positive effect that the memorandum seems to have had was to prompt the Council's chairman, the physicist and Nobel laureate Lord Rutherford, to have the Council consider whether adequate provision was being made for the encouragement and support of all types of scientific and industrial research.[69] Rutherford was particularly concerned with the difficulties with which universities and "similar" institutions were faced when they wished to develop researches in pure science demanding facilities that were costly to provide and maintain.

Three months after it had forwarded the memorandum, the Parliamentary Science Committee received from the Lord President, now Lord Halifax, a copy of the Advisory Council's generally negative report on the memorandum. The Council agreed that there was urgent need for much greater application of science by industry. They disagreed, however, with much in the memorandum because they had "had ampler opportunity of informing [themselves] at first hand about the difficulties to be overcome," and they could not accept the memorandum's conclusions because of their having been derived from what the Council regarded as incomplete or inaccurate premises.[70]

The practical objective of the DSIR, the Council asserted, was that of increasing the use of science by industry. The Department's policy had been, and was, that so long as industry showed itself willing to make effective use of science then so long would it be able to justify to the nation a program of steady development. The Council confidently anticipated that industry would become more and more research-minded, and they looked forward to the time when it would be a proved advantage to the state to find larger sums of money for industrial research. They were convinced from practical experience, and their opinion was confirmed by the advice of industrialists, that the worst way of attempting to secure increased expenditure on research by industry was to spend large sums of public money for industry's direct benefit in advance of desirable industrial support.[71]

The Advisory Council was also hostile to the proposal that the control of research money to be provided by the Exchequer no longer be the responsibility of the Lord President but of a new body that would be "autonomous" and would "undertake the regimentation of all research in the country, including University research."[72] The Council noted that the reason advanced for "replacing a Minister, *responsible to Parliament* and advised by three research Councils whose members are selected to represent knowledge and experience as distinct from interests," was the necessity of ensuring a proper distribution of money between industrial, agricultural, and medical research. But in its view, the memorandum showed no grounds on which a proposal could be submitted to Parliament "to set up, by statute or otherwise, a body such as it proposes, and to depart from the sound constitutional practice which secures Parliamentary control over public expenditure."[73]

The Parliamentary Science Committee did not feel that the Council had given adequate consideration to the memorandum and urged it in early November to reconsider the various suggestions made.[74] The Council, however, remained convinced that the means that it and its predecessors had recommended were well designed, in principle, to increase the use of science by industry in the economic and political conditions found in Britain. Furthermore, *"the provision of funds for the [DSIR] year by year in Votes maintain[ed] the interest of Parliament in an important State service."*[75] The Council was equally convinced that the proposals of the Parliamentary Science Committee were neither well calculated to attain the end the DSIR had in view nor likely to improve the coordination of state-supported research. It was the task of the Lord President, advised by the three research Councils, to secure this coordination.[76]

Disappointed, the Parliamentary Science Committee decided to take no further action for the time being. However, at the annual meeting of the council of the Association of Scientific

Workers on 26 February 1938, Bernal argued that the objective laid down in the memorandum was more important than the means by which it was to be attained.[77] He suggested that the Association make a wider examination of government research in industry, medicine, and agriculture, and then take action through the Parliamentary Science Committee. In November 1938, the Association published the memorandum and related correspondence in the *Scientific Worker* in the hope that useful discussion and action would ensue. By that time, however, the threat of war in Europe had riveted scientists' attention on the urgent question of the state's use of science in war.

A third major aspect of the first phase of the social relations of science movement was the rapid diffusion of a Marxist view of science among younger scientists who joined and quickly dominated the Association of Scientific Workers. Their forceful leader, not only at this time but throughout the course of the movement, was J. D. Bernal. Under his influence and that of other prominent socialist scientists, especially P. M. S. Blackett, a Marxist view of science inspired the activities of the Association of Scientific Workers and informed its journal, the *Scientific Worker*, down through the years of the Second World War. In 1943, for example, a pamphlet explaining how science develops, written by Blackett but published anonymously by the Association, affirmed that "while the course of the development of science is very complicated and variable in detail, *the main lines of its growth are broadly governed by urgent human needs of the time, arising out of the way in which men gain or seek to gain their living. Science advances, and within it any special branch advances, most rapidly when it is fulfilling an urgent social need.*"[78] However, it was one thing to radicalize the Association of Scientific Workers but immensely more difficult to bring about changes in the organization and financing of British science. In this regard, Bernal and the Association of Scientific Workers seized the opportunity of working through the Parliamentary Science Committee. Although their efforts were unsuccessful, the Bernal memorandum on the finance of British science represented both the most ambitious undertaking of the Parliamentary Science Committee and, at the same time, potentially the most rewarding course that Bernal could pursue in the mid-thirties in attempting to implement to some degree his ideas as to how the state should support science.

Neal Wood has written that the importance of the attendance of the Soviet delegation at the 1931 International Congress of the History of Science and Technology in London "in energizing the [social relations of science] movement cannot be overestimated."[79] But as I shall argue in the Epilogue, Wood construes the movement too narrowly, as an intellectual one involving radical scientists, and overlooks the developments involving the British Association, British Science Guild, Association of Scientific Workers, and Parliamentary Science Committee described in the preceding two chapters. The Soviet delegation provided an important inspiration for the radical scientists, but the social relations of science movement was much more than an intellectual movement among radical scientists and found its principal stimulus in the economic crisis of the day.

---

1. Radical scientists have received much attention in connection with the social relations of science movement. See the works of Neal Wood, Gary Werskey, and Robert Filner cited in the footnotes to the Introduction.

2. See *Science at the Cross Roads*, 2d ed. (London: Frank Cass & Co., 1971). This new edition of the Soviet contributions to the Congress contains an essay by Gary Werskey on the reception of the first edition in England.

Earlier sources of information on the theory and practice of Soviet science included J. G. Crowther's four articles in the *Manchester Guardian* in November 1929. These gave an account of the development of science in the USSR and of its projected development under the First Five Year Plan. According to Crowther, news of this Plan and of the planning of science and technology incorporated in it aroused increasing interest in Britain. "It was an example of socialism in action with regard to science, and it became a source of ideas for dealing with British social and scientific problems" (Crowther, *Fifty Years with Science* [London: Barrie and Jenkins, 1970], p. 84). Prior to the Congress, Crowther had also published *Science in Soviet Russia* (London: Williams & Norgate, 1930).

3. J. D. Bernal, "Science and Society," *Spectator* (July 1931), in J. D. Bernal, *The Freedom of Necessity* (London: Routledge & Kegan Paul, 1949), p. 336; Bernal is one of the five left-wing scientists (the others being J. B. S. Haldane, Lancelot Hogben, Hyman Levy, and Joseph Needham) who are the subjects of Gary Werskey's *The Visible College: The Collective Biography of British Scientific Socialists of the 1930s* (New York: Holt, Rinehart & Winston, 1978). Werskey gives a fuller account of the congress (pp. 138-47).

4. Bernal, *The Freedom of Necessity*, p. 337.

5. J. G. Crowther writes: "The most remarkable paper delivered at this Congress was that of B. Hessen.... It had never occurred to me, or to most other people, that

Newton's *Principia* had any social and economic roots. I knew from Marxist thought the principle that all science had its root in society, but I had not perceived or looked for this in any concrete and particular case, especially not in the greatest case of all: Newton's *Principia*. I had assumed in the conventional manner that it was a purely intellectual creation" (*Fifty Years With Science*, p. 79).

6. Bernal, *The Freedom of Necessity*, p. 338.

7. Ibid., p. 339.

8. Ibid.

9. Gary Werskey adds Hyman Levy, professor of mathematics at Imperial College of Science and Technology, and Joseph Needham, Sir William Dunn Reader in Biochemistry at Cambridge University: *The Visible College*, pp. 146-47.

10. J. D. Bernal to Beatrice Webb, 10 February 1939, J. D. Bernal Papers, Cambridge University Library, Box 24, Folder B 1.2.3.

11. Ibid.

12. Julian Huxley, *A Scientist among the Soviets* (New York and London: Harper & Brothers Publishers, 1932), p. 60.

13. Ibid., p. 61.

14. Ibid., p. 73.

15. Ibid., p. 74.

16. Ibid., p. 65.

17. Ibid., pp. 69-70, 78, 118.

18. Ibid., p. 131. In his book *Industry and Education in Soviet Russia* published in the same year, J. G. Crowther also declared that Britain had much to learn from the Soviets.

19. Bernal, "The Scientist and the World Today" (1933), in *The Freedom of Necessity*, pp. 339-49.

20. Ibid., pp. 347-48.

21. Ibid., pp. 348-49.

22. Bernal, "Dialectical Materialism" in *Aspects of Dialectical Materialism* (1934), reproduced in *The Freedom of Necessity*, pp. 365-87, p. 366.

23. Ibid., p. 380.

24. Bernal, "Science and Industry," in Sir Daniel Hall et al., *The Frustration of Science* (London: George Allen & Unwin, 1935), pp. 42-78.

25. Ibid., p. 78.

26. Ibid., pp. 68-69.

27. Ibid., p. 69.

28. Ibid., p. 71.

29. Hall et al., *The Frustration of Science*, p. 8.

30. *Nature* 134 (1934): p. 972.

31. Julian Huxley, *Science and Social Needs* (New York and London: Harper & Brothers Publishers, 1935).

32. Ibid., pp. 15-16, 18, 21.

33. Ibid., p. 252.

34. Ibid., p. 279. On science as a social function see also Huxley's "Science and its Relations to Social Needs," in *Scientific Progress* (New York: Macmillan Co., 1936), pp. 177-210.

35. Sir Bernard Lovell, "Patrick Maynard Stuart Blackett, Baron Blackett of Chelsea," *Biographical Memoirs of Fellows of the Royal Society* 21 (1975): 1–115, 76.

36. Huxley, *Science and Social Needs*, pp. 253, 256, 264.

37. *Nature* 134 (1934): 83.

38. Ibid., pp. 117–18.

39. *Scientific Worker* 9 (1935–37): 121. After selling over half a million copies, *Mathematics for the Million* was still in print in 1978 in the United Kingdom and United States. It was revised four times and published in at least fifteen foreign languages. There were four revisions and "about eleven" translations of *Science for the Citizen* that sold "near 200,000" copies (G. P. Wells, "Lancelot Thomas Hogben," *Biographical Memoirs of Fellows of the Royal Society* 24 [1978]: 183–221, 201).

40. Werskey, *The Visible College*, p. 235.

41. Maurice Goldsmith, *Sage: A Life of J. D. Bernal* (London: Hutchinson, 1980), p. 72.

42. *Report of the Conference on Academic Freedom, Oxford, August 1935* (Cambridge: W. Heffer & Sons, 1935), pp. 6, 5, 2, 77.

43. Ibid., p. 78.

44. Werskey, *The Visible College*, p. 235.

45. *Scientific Worker* 9 (1935–37): 4. The entire policy is given on pp. 4–7.

46. Ibid., pp. 2–3.

47. Ibid., p. 101.

48. Ibid.

49. R. W. Western, "Why the Association?", *Scientific Worker* 9 (1935–37): 138.

50. Ibid., pp. 138–39.

51. Ibid., p. 140.

52. See above, p. 65.

53. *Scientific Worker* 9 (1935–37): 173.

54. *Nature* 138 (1936): 714, 965; *Scientific Worker* 9 (1935–37): 128.

55. *Scientific Worker* 10 (1938): 110.

56. Ibid., 9 (1935–37): 173; 10 (1938): 110.

57. Sir Harry Melville, *The Department of Scientific and Industrial Research* (London: Allen & Unwin, 1962), p. 24. In the British Cabinet there were four offices— Lord President of the Council, Lord Privy Seal, Chancellor of the Duchy of Lancaster, and Paymaster-General—to which no departmental responsibilities were attached. Thus, a Prime Minister could have four persons from either House of Parliament each free to devote his time to particular problems or areas of government. As will be seen in later chapters, in addition to the Lord President, occupants of the other three offices had at various times responsibilities for specific scientific matters.

58. Ibid., pp. 26–27.

59. The academic members were Professor Alfred Fowler, Sir James Jeans, William Hobson Mills, Professor Andrew Robertson, Sir Albert Seward, Professor T. Franklin Sibly, and Professor Nevil Vincent Sidgwick; the nonacademic members were Lord Cadman, Viscount Falmouth, Sir Clement D. M. Hindley, Lord Riverdale, and Sir Harry B. Shackleton. The secretary of the Council was Sir Frank E. Smith. Rutherford died on 19 October 1937 and was succeeded by Riverdale as chariman. DSIR, *Report for the Year 1936–37* (London: H.M.S.O., 1938): 1, 6.

60. Parliamentary Science Committee, "Memorandum on the Finance of Research" (1937), *Scientific Worker* 10 (1938): 112–49; see p. 114.

61. Ibid., p. 117.

62. Ibid., pp. 120, 122, 124, 142. In an article based on the memorandum, Bernal gave a considerably lower figure for the United States, namely, three tenths of one percent of the national income: "A Policy for Scientific Research for Britain," *Nineteenth Century and After* 123 (1938): 101. The discrepancy serves as a reminder of the inadequate knowledge about science that Bernal and others were both drawing attention to and trying to change.

63. "Memorandum on the Finance of Research," p. 126.

64. Ibid.

65. Ibid., p. 128.

66. *Nature* 138 (1936): 53. For an account of *Nature*'s views on the organization of science in Britain during the period 1919–38, see: Paul Gary Werskey, "The Perennial Dilemma of Science Policy," (*Nature* 233 [1971]: 529–32).

67. Association of Scientific Workers, "Report of the Executive Committee to the Council at its half-yearly meeting on October 30, 1937," p. 1, ASW/1/2/18/19.

68. The commentary is included in the set of papers issued for the Advisory Council's meeting of 26 May 1937, DSIR 2/142.

69. Advisory Council for Scientific and Industrial Research, minutes of special meeting, 9 June 1937, DSIR 1/8.

70. "Report to the Lord President of the Council by the Advisory Council on a Memorandum by the Parliamentary Science Committee on the Finance and Development of Research" (July 1937), *Scientific Worker* 10 (1938): 145–49.

71. Ibid., pp. 146, 147.

72. Ibid., p. 148.

73. Ibid.

74. "Reply of the Parliamentary Science Committee to the Report of the Advisory Council" (November 1937), *Scientific Worker* 10 (1938): 150–51.

75. "Report to the Lord President by the Advisory Council on the further Communication dated 2nd November [1937] from the Parliamentary Science Committee" (December 1937), *Scientific Worker* 10 (1938): 152.

76. Ibid.

77. Association of Scientific Workers, Council minutes, 26 February 1938, ASW/1/2/19/5.

78. An untitled typewritten draft of the pamphlet, dated "14.9.43," is found in Blackett's Papers, Folder E. 8, at the Royal Society. Association of Scientific Workers, *The Development of Science* (London: Association of Scientific Workers, 8.11.1943), p. 3.

79. Neal Wood, *Communism and British Intellectuals* (New York: Columbia University Press, 1959), p. 124. Wood makes a similar claim on p. 137.

5

The British Association Again, and Further International Developments

The second phase of the social relations of science movement was shaped, as was the first, by an international development affecting Britain, namely, the rise of fascism in Europe. This phase is most clearly seen in the activities of the British Association which are the principal concern of these next two chapters. This chapter deals with the years 1935–37, during which the British Association's continuing concern with the social relations of science led the British Science Guild to amalgamate with it and the American Association for the Advancement of Science to initiate its own study of science's social relations. Also during these years, the persecution of scientists in Germany, German rearmament, and Italy's use of mustard gas in Ethiopia posed new questions concerning the social relations of science. Scientists of other nations, in addition to Britain and the United States, also became interested in these relations, and in 1937 the International Council of Scientific Unions created a Committee on Science and its Social Relations. All of these developments outside the British Association had their impact on activities within it.

Following the noteworthy 1934 meeting, the council of the British Association again suggested to the organizing committees of the various sections the desirability of paying attention to the social relations of science in their 1935 programs.[1] As in the previous year, the suggestion was willingly followed. Symposia were held by the agriculture section on the state control of agriculture and by the zoology section on the herring industry. Discussions were held by the economists on universities in relation to business and on future trends of scientific management in Britain, by the chemists on the chemotherapy of malaria, and by the agriculture section on sugar beet problems. Joint discussions were organized by the psychology and education sections on the place of psychology in the training and work of teachers, by the psychology and engineering sections on the applications of science in the control of road traffic, and by the economics and physiology sections on the economic aspects of diet.[2]

Despite these efforts, there was not the satisfaction that there had been with the 1934 meeting. At one session, someone had suggested that the winds at the South Pole be harnessed to produce electric power. Later, after a "wordy and futile" sectional discussion, someone else was heard to remark that the wind at the British Association should be harnessed for this purpose.[3] To Ritchie Calder, the meeting was followed by the same sense of frustration that had been felt in 1931 and 1932. The Association had generally failed to fulfill "the high hopes inspired by Sir Frederick Gowland Hopkins' leadership two years ago."[4] Calder's overall impression of the meeting was one of futility and wasted effort. There had been "no real drive, no 'scientific crusade,' no sense of coordinated effort, in spite of the evidence of good intentions."[5]

In a *Nature* editorial, Rainald Brightman also expressed misgivings about the various discussions conducted at the 1935 meeting. These, he said, in a telling criticism, "are commonly regarded as contributions to the development of social science. Granted that they represent contributions to our body of organised knowledge bearing on the problems which confront an organised society, how far do they contribute to the solutions of problems as visualized by those who are seeking to attack such problems systematically by scientific methods? Can such haphazard discussions lead to the evolution of either policy or technique permitting of the solution of social problems in anything like the way in which problems of physical science are solved?"[6] Brightman indirectly answered these questions in the negative, but went on to say charitably that scientists, having recognized their responsibility to give a lead in social science to administrators and politicians, were now attempting to integrate various branches of research. "One of the most valuable services rendered by the meetings of the British Association is its educational work in assisting that integration, and thus promoting a broad point of view in which the wiser direction of available scientific effort upon national and social problems becomes possible."[7]

There were some members at the British Association meeting who thought that even more could be done. A resolution advanced by the economics section and supported by the psychology section recommended that the Association indicate the importance it placed on the development of the social sciences by appointing a third general secretary who would be associated primarily with the social sciences. Such an action, it was argued, would inform the public that the Association attributed as much importance to the social sciences as it did to the mathematical, physical, and biological sciences. It would also have been a stimulus to the serious, and not merely amateur, study of the social relations of science. A committee was appointed by the council of the Association to consider the recommendation.[8]

On the report of this body, however, the council resolved that the appointment of a third general secretary should not be recommended to the general committee. Nevertheless, the council formulated an alternative recommendation that was put into effect with the help of the organizing sectional committees. Papers listed in the program for the annual meeting and involving the relations between science and the community would henceforth be distinguished by placing them in a separate group under an appropriate heading. The council curiously believed that this procedure would provide, without involving any violent reform of the program, the necessary evidence that the Association was discharging its function of "obtaining a more general interest for the objects of science."[9] The council further recommended that at least one discussion at each annual meeting should deal with "the application of science to social problems." That, though vague, was at least a new twist. For in 1934 and 1935, the emphasis had been on the social impact of science.

These innovations were put into practice at the hightly publicized 1936 annual meeting when to show its concern for the life of the community, the Association met in the common man's seaside resort, Blackpool. As the *Manchester Guardian* reported, the teacher had descended "to the hunts of the multitude."[10] Josiah Stamp, the only businessman or economist to have been elected to lead the Association, gave his presidential address in the Tower Ballroom. Many in his audience wore stiff shirts and white waistcoats, the traditional way of indicating membership in the Association. In a new departure, the official procession to church on Sunday was led by a Salvation Army band.[11] Prominent Association members showed that they were really ordinary individuals by indulging in such activities as eating ice-cream cones and riding rollercoasters. But others missed familiar comforts. H. G. Wells, who had come to his first meeting in a long time because of the Association's new interest in the social relations of science, found neither his hotel food nor wine to his liking; James Bryant Conant, the visiting president of Harvard University, found "the arrangements in Blackpool so uncomfortable that he left almost immediately."[12]

Those who remained found, as was now to be expected, much on the social relations of science. Discussions were held on chemistry and the community, chemistry and food science, the cultural and social values of science, agriculture and national nutrition, the psychology of mass entertainment, the reform of the national examination system, the poultry industry, traffic safety, and the strain of modern civilization. These and various addresses, such as that on the engineer and the nation, all indicated to Rainald Brightman, writing in a *Nature* editorial, the wide areas over which scientists were seeking solutions to problems of social well-being and interpreting their findings to the general community.[13] Once again, there was satisfaction with the program. The new emphasis of the 1934 meeting was becoming a permanent element of the annual programs. At the 1937 meeting in Nottingham, most sections were to have at least one lecture on the subject of the effects of scientific research on the public welfare.[14]

These developments within the British Association were watched with interest by the British Science Guild, which in 1935 formed a committee, consisting of Sir Richard Gregory, Sir Albert Howard, Commander L. C. Bernacchi, and Professor B. W. Holman, to decide the future of the Guild—many of its original members had died and it was experiencing difficulties recruiting new ones.[15] The Guild "still did good work, by arranging meetings and lectures, and popularizing science: it had substantial assets and was supported by Lady Lockyer in memory of, and in justice to, her late husband."[16] Gregory and Howard were its strongest supporters. On 23 October 1935, Holman told the executive committee of the Association of Scientific Workers, of which he was still a member, about the Guild's new committee and suggested that the Guild and Association amalgamate. The executive committee considered that impracticable, but did suggest a working arrangement between the two organizations, promising to give to it all possible assistance.[17]

The Guild's council became convinced that the particular purposes of the Guild were now being effectively achieved by the British Association and that the incorporation of the two bodies was desirable. The Guild, therefore, proposed to the Association that they amalgamate. The council of the Association arranged for a joint committee consisting of the president and treasurer of the Association, W. W. Watts, Emeritus Professor of Geology, Imperial College of Science and Technology, and P. G. H. Boswell, and Gregory and Howard from the Guild, to consider the proposal. Having received its report, the council resolved in March 1936 to recommend to the general committee at the Blackpool meeting the incorporation of the Guild with the Association.[18] Aware that the recommendation had the approval of the Guild, the general committee adopted it unanimously and with acclamation on 9 September 1936.[19] This action, which doubtless would have greatly pleased Sir Norman Lockyer, was largely due to the determined efforts of his protégé, Gregory.[20] It provided a further reminder of the changes that had occurred within the British Association.

The Guild brought its assets of some £4,355, 62 life fellows, 273 life members, and 242 subscribers to the Association.[21] One intention was that the Guild should in some sense continue to exist within the Association. A special committee of three former Guild and three Association members was charged with keeping the Association's council informed on matters concerning the application of scientific methods and results to social problems and public affairs. In addition, the committee was responsible for continuing the Guild's annual Norman Lockyer and Alexander Pedler lectures, also transferred to the Association. The purpose of each lecture had been to publicize the applications of science. The Lockyer lecture, given annually in London since 1929, was to continue to be delivered annually on some topic connected with the charge of the new committee. The Pedler lecture, given yearly in a different provincial city from 1929, was now to be given yearly before one of the corresponding societies of the British Association.

Under the influence of Gregory and others, consideration of the social relations of science continued to be a prominent feature of the British Association's annual meetings. By 1938, however, there was concern not merely with the social, but also with the international, relations of science. The trend in this new direction was already apparent at the Association's 1936 meeting. It had its origins in the alarm created in Britain by certain actions on the part of the Fascist nations, Italy and Germany.

In October 1935, Italian forces attacked Ethiopia, and by May 1936 that country was defeated and under Italian control. During the conflict, the Italians used aircraft to bomb and spray Ethiopian civilians with mustard gas.[22] The effect of these actions in Europe was to confirm a growing fear. During the summer of 1935, it had been reported that the potential for air attacks on the civilian populations of cities was perturbing many Europeans. It was generally conceded by those most competent to judge that there was no effective means of defense against air attack—in Stanley Baldwin's famous remark, the bomber would always get through—and so any country that was attacked would be forced to retaliate by attacking the cities of the aggressor. Civilian populations on both sides would, therefore, be subjected to explosive and incendiary bombs and gas attacks. It was true that the use of gas in war had been prohibited by the Geneva Gas Protocol of 1925, to which most of the world's principal powers had subscribed. However, in view of the failure to achieve general disarmament there was less confidence in the Protocol in 1935 than there had been in 1925.[23] Unfortunately, this loss of confidence came to be justified through Italy's use of mustard gas in Ethiopia. To Rainald Brightman it seemed that even the dullest individual could not fail to see what was involved in Italy's cynical disregard of international conventions in using poison gas against a nation that possessed no means either of defense or of reprisal in kind. This destruction of faith in international pledges was accentuated, he added, by the reluctance of powerful members of the League of Nations to restrain Italy. However, "the alternative to collective security resolutely enforced is so terrible that it would be rash to predict that any intelligent people will indefinitely allow their rulers to repudiate or disregard their international obligations and plunge all alike into a catastrophe from which there may be no recovery."[24] Significantly, Brightman urged scientists to consider whether there was some decisive contribution that they might make collectively to remedy this "condition of our so-called civilization." Here then was another problem whose contemplation would also assist in defining the scientist's social role.

The problem was further accentuated by Germany's actions. In March 1936, Hitler reoccupied the Rhineland in violation of both the Treaty of Versailles and the Locarno Pact. This move greatly strengthened his military power, and during the ensuing months the question of the extent of Germany's rearmament occupied a prominent place in British Parliamentary debates and in the public mind.[25]

Prior to these developments, there had been concern in British scientific circles about other occurrences within Nazi Germany. In his Huxley Memorial Lecture in November 1933, A. V. Hill condemned the violations of freedom of thought and research.[26] A year later, P. M. S. Blackett jotted down notes on the state of German science. The extent to which science was being used for war preparations was very great, many eminent scientists had left Germany, science was being used for anti-working-class activities, and scientific fact was being deliberately distorted to accord with Nazi teachings.[27] Still later, *Nature* explained how the Nuremberg legislation of September 1935, by confining Reich citizenship to "Aryans" and those who accepted the National Socialist conception of the State, had deprived Jews and "non-Aryans" of all citizens' rights including the right to hold office and the right to make a living through the practice of a profession, trade, or business. In opposing these measures, *Nature* pointed out that the struggle for the freedom of the individual irrespective of race, creed, or conviction, was of vital importance to scientists. A political faction that sought to control and mold the religious beliefs of its countrymen in order to support its own political views would not, it warned, lightly allow scientific research and the advancement of knowledge to proceed in search of truth without regard to extraneous circumstances.[28]

Shortly thereafter, an appeal over Lord Rutherford's name was made for support in forming the Society for the Protection of Science and Learning. The Society was to act as a permanent successor to the Academic Assistance Council which during the preceding three years had given assistance to German university professors displaced for political and 'racial' reasons and to refugee scholars from other countries.[29] Of all of these, 363 had been found permanent positions and 324 were being temporarily maintained as research guests. Recent developments in Germany, especially since the publication of the Nuremberg legislation, had convinced Rutherford of the continuing need for an assistance organization and of the fact that the Council could no longer regard its work as purely of a temporary emergency nature.[30]

The international developments touched Britain directly, leading her to rearm—an effect that thrust another formidable problem before British scientists, namely, should they or should they not contribute their knowledge to military use? In May 1936, a group of twenty-two Cambridge scientists wrote a discerning letter on the subject to *Nature*. It helped to clarify one facet of the question of scientific responsibility.

All scientists who wished for peace, said the letter, must strive for certain objectives such as the maintenance of the international character of science and the safeguarding of the public from scaremongering or scientifically inaccurate statements.[31] Beyond this, however, scientists were divided as to how peace was to be attained. First, there were those who thought that although war was undesirable it was still the final support of justice, and hence in the long run it might lead to peace. Such scientists would not be opposed to creating strong armaments or to conducting war research, but they would insist that at the same time the fullest support be given to the principle of collective security. Second, there were those who regarded all war as a barbarous and destructivce activity, a means that no end could justify, and who considered that aggression would be stopped by effective and whole-hearted action of a nonmilitary character by the nations of the world. They would refuse to conduct war-directed research and would try to prevent such research from being carried on in universities. They would demand that any money spent on war research should be for defensive purposes only and that the results of the research should be published. Further, they would organize themselves to resist the attempts of a war-making government in any country to coerce them into helping a war effort. Third, there were the complete pacifists, who differed as far as their practical activities were concerned in their objection to sanctions of any kind, but who would, as scientists, join in any organization to resist use being made of their services in war. Last, there were those who thought that in the long run war was inevitable under capitalism and who were working for the establishment of world socialism. Their immediate policy was to use any means to prevent the outbreak of war, and they would therefore be willing to join with those of other opinions in various practical ways.

Later that summer, sentiments common to the second and third groups were expressed also in Oxford. There the British Medical Association considered the question of chemical warfare and adopted the following resolution: "That this meeting condemns unreservedly the use of poison gas in warfare as inhuman in its results and degrading to civilization, and relies upon the Council to do everything in its power with a view to securing the co-operation of the medical profession in all countries in order to prohibit the use of poison gas."[32] During discussion of the resolution, an appeal was made to scientists in general, as well as to physicians, to protest collectively against the destruction of civilian populations by poison gas and to try to secure international cooperation in the matter. In reporting this, *Nature* urged that though politicians might consider it impracticable to do anything to prohibit the destruction of human life by indiscriminate chemical warfare, nevertheless scientists should let it be known that they dissociated themselves from such activity. The journal hoped that the lead given by the British Medical Association would be followed by the British Association at its forthcoming (1936) annual meeting.[33]

One of a number of Americans drawn to this meeting because of the Association's concern with the social relations of science was the science correspondent of the *New York Times*. He reported that, unlike the Association's meetings of the previous ten years, the 1936 one was "restless even rebellious"—"Fascism and the dread of war have caused chemists, biologists, and engineers to break the bounds of scientific reserve."[34] To the *Manchester Guardian*, the most remarkable thing about the meeting was the agreement expressed on two points: "indignation at the prostitution of scientific inquiry to war and nationalism, [and] apprehension over the spread of political systems which reduce science itself to servitude."[35] Rainald Brightman observed in *Nature* that there had never been more general alarm about the possible applications of science to war or greater willingness to endorse the view of the late Sir Alfred Ewing that man knew how to command nature without knowing how to command himself.[36] He cautioned that in the growing preparations for war it was easy for the possibilities of a higher standard of living, which science could provide, to be overlooked and for the scientist to become associated instead with the perversion of his knowledge for destructive purposes. Furthermore, warned Brightman, the intensification of preparations for self-defense had tended in Britain to strengthen the fetters on freedom of investigation and exposition that dictatorships in many countries had already placed on both industrial and academic workers.

The outstanding session at the British Association meeting was perhaps the discussion in the education section of the social and cultural values of science. At a time when science was under attack, it was only natural that scientists should emphasize such values. The principal theme of all of the speakers was liberalism—with all deploring the decline of democracy and liberal culture. One of them, Gregory, assailed the "insane mis-use" of science involved in air and gas warfare. In a moving address, Sir Daniel Hall, Director of the John Innes Horticultural Institution, spoke in support of liberty of thought. He said that if scientists did not take heed of the degradation of their discoveries in propaganda, war, and other antisocial activities, they would soon discover they were slaves, and when that happened the motive and fascination of scientific work would disappear. Yet, interestingly, he did not believe that scientific organizations such as the British Association and Royal Society should protest on behalf of scientists against the abuse of science. Rather, he thought that it would be more satisfactory to create an institute for the investigation of the influences of science on society.[37]

That something had to be done was also clear to others. In his presidential address to the chemical section, J. C. Phillip, Professor of Physical Chemistry at the Imperial College of Science, urged his fellow chemists and other scientists to throw "their weight into the scale against the tendencies which are dragging science and civilization down and debasing our heritage of intellectual and spiritual values."[38] The visiting president of the American Association for the Advancement of Science, Edward G. Conklin, Professor of Zoology at Princeton University, was eager to bring about a union of English-speaking scientists as a first step toward a "Science League of Nations" that would, he hoped, safeguard the international freedom of science, examine major social problems, and pool its discoveries for the good of humanity. Although he discussed the idea with Gregory, nothing came of it at the time.[39] Nevertheless, Conklin greatly appreciated what he had found at the British Association meeting: "When the whole foundations of science are being undermined by extreme nationalism, when scientists are being disciplined and have to click their heels to the orders of a political sergeant-major, the British Association has taken its stand for the international democracy of science."[40]

There were some scientists, however, who thought it of little value for scientists to protest against the use of science in war. But in a *Nature* editorial, Rainald Brightman reminded them of the necessity of science delivering its spiritual message regardless of whether or not it was heeded and of scientists promoting the educational work that was the essential condition for the creation of a better world order.[41] These indeed were the directions that the British Association would shortly take. Already in 1936, Brightman found that scientists were willing to study these matters intensively and prepared either to develop an organization through which to follow an effective common policy or to secure a reorientation of existing organizations that would be equally effective. He hoped that *rapprochement* between the British and American Associations for some such purposes as these might not be impossible or impracticable.

Meanwhile, the British Association finally began to consider whether it should create a special section for social science. The alternative, according to that year's president, Sir Josiah Stamp, was to increase the number of intersectional discussions on, and to encourage every section to give increasing attention to, the social relations of science.[42] In connection with other developments, which must now be considered, these various new currents of thought would in 1938 lead to significant innovations in both the Association's outlook and structure.

Concern for the social relations of science was not solely a British phenomenon.[43] American and other European scientists were likewise concerned, and at a meeting of the Interantional Council of Scientific Unions in London in April 1937, Professor J. M. Burgers introduced a resolution on the subject on behalf of the Amsterdam Royal Academy of Sciences.[44] He proposed that the International Council appoint a committee to coordinate the views expressed in recent years on the social responsibilities of scientists in regard to the dangers menacing civilization's future.[45] Such views had been voiced at meetings of, in addition to the British Association, the World Power Conference, the Society of Chemical Industry, and the Association of Scientific Workers. They had also appeared in the reports of the Carnegie Institution of Washington and the Carnegie Endowment for International Peace. With nations rearming and malaise growing in Europe, the problems associated with these views were becoming more urgent.

The Amsterdam proposal occasioned a lively discussion involving eighteen participants at the International Council's meeting. A number of the speakers declared themselves unable to vote for the proposal as it stood without consulting with their parent organizations. Consequently, a restricted field of activity for the proposed comittee was the most that could secure general support. The matter was then referred to a small committee representative of the various views expressed in the discussion. This committee subsequently reported that the appointment of a committee with the full powers proposed by the Amsterdam Royal Academy lay outside the objectives of the International Council, which excluded economic and social matters.[46] At a subsequent meeting of the Council, the Amsterdam motion was withdrawn in favor of one to appoint a committee with the following terms of reference:

> The Committee, at suitable intervals, should prepare a survey of the most important results obtained and of the directions of progress that are opening and of points of view brought forward in the physical, chemical and biological sciences with reference to:
> (1) their interconnections and the development of the scientific picture of the world in general;
> (2) the practical application of scientific results in the life of the community.[47]

The terms of reference also clearly stipulated that the work of the Committee was to be "limited strictly to scientific activity." Thus, the question of the scientists' social responsibility was to be avoided. As a British member of the Committee noted, its work was to be limited to the scientific aspect of the original proposal, and so the more political side might be taken up elsewhere.[48]

The Committee, whose full name was the Committee on Science and its Social Relations, was appointed in July 1937 with a membership of seven: two members from each of Britain and France, and one member from each of Denmark, Czechoslovakia, and the Netherlands. The Committee's president was F. J. M. Stratton, Professor of Astrophysics at Cambridge, who had been involved in the changes within the British Association during 1933-34; the vice-president, S. Chapman, Chief Professor of Mathematics at Imperial College of Science and Technology; and the secretary, Professor Burgers, who had presented the original resolution.[49]

As a development of its original terms of reference, the Committee on Science was soon to adopt a "more precise" statement of its objectives, namely, "to consider the progress, interconnections and new directions of advance in the mechanical, physical, chemical, and biological sciences, especially in order to survey, at suitable intervals, *and to promote, thought* upon the development of the scientific world picture, and upon the *social significance* of the applications of science."[50] (Italics added.) Thus, the Committee itself decided to promote thought on the social significance of the applications of science. This brought it closer in intent to the committee originally proposed by Burgers. At least now at an international level scientists would observe and promote thought upon the social effects of science, even though they would not officially concern themselves with what their responsibilities might be regarding those effects.

The Committee on Science set about collecting recently published materials relating to its objectives with the intention of preparing a report and bibliographies. The subject matter of the report was to be considered under four headings: outstanding developments and problems in scientific work, new applications of science in human society, interpretative work on the world picture as given by science, and thoughts on the social relations of science and the effects of its applications.[51] Data for the report were to be obtained from three sources: national correspondents appointed by national academies or research councils who would provide information on national aspects of the social relations of science, correspondents appointed by international organizations dealing with the various sciences who would report from the point of view of their particular sciences, and individuals having a special interest in, or knowledge of, these matters.[52] It was hoped that the report would be presented at the next meeting of the International Council in 1940 and subsequently published and widely circulated.[53]

Meanwhile, international cooperation of a different sort was beginning to be sought elsewhere. As mentioned previously, President Conklin of the American Association for the Advancement of Science had suggested at the Blackpool meeting of the British Association in 1936 that some sort of union be created betwen the British and American Associations. That meeting had attracted representatives of the American press as well as American scientists, among them the science correspondent of the *New York Times*, Waldemar Kaempffert. Following conversations with Conklin, Kaempffert and Ritchie Calder proposed to the president of the British Association, Sir Josiah Stamp, that the presidents of the British and American Associations should jointly issue "a Magna Charta, a Declaration of Independence, of Science, proclaiming that freedom of research and of exchange of knowledge was essential, that Science sought the common good of all mankind, [and] that 'national science' was a contradiction in terms."[54] But nothing came of this proposal at the time.

Across the Atlantic that same year, leading scholars from around the world gathered at Harvard University for the celebration of the institution's tercentenary. At this, as at the earlier British Association meeting in Blackpool, there was concern about the threat that the misuse of science posed to civilization. It was asked how the world could escape destruction at the hands of its own science and technology. Scientists were "regarded as being responsible for a world in which little children drill in gas masks and the populations of great cities have intensive training in hiding in cellars."[55] In this atmosphere, certain remarks by the prominent French medieval scholar, Étienne Gilson, led to an unscheduled symposium at the tercentenary authorized by Harvard's President Conant. Gilson had urged university faculties, scientists, artists, and philosophers to assume responsibility for teaching that "there is a spiritual order of realities whose absolute right it is to judge even the State, and eventually to free us from its oppression."[56] At the symposium, members of the National Association of Scientific Writers asked a group of leading scholars about the advisability of forming a permanent "court of wisdom" in which the world's foremost scholars would give mankind the benefit of their collective wisdom on world affairs. The scholars welcomed the idea, if only it could be realized in practice.[57] That, of course, was the difficulty, and at the time nothing came of the suggestion.

Neither was any action taken a year later at the 1937 British Association meeting to lessen the "threat to civilization." So on the last day of that gathering, the gadfly Ritchie Calder published in the *Daily Herald* an open letter addressed to Lord Rayleigh who was to be the Association's president during the following year. Rayleigh was asked whether he would speak on behalf of all the scientists who had been silenced by tyranny, declare that science, which could do so much to advance civilization, would refuse to be an instrument for its destruction, and reaffirm "the transcendental truth of science—that it seeks knowledge for the common good of all mankind."[58] Calder reminded Rayleigh that the council of the British Association had taken the initiative at that year's meeting in "showing the nations of the world that they are members of a great commonwealth and in furthering the cause of international peace."[59] He was referring to the fact that the council had decided to send to India a deputation of scientists headed by Lord Rutherford to sit in joint session with the Indian Science Congress and thereby to serve as a model for similar international conventions. Calder then asked Rayleigh whether he would give effect to the proposal that Calder and Kaemppfert had made to Stamp in the previous year, and, in addition, implement their further proposal that the British and American Associations should form "the nucleus of a democratic 'World Association of Science' concerned with the solution of international social problems."[60]

Rayleigh did not reply, but a year later in his presidential address to the British Association he spoke to some of the issues raised by Calder. In the meantime, Calder's letter bore fruit elsewhere. It was the subject, doubtless with Kaempffert's help, of an approving editorial in the *New York Times* in October 1937. According to the *Times*, Calder had not exaggerated. His World Association—"an organization which shall indicate how the objective attitude of the laboratory may be applied in governing a people, in breaking down prejudices, [and] in solving problems that mean progress not in one country alone but the world over"—was needed to save the world and, thereby, science.[61] Although the British Association had taken no action on the proposal, the *Times* nevertheless curiously asked whether the American Association would heed "the appeal" of its British counterpart. "There was never," it stressed, "a time when science had so vital a message to deliver, so high a social mission to perform."[62] Science, the paper continued, achieved the only true internationality the world had ever known, and it afforded striking evidence that men could sink both their differences of opinion and their passions and work for a common cause.

Within a week, this *Times* editorial was reproduced verbatim in *Science*, the weekly journal of the American Association.[63] The Association was by then in the process of following the British Association's lead in paying attention to the social relations of science, and the following week's issue of *Science* announced and briefly described the scope and purposes of five conferences to be held on the subject of "Science and Society." The first of these would be held at the forthcoming December 1937 meeting of the American Association in Indianapolis, Indiana, and the others at successive meetings. These conferences were organized by the Association's section on the social and economic sciences under the direction of Harold G. Moulton, president of the Brookings Institution.[64]

One purpose of the conferences was to investigate and describe in a systematic and comprehensive manner the effects of science and its applications upon individuals and society. It was explained that science was by far the most important influence to which man had ever been subjected. It had transformed his environment and outlook and had enormously increased his ability to satisfy his physical needs. In his economic and social relations, it had changed him from being a largely self-sufficient individual into being one essentially dependent on the remainder of the world. However, it had also created problems. The effects of scientific discovery on the processes of wealth production had at times caused serious economic and social maladjustments. Also, there were fears that science and technology might become the masters, rather than remain the servants, of man and even cause the destruction of society.

A second purpose of the conferences, which had had no counterpart at the British Association, was to indicate the ways in which economic, social, and political institutions affect scientific development. It was pointed out in this conneciton that changing economic conditions, as well as imperfections in the operation of the economic system, react upon the development of science. It was further explained that confused and conflicting views with respect to the present and future state of science and its applications pervaded contemporary public discussions and led to governmental controls that were often inimical to both science and society.[65]

The first conference consisted of five daily sessions at the December 1937 meeting of the American Association on the subject of "Fundamental Resources as affected by Science," at the first of which Moulton spoke on the general topic of the mutual relations of science and society.[66] The *New York Times* gave much attention to this aspect of the meeting. It argued that the time seemed ripe to realize Calder's proposal for a Magna Charta—"a declaration of independence which shall state the function of science in modern society and the need of freedom of research, freedom of theorizing, freedom of discussion."[67] It urged that if the American and British Associations would together "formulate and announce the simple principles that guide scientific men and proclaim the might and sanctity of scientific thought, the world [was] bound to listen and to profit." This was because "scientific men in all countries are spiritually welded together by a community of interest and an idealism which contrast markedly with fascistic nationalism and communistic class hatred. They set an inspiring example of devotion to a cause that knows no country, no creed, no race. They prove that collective thinking for the common good is possible."[68]

Having been doubtless both flattered and encouraged by this, the American Association two days later took, in the words of the *Times*, the first concrete steps toward the creation of a "Court of Wisdom" in which scientists of the freedom-loving nations would unite to promote peace and intellectual freedom. In contrast to the British Association's inaction on Calder's and Kaempffert's proposal, the council of the American Association had unanimously adopted a resolution to extend to the British Association and all other organizations with similar aims an invitation to cooperate for the preservation of the intellectual, spiritual, political, and moral values of civilization "so that science [might] continue to advance and to spread more abundantly its benefits to all mankind."[69] The American Association's permanent secretary, F. R. Moulton, was instructed to explore with the British Association the desirability of inviting its members to join the American Association in order that the scientists of Britain and America might act as a "united world intellect." Such a union, it was naively hoped, would exert a moral force on the peoples of the world similar in influence to the moral force that supposedly had been exerted by the universities in Western Europe during the later Middle Ages.

The *New York Times* was of course delighted with the "momentous step" taken by the council of the American Association—"No more momentous decision," it exaggerated, "has ever been made by organized science in any country."[70] If the national scientific societies in democratic countries would accept the invitation, it continued, scientists would then take their places beside the priests of all faiths who had braved persecution and imprisonment rather than sacrifice freedom of conscience. To the *Times*, religion and science, once thought to have nothing in common, now seemed destined to present a united front against tyranny!

Other newspapers, and of course members of the American Association, also paid much attention to the resolution, which read as follows:

> Whereas, Science and its applications are not only transforming the physical and mental environment of men, but are adding greatly to the complexities of their social, economic and political relations among them; and
> Whereas, Science is wholly independent of national boundaries and races and creeds and can flourish permanently only where there is peace and intellectual freedom; now, therefore, be it
> *Resolved* by the council on this thirtieth day of December, 1937, that the American Association for the Advancement of Science makes as one of its objectives an examination of the profound effects of science upon society; and that the association extends to its prototype, the British Association for the Advancement of Science, and to all other scientific organizations with similar aims throughout the world, an invitation to cooperate, not only in advancing the interests of science, but also in promoting peace among nations and intellectual freedom in order that science may continue to advance and spread more abundantly its benefits to all mankind.[71]

In Britain, several individuals actively involved in the study of the social relations of science expressed views on the American Association's actions. Lancelot Hogben was very enthusiastic about the results of the Indianapolis meeting and congratulated the *New York Times* on the part it had played. To Sir John Orr, who was doing much to bring about an improvement in the average British diet, the American Association's decision was "one of very great importance." Though reluctant to comment, Julian Huxley did however say that "any proposal for unifying scientific thought and possible action on an international basis is bound to be a good thing."[72]

Nature carried its first news of the American Association's "momentous announcement," which it characterized as a challenge to science and society, on 22 January 1938. The informed journal thought it unlikely that the British Association would disregard the invitation to cooperate in forming the nucleus of what would be a World Association for the Advancement of Science and Society—an international 'brains trust'—since, as *Nature* ungraciously observed, it had been the success of the British Association's Blackpool meeting and its concern for social problems that had inspired the American Association's action.[73] The American Association's project was an ambitious one, added *Nature*, but the ideals it propounded were those of science itself. In a world of misrepresentation and dangerous prejudice, objectivity was sorely needed.

The American Association's invitation was soon engaging the earnest attention of the general secretaries and general treasurer of the British Association and was brought before the British Association council. The American Association was subsequently somewhat cooly informed, however, that the British Association looked forward to discussing the project further with an American delegation that was to attend the British Association's 1938 meeting in Cambridge. In fact, British Association leaders were less than enthusiastic about the proposal. In the meantime—before the end of March 1938—a scheme in rough outline, now unfortunately lost, was sent by the British Association to F. R. Moulton in the hope that it might prove possible, either on the basis of that scheme or some other, to lay practical proposals before the governing bodies of both associations at an early date.[74] While these negotiations continued, other developments of greater consequence to the British Association were under way. They form the subject of the next chapter.

The new attention paid to the social relations of science by the British Association in 1934 continued to broaden and intensify after that year's meeting, and even though there were doubts about the value of the Association's contributions to the study of those relations, the Association's efforts nevertheless stimulated the interest of other organizations at home and abroad. The British Science Guild, which had been formed at the beginning of the century because the Association had chosen to ignore the social relations of science, readily allowed itself to be absorbed into the Association in 1936. The American Association for the Advancement of Science, impressed by the British Association's concern with the social relations of science, energetically began its own public consideration of those relations in 1937 and soon proposed the creation of an international organization of scientists for the preservation of the intellectual, spiritual, political, and moral values of civilization. There were several reasons for the American proposal. Scientists were disturbed by the persecution of their fellow scientists in Nazi Germany. Also, Italy's use of mustard gas in Ethiopia during the winter of 1935–36, followed by Germany's rearmament, left many scientists and civilians troubled by the question of the use of science in warfare. Scientists of nations other than Britain and the United States were drawn to the study of the social relations of science, and in 1937 the International Council of Scientific Unions created a Committee on Science and its Social Relations. The combined effect of all of these developments intensified concern with the social relations of science within the British Association.

1. British Association, Council minutes, 7 December 1934.
2. *Nature* 136 (1935): 409.
3. *Daily Herald*, 11 September 1935, p. 8.
4. Ibid.
5. Ibid.
6. *Nature* 136 (1935): 410.
7. Ibid.
8. *Report of the British Association for the Advancement of Science* (1935), p. xlviii; ibid., (1936), p. xxxiii.

9. *Report of the British Association for the Advancement of Science* (1936), p. xxxiii; British Association, Council minutes, 6 March 1936.

10. *Manchester Guardian*, 9 September 1936, p. 4.

11. J. Harry Jones, *Josiah Stamp: Public Servant* (London: Sir Isaac Pitman & Sons, 1964), pp. 276-78.

12. J. G. Crowther, *Fifty Years with Science* (London: Barrie & Jenkins, 1970), pp. 168, 170.

13. *Nature* 138 (1936): 417.

14. *Scientific Worker* 9 (1937): 213.

15. Association of Scientific Workers, Executive Committee minutes, 23 October 1935, ASW/1/2/16/14; "A Narrative written by P. G. H. Boswell for his wife," (typescript, 1942-ca. 1948), pp. 241, 249. The narrative is in the archives of Liverpool University.

16. "A Narrative written by P. G. H. Boswell for his wife," pp. 241, 249.

17. Association of Scientific Workers, Executive Committee minutes, 23 October 1935, ASW/1/2/16/14.

18. British Association, Council minutes, 6 December 1935; ibid, 6 March 1936.

19. British Association, General Committee minutes, first meeting, Blackpool, 9 September 1936; British Science Guild, "Incorporation with British Association for the Advancement of Science," (1936), p. 1.

20. *Manchester Guardian*, 9 September 1936, p. 4.

21. W. H. G. Armytage, *Sir Richard Gregory* (London: Macmillan & Co., 1957), pp. 102-3, 134-35.

22. *Nature* 137 (1936): 757; Richard Greenfield, *Ethiopia: A New Political History* (New York: Frederick A. Praeger, Publishers, 1965), p. 209.

23. *Nature* 136 (1935): 170.

24. *Nature* 137 (1936): 758.

25. Ibid., p. 978.

26. The lecture is reprinted in A. V. Hill, *The Ethical Dilemma of Science and Other Writings* (New York: The Rockefeller Institute Press, 1960), pp. 205-21.

27. P. M. S. Blackett, "Notes, December 10, 1934, on State of Science in Germany," P. M. S. Blackett Papers, The Royal Society, H2.

28. *Nature* 137 (1936): 16.

29. The Association of Scientific Workers had endeavored to cooperate with the British Science Guild and the Association of University Teachers in approaching the Academic Assistance Council and the Government with the object of obtaining endowed chairs in British universities for "certain displaced German professors" (Association of Scientific Workers, "Annual Report of the Executive Committee," 24 February 1934. ASW/1/1/16/1).

30. *Nature* 137 (1936): 486.

31. Ibid., p. 829. For a second letter on the same subject from a group of scientists at the University of Cape Town, South Africa, see *Nature* 138 (1936): 80.

32. *Nature* 138 (1936): 155.

33. Ibid.

34. *New York Times*, 13 September 1936, section 2, p. 1.

35. *Manchester Guardian*, 17 September 1936, p. 10.

36. *Nature* 138 (1936): 417-19.

37. *Manchester Guardian*, 17 September 1936, p. 5; ibid., 11 September 1936, p. 5; ibid., 17 September 1936, p. 10. On 29 October 1936, the executive committee of the Association of Scientific Workers adopted the resolution: "That this meeting is in complete agreement with the vigourous attack made by Sir Richard Gregory at the meeting of the British Association against the Prostitution of Science and urges the Council to study the text of his speech. It also draws attention to a leader in the *Daily Herald* [See n. 58, below.] regarding the foundation of the International Association for the Advancement of Science and urges the Council to do all in its power to implement the development of such a body" (ASW/1/2/17/13/ii); ibid., 11 September 1936, p. 5. The speeches of Gregory and Hall were among six from the 1936 meeting published in Sir John Boyd Orr et al., *What Science Stands For* (London: George Allen & Unwin, 1937).

38. *Nature* 138 (1936): 498.

39. W. H. G. Armytage, *Sir Richard Gregory*, p. 165.

40. *Daily Herald*, 17 September 1936, p. 6.

41. *Nature* 138 (1936): 698-99.

42. *Daily Herald*, 17 September 1936, p. 6; *Manchester Guardian*, 17 September 1936, p. 5.

43. See chapter 2, n. 9.

44. For information on the ICSU see *Nature* 139 (1937): 697-98; ibid., 140 (1937): 337-38.

45. *Nature* 139 (1937): 689, 697-98.

46. Ibid., 142 (1938): 279.

47. Ibid., 139 (1937): 869-70.

48. F. J. M. Stratton, "International Co-operation in Science," *Nature* 140 (1937): 337-38; p. 338.

49. Ibid., pp. 358, 983.

50. Ibid.

51. Ibid.

52. The Royal Society of London nominated Professor F. E. Weiss as its correspondent and created a subcommittee representing the various branches of science to assist him (*Nature* 142 [1938]: 278; ibid., 141 [1938]: 723). For further details on the proposed collection of information, see "International Committee on Social Relations of Science," *Nature* 142 (1938): 278-79.

53. The work of the Committee was in fact interrupted by World War II. At its first postwar meeting (in London, 11-12 September 1947), the Committee decided to resume its work (*Nature* 160 [1947]: 742-43). See also, *Advancement of Science* 3 (1946): 350.

54. *Daily Herald*, 8 September 1937, p. 8.

55. *New York Times*, 13 September 1936, section 4, p. 5.

56. Ibid., 27 December 1937, pp. 1, 8.

57. Ibid.

58. *Daily Herald*, 8 September 1937, p. 8.

59. Ibid.

60. Ibid.

61. *New York Times*, 17 October 1937, p. 8.

62. Ibid.
63. *Science* 86 (1937): 375–76.
64. *New York Times*, 26 December 1937, p. 24.
65. *Science* 86 (1937): 387–91.
66. *New York Times*, 26 December 1937, p. 24. The other four conferences were scheduled for the June and December meetings of 1938 and 1939. Their subjects would be, in chronological order: standards of living as affected by science, the economic system in relation to scientific progress, governmental policies in relation to scientific progress, and science and human beings.
67. Ibid., 28 December 1937, p. 20.
68. Ibid.
69. Ibid., 31 December 1937, p. 5.
70. Ibid., 1 January 1938, p. 18.
71. *Science* 87 (1938): 10.
72. *New York Times*, 20 January 1938, p. 2; 30 January 1938, p. 26.
73. *Nature* 141 (1938): 150. *Nature*'s observation, although it went unchallenged by the AAAS, was not entirely true. The AAAS was slower than the BA to concern itself with the effects of science on society, but it had led the BA in expressing concern about encroachments on scientific freedom. In 1933, the AAAS adopted "A declaration of intellectual freedom" which it reaffirmed in 1937 and which read:

> The American Association for the Advancement of Science feels grave concern over persistent and threatening inroads upon intellectual freedom which have been made in recent times in many parts of the world.
>
> Our existing liberties have been won through ages of struggle and at enormous cost. If these are lost or seriously impaired there can be no hope of continued progress in science, of justice in government, of international or domestic peace or even of lasting material well-being.
>
> We regard the suppression of independent thought and of its free expressn as a major crime against civilization itself. Yet oppression of this sort has been inflicted upon investigators, scholars, teachers and professional men in many ways, whether by governmental action, administrative coercion or extra-legal violence.
>
> We feel it our duty to denounce all such actions as intolerable forms of tyranny.
>
> There can be no compromise on this issue, for even the commonwealth of learning can not endure "half slave and half free." By our life and training as scientists and by our heritage as Americans we must stand for freedom (*Science* 87 [1938]: 10).

74. *Science* 87 (1938): 368.

6

The British Association's Division for the Social and International Relations of Science

Within the British Association, two major developments in the social relations of science movement occurred in 1938. First, an answer was given to the troubling question concerning the scientist's responsibility regarding the uses of science, namely, that all members of society and not only scientists bore responsibility for such uses. Second, in an action unprecedented in its history, the Association formed a Division for the Social and International Relations of Science in the belief that the study of these relations was just as important as the scientist's traditional study of the natural world. These developments, together with the work of the Division down to 1945, covering its most important period, are the chief subjects of this chapter.

The developments within the British Association now to be described were hardly surprising, given the times. During 1937, the Association of Scientific Workers arranged a series of five public lectures on science and society that were so successful that the Association decided to make such lectures a permanent feature of its work.[1] The adoption by the American Association for the Advancement of Science of a course of action that the British Association's leaders had not chosen to follow promoted even further introspection within the British Association in early 1938. Also, there was a great desire on the part of many scientists, particularly chemists, to do something positive in counteraction to the misuse of science in war. In addition, ever since its 1936 meeting there had been talk within the British Association of creating a new section devoted to the social sciences. Finally, following the Association's annual meeting in 1937, the *Times*, declaring that scientific progress clearly affected politics, economics, and even the general fabric of social and national organization, had posed the questions: "Is it the business of a body like the British Association to discuss such impacts and implications of science? Should it discuss them at all, or should it go further and deliberately organise its discussion of them?" Noting that opinion within the Association was divided on these questions, the *Times* suggested that if study of the social effects of science was to be conducted on a large scale and in an organized way, then "either a new section of sociology must be erected, or a 'general section' must be formed in which matters affecting science as a whole can be treated." Alternatively, it added, "a new body can be organised where such questions can be discussed, leaving the British Association to pursue traditional courses."[2]

In these circumstances, Sir Richard Gregory observed in *Nature* in April 1938 that the report of the International Council of Scientific Unions' Committee on Science and its Social Relations to be issued in 1940, and perhaps triennially thereafter, was far from sufficient to meet the need for thought, discussion, and publication on the social relations of science.[3] Also, the provision for such discussion at British Association meetings was inadequate. Consequently, explained Gregory, some were suggesting that there should be a society for the study of the social relations of science. Its membership should not be confined to scientists, though they should play the principal part in it. It should, said Gregory, be a society for the advancement of knowledge, not a propagandist body for the advancement of science like the former British Science Guild, or for the advancement of professional scientific interests like the Association of Scientific Workers. It should perform functions not performed by such bodies as Political and Economic Planning and the Engineers' Study Group on Economics, both of which were doing valuable work on the social relations of science.⁴ The society should receive, read, discuss, and publish papers submitted to it and perhaps arrange symposia. Its area of concern would inevitably involve questions of political importance, but its attitude toward papers submitted to it should be scientific and objective; it should not in general express corporate opinions, but leave readers to accept or reject the statements in its publications as they judged appropriate.⁵

Advance proofs of the *Nature* editorial carrying these views were sent to prominent scientists and nonscientists, and their reactions solicited. Forty replies were received which were published in *Nature* together with the editorial.⁶ Some respondents doubted that a new scientific society was needed and suggested that the desired objectives might be obtained through cooperating with existing organizations. However, there was general agreement that steps should be taken to further the study of the social relations of science so that scientists and society might become more conscious of their common interests and responsibilities.

Gregory had arranged that Allan Ferguson, general secretary of the British Association, should write a leading article for *Nature* commenting on the various views expressed in the responses. On 7 April, he wrote to Ferguson:

> You know yourself that I am not actually committed to the formation of a new society, but I should certainly support it if there is a decided opinion among scientific workers in favor of it. A meeting is to be held at the Royal Society in May, possibly with Sir William Bragg [the Society's president] in the Chair, for the discussion of the matter, and the proposal will then be made that a new society be founded either through the British Association or otherwise. I think it would not be the best of policies to stress the British Association avenue so obviously as to suggest the author was officially connected with it, but you are such a sensible and reasonable chap that I am sure you will put the points forward as any scientific man would wish them to be presented.⁷

Five days later Gregory wrote again to Ferguson:

> I have not committed myself or *Nature* to the idea that a new society is the only way of getting the work done. All I am keen about is that biologists, chemists and other workers in the realm of the

natural sciences should by virtue of their particular knowledge and methods of investigation get closer contact with workers in the field of social science, even though these fields have political aspects. I doubt very much whether the B. A. council would go very far in this direction, but I do believe with you that as it has already accepted the field as appropriate to its work, it is worth while suggesting that it might appropriately extend its activities in this direction by fathering a section or other organisation which could deal with papers throughout the year instead of limiting them to the annual meeting.[8]

The next step was the calling of a representative conference to appoint a committee to explore a possible constitution for the proposed organization and also its relationships to existing scientific organizations.[9] Prior to the deliberations of that committee, and one week after the initial leader had been published, Ferguson's anonymous editorial suggested that before an attempt was made to constitute a new society the British Association might be invited to undertake the task envisioned for the proposed body.[10] The Association, said Ferguson, might very well consider arranging discussions on aspects of the social relations of science to be held in London or elsewhere at regular intervals between annual meetings. In that case, it would be better to have within the Association a new department, rather than a new section or subsection, for the social relations of science. Were the Association to accept the responsibility, its annual reports would have to be supplemented by a new periodical devoted primarily to the advancement of knowledge of the mutual impact of science and society.[11] However, concluded Ferguson, it might well be that serious questions of finance and policy would make it too difficult for the British Association to undertake the task.

An informal meeting of specially interested persons called by Gregory and Professor Sydney Chapman and including the British Association's treasurer P. G. H. Boswell, Sir William Bragg, H. G. Wells, and W. A. Wooster, was held on 1 June.[12] Wooster represented the Association of Scientific Workers of which he was general secretary. The Association had written earlier to *Nature* in support of the creation of an independent body to study the social relations of science,[13] and its executive committee had instructed Wooster that he should be guided by that view.[14] However, at the meeting it was stressed that there was nothing in the constitution of the British Association to forbid the establishment of machinery within the Association for the study of the social relations of science. The group agreed to submit a resolution to the council of the Association inviting the Association to establish a special department that would consider the social and international relations of science by means of inquiry, publication, and the holding of meetings not necessarily confined to the annual meetings of the Association.[15] The international relations of science were specified in the resolution "primarily because of the deep interest" of the American Association for the Advancement of Science in the subject and also because of the negotiations underway between the British and American Associations.

Within the British Association about this time, the general officers (P. G. H. Boswell, F. T. Brooks, Professor of Botany at Cambridge, and Allan Ferguson, Assistant Professor of Physics, Queen Mary College, London) commended to the council the correspondence and leading articles recently appearing in *Nature*.[16] They pointed out that there was nothing in the Association's constitution to prevent it from forming a new department as had been suggested. As it would be the funciton of the general committee to create a new department were such a course considered desirable, the general officers suggested to the council that the matter be brought to the attention of the general committee in advance of that year's meeting, to be held in Cambridge, with or without a specific recommendation from the council.

The council, however, did resolve to recommend to the general committee that it "take action in the direction" of founding a new department or organization other than a section to undertake the proposed task and it appointed a committee to formulate a scheme and report to the general committee. It is hardly a coincidence that in addition to Chapman, A. V. Hill, Foulerton Research Professor at the Royal Society, Sir Thomas Holland, Principal and Vice-Chancellor of the University of Edinburgh, and the president and general officers of the Association, this committee also included Gregory.[17] As with earlier developments in the decade, Gregory and *Nature* were here again most influential.

The committee subsequently formulated a "Proposal for the establishment of a Division to deal with the Social and International Relations of Science." The term *division*, suggested by Boswell, was chosen to differentiate the proposed organization from the traditional unit of the Association, the section. The division's principal purpose would be "to further the objective study of the social relations of science." In other words, it would concern itself with the effects of scientific advance on the well-being of the community and, reciprocally, with the effects of social conditions upon the development of science. Thus, its goals were similar to those of the series of conferences embarked upon by the American Association for the Advancement of Science. The division would have an executive committee nominated annually by the council and appointed by the general committee.[18] The proposal was discussed by the general committee at the annual meeting in Cambridge on 17 August 1938 and adopted.[19] The British Association's unique Division for the Social and International Relations of Science thus came into being. Gregory, "with a vision as great as his social conscience, had worked for and achieved its inception."[20]

The Division was welcomed "very cordially" by the Association of Scientific Workers, which wished it success in the task it had undertaken, "a task for which the time is fully ripe," and suggested a list of subjects for its attention.[21] Later, the Association's journal, the *Scientific Worker*, would describe the creation of the Division as "a very important step" which "comes very near one of the objects of our own Association," namely, "to secure the wider application of science and scientific method for the welfare of society."[22] At that time, the Association's president, Professor F. G. Donnan, and several of its vice-presidents and members sat on the general committee of the Division.

The reason for including the international relations of science, it is recalled, was the American Association's interest in the subject. Earlier, at its Ottawa meeting in June 1938, the American Association had appointed a committee to confer with representatives of the British Association at the Cambridge meeting regarding the possibility of closer relationships between the two Associations. The members of this committee were the American Association's new president, George D. Birkhoff, Professor of Mathematics at Harvard University; its two vice-presidents, Harold G. Moulton, president of the Brookings Institution, and Herbert E. Ives, a physicist at the Bell Telephone Laboratories; and it permanent secretary, F. R. Moulton. Prior to the Cambridge meeting, they had lengthy discussions with representatives of the British Association at the country estate of the Association's president, Lord Rayleigh.[23] In addition to Rayleigh, the British Association's representatives were its general secretaries, F. T. Brooks and Allan Ferguson; its secretary, O. J. R. Howarth; and, of course, Gregory. These conferences led to the passing of two resolutions by the council and general committee of the British Association. According to the first, each Association was to invite, on alternate years, a distinguished representative of the other Association to deliver a principal address at its annual meeting. There would be no limitations on the subjects of the addresses, which it was assumed would be of the general quality of those given at meetings of both Associations. Writing later in *Science*, F. R. Moulton added that it would, of course, be gratifying if from time to time the subjects chosen were of international interest. But whether or not such subjects were selected it was assumed that having visiting speakers would add to the attractions of the meetings and that the publicity given to their addresses in newspapers would promote cordial international relations. It was hoped that this arrangement would prove so successful that it would lead to similar arrangements with scientific organizations in other countries.[24] Under the terms of the second resolution, each Association would elect as honorary members the principal administrative officers of the other Association. The purpose of this arrangement was again to promote cordial international relations and to familiarize the officers of each Association with the work of the other.[25]

F. R. Moulton described the resolutions and the decision to create the Division as actions that promised to have important effects on the British Association's future activities and influence. He doubted whether decisions of equal importance had been taken previously by the Association.[26] Nevertheless, he and other Americans cannot have been entirely happy with the resolutions. For the grand American vision of a moral force generated through international scientific cooperation was hardly advanced. The *New York Times* sadly reported that for the time being no steps were to be taken to form a world association of scientific organizations in defense of freedom of thought, although it was believed that such an association might well grow out of the new relationships. "More militant members" of the British Association were disappointed that the Association had adopted "so conservative" a policy of cooperation, but they were pleased that at least the first step had been taken to, as the *New York Times* obscurely put it, interpret science internationally.[27]

Yet it was a small step compared with what F. R. Moulton had had in mind in April 1938. He had envisioned holding an international conference in London during the coming summer that would produce a set of fundamental ethical principles and formulate inviolable methods of international intercourse and cooperation among scientists.[28] But none of this happened.

Moulton, nevertheless, could doubtless share the optimistic view of the irrepressible Ritchie Calder who wrote that the new Division could, and must eventually, become a "World Association of Science." Calder explained that American journalists were to campaign for a similar division in the United States, and he confidently anticipated that little difficulty would be experienced in also creating divisions in France, Scandinavia, and the Netherlands. A federation of such working units would produce a world association.[29]

In regard to the creation of the Divison, Calder commented that the Association's general officers deserved the fullest credit for channeling the movement into the British Association, "through a Council which still [included] a lot of Professor Blimps," and then through the general committee. Their success, he imagined, astonished even themselves. Indeed, many members were amazed at the remarkable unanimity in support of the Division. Individuals with opposing political views were equally strongly in favor of it. No activity of the British Association in the twentieth century had inspired so much enthusiasm among diverse personalities.[30] In a second editorial in *Nature*, Allan Ferguson described the 1938 meeting as one of the most momentous in the British Association's history.[31]

The Division was a child of its times, in part the response of British scientists to the misuse of science in war. Appalled at what had happened in Ethiopia, the members of the chemistry section of the British Association had requested the council in 1936 that all possible publicity be given to the following statements:

> 1. The extent to which Chemistry is applied for beneficient purposes in connection with the industry of the British nation and the health of its citizens, is enormously greater than the scope of its employment for purposes of warfare.
> 2. Whilst the individual must remain free to determine his own action in relation to national defence, chemists as a body view with grave concern the increasing use of science for destructive ends.[32]

The council, however, decided to take no action.[33] In the Spanish Civil War, German and Italian planes bombed several cities, including Madrid and Barcelona.[34] These unsettling events were reported even in the British science press, and British scientists learned of them with horror.[35] During the years 1936-38, J. B. S. Haldane spent nearly three months in Republican Spain and was present during a number of air raids. Believing that the lessons of the war were "quite literally matters of life and death to the British public,"[36] he published a book on air raid precautions in 1938. In it he wrote: "Air raids are not only wrong. They are loathsome and disgusting. If you had ever seen a child smashed by a bomb into something like a mixture of dirty rags and cat's meat you would realize this fact as intensely as I do. And I sympathize with the attitude of those who feel that the whole business is so horrible that they will have nothing to do with it."[37]

Although not directly involved in these developments, British scientists nevertheless could not escape a sense of guilt and frustration. The former Prime Minister, Ramsay MacDonald, was well aware of their distress, and he tried to alleviate it when he gave the British Association's first Radford Mather lecture at the Royal Institution in October 1937.[38] Speaking of science in relation to war, he said:

> Science increases power which can be applied both to life and death. The men who have made air forces possible, for example, have also created civil air fleets, and if the communities cannot make and keep peace, or if they are so blind as to follow the aggressive actions of their rulers, democratic or dictatorial, the consequences are theirs. If

peace is not secured by, say, diplomacy and the will not of one but of all nations, it is both a false judgement and a cowardly one to blame the scientific engineer and worker. The action of the farmer in growing corn and food for war is exactly of the same kind as the engineer who makes flying engines. Peace or war is not the responsibility of scientists as *scientists*, except in very special cases, so long (and it will always be) as the discoveries which increase our peaceful and beneficial resources can be used for war machinery.[39]

During 1937 and 1938 a principal activity of the Scientists' Group of the Left Book Club was the collecting and arranging of materials, including pictures, graphs, and diagrams, illustrating the frustration of science.[40] In 1938, these were presented in an exhibition, the preliminary display being opened in London in May by P. M. S. Blackett and Hyman Levy.[41] Levy was also scheduled to speak at a second display of the exhibition at the British Association meeting in Cambridge in August. Unable to be there, Levy asked Bernal to read the following statement, which is of interest not only because it is in agreement with MacDonald's views:

> The material at this exhibition presents in sharp form the problem that scientists are being called upon to face at this meeting of the British Association, viz. the contradiction between the potentiality of plenty and the actuality of poverty, between the possibilities of construction and the fact of destruction. Many of our scientific spokesmen are hopelessly ignorant of the nature of this problem. This is manifested by the defence they are putting up against a charge no sensible person makes, viz. that scientists are, in some sense, responsible for the misuse of science, and that by their actions alone they can ward off war. That is a shallow analysis. As I see it the position is this. No man can separate his work from its ultimate social application. If, therefore, he sees the fruits of his labour rendered nugatory, and even being turned to create a situation in which scientific knowledge cannot flourish, there rests on his shoulders the responsibility not only of raising his voice in protest, but of discovering precisely how such a situation has emerged. No scientific man who desires to see beyond the end of his nose can now escape the necessity for analysis, in an objective scientific manner, of the nature of the changes that are so rapidly coming over society. The charge that can really be laid at the door of scientists is that they have left this analysis so late that the world is tumbling about their ears before they have wakened up to its necessity. But merely to

analyze it is not sufficient. There are plenty of problems that can be studied behind closed doors, without adding to their number. It is his business now, when the common man is ready to listen, to go forth and tell the world about it, to explain what is happening to his work and why; and finally to draw the proper political moral. For scientific men at this juncture to remain non-political is at least to connive at a scientific crime. Against such the charge that scientific spokesmen are busy rebutting becomes, in fact, true. They will have helped to encompass the destruction of large numbers of innocent people.[42]

After reading Levy's statement Bernal spoke to a packed meeting on the subject of the progress of scientific discovery in relation to social development. A vigorous discussion ensued.[43]

As President of the British Association for 1938, Rayleigh must have been in the audience at MacDonald's Radford Mather lecture, and as he listened he may have recalled Calder's open letter in the *Daily Herald*.[44] He gave his own views on the subject in his presidential address to the British Association in Cambridge. The second part of this was entitled "Science and Warfare," in which he showed through several examples how one often cannot foresee the future uses of scientific discoveries. He concluded that "the application of fundamental discoveries in science to purposes of war is altogether too remote for it to be possible to control such discoveries at the source."[45] He continued:

> For good or ill, the urge to explore the unknown is deep in the nature of some of us, and it will not be deterred by possible contingent results, which may not be, and generally are not, fully apparent until long after the death of the explorer. The world is ready to accept the gifts of science, and to use them for its own purposes. It is difficult to see any sign that it is ready to accept the advice of scientific men as to what the uses should be.
>
> Can we do nothing? Frankly, I doubt whether we can do much, but there is one thing that may be attempted. The British Association has under consideration a division for the study of the social relations of science which will attempt to bring the steady light of scientific truth to bear on vexed questions.[46]

Rayleigh's views found wide support. In a *Nature* editorial, Rainald Brightman observed that his remarks should have made it clear beyond question that the scientific discoveries that had been used in warfare had been made in no nefarious quest, and had indeed in some instances been laid aside by scientists as of no practical value. "Dismissing, therefore, the idea that scientific men are specially responsible for the application of fundamental discoveries of science to purposes of war, we can face the essential and wider problem of assisting a world anxious to make wiser use of the knowledge and powers which science can bring."[47] This conclusion was one of the most important of the entire social relations of science movement. Its appearance marked a turning point in the scientist's thought about his responsibilities for the uses of science. Prior to 1931, he had enthusiastically worked for the advancement of science, but from 1931 to 1938 he was frustrated because, partly in his own mind and partly in the minds of others, it was felt he was responsible for the misuses of science. In 1938, however, the burden of responsibility for the uses of science was moved from the scientist's to society's shoulders, and in that new situation the scientist sought and found, through the Division, a role that largely restored his peace of mind.

Parts of Rayleigh's address were reproduced in *Discovery*, whose editor, C. P. Snow, admirably summarized the views of contemporary scientists in regard to war.[48] First, science was a nonmoral activity. How its discoveries were used depended on society and not on scientists. Second, science could no more prevent itself making possible, for example, the bombing airplane than the ordinary automobile. Third, the scientific means of making war had been grossly exaggerated. Finally, any utopian solution in which it was hoped that scientists would not cooperate in war could be rejected as a dream. However, that did not eliminate the responsibility of scientists—as members of the community, and specifically as members of one of the most influential professional groups in the world—in trying to prevent war.[49]

Snow explained that the last point had been accepted wholeheartedly by many scientists. Living in a world of crisis, scientists had been compelled to learn that war was a symptom of society's sickness, not a single phenomenon on its own account. To understand the causes of war, in the hope of preventing it, society itself had first to be understood. In particular, the change in the world over the last 150 years, during which applied science had enabled the whole scale of industry and organization to be dramatically altered, had to be understood. Until this was accomplished and until the powers of applied science were directed consciously to the benefit of the peoples of the world, human life would be a precarious business at best. Few, said Snow, had realized the urgency of this problem so clearly as Gregory, whose "gospel" was: "We must *know* as much of the social relations of science as we do of science itself. Since the industrial revolution, science has been changing the world; we must understand exactly how the change is working if we are to keep our hope for the human race. And we must understand as coolly and as dispassionately as scientists aspire to think in their own fields."[50] With the creation of the British Association Division, Gregory had persuaded organized science to follow him. J. D. Bernal was like most other scientists enthusiastic about the Division: "An attempt will be made to find out in an accurate and practical way what are the actual results of applied science in the contemporary world and what they might be if science were applied in a rational and ordered way for human welfare. Armed with this knowledge the peoples of the world will for the first time be in a position to judge what are the real possibilities for human advancement and how they are being frustrated by stupidity and greed."[51]

During the Cambridge meeting, the general committee, having approved the creation of the Division, also appointed a nuclear group to the Division's committee with power to recommend additional names to the council for appointment to the committee.[52] This group was charged with drawing up the permanent constitution of the Division, creating its machinery, suggesting priorities among subjects of inquiry, and reporting on all of these to the council by November 1938.[53]

When the group met in October, Gregory was made chairman of the committee.[54] An executive subcommittee was appointed consisting of Gregory, Ritchie Calder, A. M. Carr-Saunders, Director of the London School of Economics, Julian Huxley, Hyman Levy, Sir John Orr, Director of the Rowett Research Institute and of the Imperial Bureau of Animal Nutrition, and the Association's general officers.[55] The full Division committee was appointed in early November, and by the end of the month it had held its first meeting and created several subcommittees on which progressive individuals were well represented.[56] Calder and Levy were members of the subcommittee charged with arranging meetings of the Division. Calder and Orr sat on the subcommittee on nutrition and agriculture, and also sat on the subcommittee charged with considering means of studying the economic requirements of nations in relation to their sources of raw materials, populations, standards of living, and industrial development. P. M. S. Blackett and C. H. Desch, superintendent of the Metallurgy Department of the National Physical Laboratory, were on the subcommittee concerned with the influence of scientific and technological developments on the relative importance of different industries and on the volume of employment. J. D. Bernal, Huxley and Orr were members of the subcommittee that was to report on the desirability of supplementing existing national research organizations whether in normal circumstances or in times of emergency.[57]

The most significant activity of the Division was the holding of public conferences usually organized by a subcommittee. As Gregory was to explain in 1943: "Scientific societies and technical institutions exist to receive and discuss contributions to natural knowledge; the main function of the Division is to be the liaison between such bodies and the public generally, not so much by public lectures of a descriptive kind as by public conferences which present objective realities of science in their contacts with progressive life and needs."[58] Gregory strongly believed that the British Association had "a distinctive part to play in educating the public as to the meaning of the spirit and service of science in modern life."[59] Thus, the Division's primary thrust came to be educational—it would help educate the public in the beneficial uses of science.

This was a significant departure from the original intention of objectively studying the social relations of science. Nevertheless, it was quite in keeping with the direction the British Association's thought had taken in 1938. Society—or, more accurately, society's elected representatives—and not scientists, determined whether science would be used beneficially or destructively. Yet with respect to other citizens, scientists were in a unique position—they knew science, and therefore were better able to foresee its potential uses. Thus, they had the ability and therefore, it was assumed, the responsibility to inform the rest of society of potential beneficial uses and to warn it of hazardous ones. As Gregory wrote to King George VI in 1946: "It is a primary objective of the BA to help towards the material and spiritual advancement of the human race by promoting general understanding of the right uses of scientific knowledge, in order that this may be neither neglected nor misapplied."[60] This was the role the Division played, or at least attempted to play, for it is difficult to judge how effective its several conferences were in educating the public. In any event, under Gregory the Division nurtured rather than objectively studied, as had been intended, the social relations of science.

Gregory's influence was great. The British Association's annual meeting of September 1939 was brought to a premature close because of the outbreak of World War II.[61] No further annual meetings were held until 1946, so the principal wartime activities of the Association were the several Division conferences. As an important coincidence, Gregory was elected president of the Association at the truncated 1939 meeting. He took office in January 1940 and continued throughout the war as both president of the Association and chairman of the Division committee. He was thus in a supreme position to make the most of the Division's potential to promote the social relations of science as he interpreted them.

The principal means would be the public conference of one, two, or three days duration. The Division had the power to arrange functions in places other than those of annual meetings of the Association and independently of them. This had been granted to enable the Association to cooperate with local scientific organizations in places not large enough to host a full annual meeting.[62] With each of the conferences, the hope was that their proceedings would, with the help of the press, stimulate influential laymen—members of Parliament, in particular—and scientists to appropriate action.

The initial Division conference, the first of three prewar ones, was held at the University of Reading on 28 March 1939. This was organized by the subcommittee on nutrition and agriculture, one of whose tasks it was to assist in the coordination and presentation of work already done by the British Association in these related areas. The committee decided that the nutritional and related aspects of milk was a subject requiring immediate attention and made it the topic of the conference. As was to be the practice at all of the conferences, invited papers on various aspects of the subject were given by qualified individuals and then discussed by those in attendance. Topics covered at Reading included the deficiencies in contemporary diet and the implications of the compulsory pasteurization of milk. The conference was attended by some two hundred persons, including representatives appointed by the Milk Marketing Board, Ministry of Agriculture, Cooperative Societies, Associations of Medical Officers of Health, farmers' unions, and dairy firms.

Nutrition was a much discussed and debated subject in Britain during the 1930s. It was one of the areas in which the British Association and individual scientists, notably Sir John Orr, had been attempting to educate the public and arouse the government to action even prior to the creation of the Division. So although it was not a new subject in 1939, it was nevertheless one in which much progress was still possible since there continued to be opposition to such measures as the pasteurization of milk. At Reading, this "highly controversial subject of pasteurization received much attention, and rightly so," reported *Nature*, "for it is with such problems, involving scientific, economic and even political issues, that the Division may be expected to deal."[63] To the journal, one of the prime reasons for the existence of the Division—namely, the need in controversial issues involving science for propaganda based on scientific fact rather than subjective opinion—was well exemplified by this first conference.

The participants approved the following resolution and forwarded it to the council of the British Association, to which the Division was responsible, for its consideration and possible action:

> In view of the proved danger of the spread of epidemic and other diseases by the consumption of raw milk, of the efficiency of controlled pasteurization in abolishing this danger and of the slight damage to the nutritive and other properties of milk caused by effective pasteurization, it is essential for the national health that it be made compulsory in all urban areas with a population of over 20,000 or more to pasteurize effectively all milk before sale to consumers to ensure its safety and to assist in securing that increase in the *per capita* consumption of liquid milk which is essential for improvement in the national level of nutrition.[64]

The council cautiously decided that before further action was taken on the resolution a factual report on existing knowledge of the subject should be compiled. The Division's subcommittee on nutrition and agriculture was empowered by the council to proceed in the matter.[65]

Not every conference produced a resolution as a stimulus to further action. It was usually considered sufficient for the Division to cast its bread upon the waters. The expectation was that the views of such a prominent body as the British Association would naturally come to the attention of those who would benefit from them. This philosophy was expressed explicitly in connection with a conference on university education: "The Association aims, among other things, to assist progress by discussion and report, and it is hoped that both the occasion and the reports [on university education] may have contributed usefully to current thinking on what a university is and does, or should do within its own walls and outside them both regionally and internationally. *Action now rests with those directly concerned.*"[66] (Italics added) In this case at least, editorial comment in newspapers and periodicals recognized the conference as a commendable service in the public interest.[67]

Resolutions could have been addressed to the government, but none ever was. The Division, or rather the British Association's council that oversaw it, would go only so far, and the rest was left to fate. The Association's long established practice not to become directly involved in governmental matters was continued.

The second and third prewar conferences occasioned no controversy and produced no resolutions. The second was held at the Royal Institution in London on 25 May 1939 with some two hundred and fifty persons participating.[68] The objectives of this conference were to show how science and society were "out of gear" and to indicate the nature of the task that the Division had undertaken in trying to assist in bringing about an adjustment.[69] From the chair, Gregory said that the Division had plenty to do if only to repudiate the view that science must be held responsible for the economic disturbances and for the destructive forces that were now threatening what ought to have been a world in which there was economic and other stability.[70] There were two papers: one by the political scientist Professor

Ernest Barker on the social implications of science, the other by Sir Daniel Hall on the various ways in which the application of science to British agriculture was being impeded. It had originally been suggested that Bernal, whose now famous book, *The Social Function of Science*, had recently appeared, was the best qualified to give the first paper.[71] But Gregory wrote to O. J. R. Howarth, secretary of the British Association: "You will agree with me that we do not want to give the impression at this stage that the Division represents only the extreme Left-wing, and I am afraid that this would be so if we get Bernal to give the address, though he has given as much attention to the subject as anyone."[72] Howarth himself had informed Hall that his paper would be acceptable only if it avoided "entering upon the field of politics. . . . I am clear from what I have heard in the [Divisional] Committee that our attitude I think very properly is to be that of the brothers in Princess Ida: 'Politics we bar, they are not our bent.'"[73] Later the same year, however, Gregory wrote to Bernal urging him to express his views on international intellectual cooperation at the forthcoming British Association meeting in Dundee. Gregory hoped he would consent, "for the last thing I want is for the new Division to be regarded as academic or preservative."[74]

The third prewar conference was held in Manchester just over a month later, on 29 June, in cooperation with the Manchester Literary and Philosophical Society. The cooperation resulted from the Division's policy of establishing contact with appropriate organizations. To that end a statement of the Division's objectives, together with a request for information on any work being done in the field of the social relations of science, had been circulated to some 350 associations, institutions, and learned societies in the British Isles and abroad.[75] Among other things, the communications led to cooperation with Political and Economic Planning in the preparation of a report on the organization of research in Britain. Also, the Parliamentary Science Committee co-opted Gregory, as Chairman of the Division committee, to its executive committee—an honor previously restricted to members of Parliament.[76]

The Manchester conference clearly indicated that educational work would be a principal activity of the Division. At its opening session, Gregory explained that the prelude to effective action regarding the social implications of science was the promotion of clear thinking about the current uses and misuses of science and the fuller uses that could be made of it for promoting general social welfare. To him, such educational activity was also of importance in securing the widespread support that was equally essential if scientific investigation was to be conducted on an adequate scale in the field of the social relations of science. Study of the social relations of science had been the declared goal of the Division, but already it was in second place with respect to the active educational role that would become the Division's principal concern. That a change of direction was underway was confirmed by Hyman Levy's remarks. In speaking of the social relations of science in the same session, he pointed out that the Division was not solely concerned with the education of scientific and public opinion in the social consequences of scientific discovery. It was, he said, equally concerned with the elaboration of the technique for the investigation of social change.[77] But this concern was not to be a primary one for the Division, although a committee was set up to investigate and report on the subject, which it did.

The second and third sessions were given to addresses on the influence of science on, appropriately for Manchester, the old cotton and new plastics industries, respectively. The intention was to make a parallel study of the effects of science on each.

In assessing the Manchester conference, Rainald Brightman stated in a *Nature* leader that it had succeeded at least in providing impressive illustrations of the opportunities to apply scientific method not merely in industry but also in the solution of problems of social change caused by industrial development.[78] It might indeed be hoped, continued Brightman, that the conference would encourage many more scientists to seek to bring to the solution of such problems the same attitudes of mind and principles that they employed in their research and also stimulate both the formulation of specific projects for investigation and the elaboration of the requisite techniques and principles of investigation.

The Division had planned three sessions for the British Association's 1939 annual meeting in Dundee, but due to the premature termination of the meeting because of the outbreak of war only one of the three sessions was held. This session discussed a preliminary report by Julian Huxley on the joint inquiry by the Division and Political and Economic Planning (PEP) into the organization of British scientific research.[79] The detailed supervision of the inquiry had been entrusted to PEP which had begun with an analysis of academic research, the subject of Huxley's report.[80]

It was not until the following April that the Division announced its next conference, planned for July 1940. The general theme of this meeting was to have been science in its national and international aspects, with particular attention being focused on the areas of international intellectual cooperation, national resources and national needs, social aspects of human nutrition, and scientific discovery in relation to progressive industry. However, the conference had to be cancelled, no doubt because of Britain's seriously deteriorating war situation in the spring and summer of 1940.[81]

It was more than a year later, but in a much improved war situation, that the Division held its next conference in London during 26–28 September 1941 on the grand subject of "Science and World Order" conceived by Gregory. A committee consisting of Ritchie Calder, J. G. Crowther, C. H. Desch, and Julian Huxley (chairman) arranged the program. Of all the conferences sponsored by the Division, this one was by far the most important and received the greatest attention, due largely to the nature of its organization. This in turn reflected the thinking of the Division and in particular of Gregory, who made some significant observations in an article published prior to the conference.

When the article appeared, the resources of science were being used, as Gregory observed, "to devastate the civilized world." Furthermore, no limit could be seen to science's powers and no end to the horrors they created when exercised without regard for its sanctity. Although, said Gregory, in repeating Rayleigh's earlier distinction, scientists were responsible for discovering these powers, nevertheless it was communities and governments and not scientists that decided how they were used. "Whether scientific knowledge is used for social betterment or to make civilization a mockery depends upon statesmen and not upon men of science, who, however, alone understand its possibilities."[82] Thus, more forcefully than Rayleigh, Gregory placed the social responsibility for science in the hands of statesmen. However, as at the creation of the Division in 1938, it was assumed that the scientist, because of his special knowledge, had a responsibility to see that the best uses were made of science. He had to do this by educating the public, and especially its representatives in government. Thus, the educational efforts of an organization such as the Division should be directed more at statesmen. As Gregory said: "It may not be necessary to have intimate acquaintance with [scientific] knowledge in order to anticipate effects of its applications, but it is obviously desirable for statesmen and administrators to have full appreciation of its powers."[83] To educate statesmen about science's powers became the principal goal of the Division. Gregory claimed, and others agreed,[84] that the conference on science and world order was "the first occasion upon which representatives of science, administration and government met together to consider problems of the adjustment of progressive scientific knowledge to social action."[85] Thus, rather than objectively studying the social relations of science in general, the original goal, the Division under Gregory turned from 1941 to fostering the relations between science and government by educating statesmen. The war provided the Division with a greater opportunity for doing so than it otherwise would have had. However, the war itself was more effective than the Division in teaching politicians to appreciate the powers of science.

The conference on science and world order was deliberately designed to deal with the relations of science to government and other agencies concerned with constructive planning. It was also intended to be international in spirit and to focus on world resources and human needs generally. Distinguished ambassadors and other government leaders were invited to participate. Scientists from twenty-two nations also attended.[86] The hope was that the conference would promote social and international contacts of far-reaching consequences.[87]

By the time of the conference, when the British Association was endeavoring to educate the government from the outside, some scientists in Britain were attempting to educate and advise the government from the inside. These were the president and two secretaries of the Royal Society who were also members of the Scientific Advisory Committee to the War Cabinet. The

Committee had been created reluctantly by the government in September 1940.[88] However, a new government attitude was revealed at a luncheon given at the Savoy Hotel for the delegates to the conference on science and world order. The British Secretary of State for Foreign Affairs, Anthony Eden, echoed Gregory's views in a remark that *Nature* was to make much of, namely: "If after the War we are to remove the fear of want as well as of war, science and statecraft must work together."[89] Later, another Cabinet member—Herbert Morrison, Home Secretary and Minister of Home Security—spoke at the session on Science and Human Needs.[90] He expressed an appreciation of what science had done, and could do, to meet such needs. Churchill had been invited to the opening of the conference but had not attended. However, he wrote to Gregory in what must have been encouraging terms:

> One of our objects in fighting this war is to maintain the right of free discussion and the interchange of ideas. In contrast to the intellectual darkness which is descending on Germany, the freedom that our scientists enjoy is a valuable weapon to us, for superiority in scientific development is a vital factor in the preparation of victory.... It will take a long time for the civilized powers to repair the trail of material and moral havoc which the Germans leave behind them. It will require all the resources of science. But I look forward to the day when the scientists of every nation can devote all their energies to the common task, and I wish you every success in the work you are undertaking now.[91]

In addition to the session on science and human needs chaired by the United States Ambassador to Britain, John Gilbert Winant, there were five other sessions, three of which were also chaired by statesmen. The Soviet Ambassador to Britain, Ivan Mikhailovich Maisky, chaired the session on "Science and World Planning;" the president of Czechoslovakia, Eduard Beneš, the session on "Science and Technological Advance;" and the Chinese Ambassador, Wellington Koo, the session on "Science and Post-War Relief." The opening session, "Science in Government," was chaired by Gregory; and the closing one, "Science and the World Mind," by H. G. Wells.[92] Each of the session subjects was geared to the general purpose of the conference, namely, "the discussion of the place which Science should find in World Order, its relations with the democratic State, its contributions to the relief of human needs and suffering, its potentialities in connection with the just distribution of world resources, and the influence which by its example it should bring to bear in directing the minds of men toward peaceful collaboration in future for the common welfare."[93] In all, some seventy-five papers were presented by British and foreign delegates during the three days of the conference, which attracted more attention from the press and public than was usually given to scientific events and discussions, including Division conferences.[94] In that sense, the conference was a huge success.

Among participants, however, there were mixed views as to the value of the conference. According to *Nature*, it left behind a jumbled impression of brilliant flashes of intellect and dark patches of uncoordinated effort. There had been much lack of unity and proper relationship among the too numerous papers, insufficient drive toward results to be achieved, and lack of clear vision of the potentialities of the situation. Some claimed the conference to have been nothing but a sterile hybrid between, on the one hand, the free-platform attitude of regular British Association meetings and, on the other, the purposeful social drive that currently inspired many younger scientists. However, others optimistically held that in spite of all the shortcomings one could see in the conference "the amoebic beginning of a world mind, as yet halting and incoherent, but full of promise for the future."[95] Later, *Nature* claimed that the conference had undoubtedly done much for scientists by clarifying their social responsibilities.[96] And later still, it judged the value of the conference to have lain largely in its emphasis on the increasingly closer relations developing between science and government. Indeed, the conference was considered significant for the evidence it provided that the fundamental problem in the area of the social relations of science was that of securing the proper relations between science and government.[97] In keeping with this, the Division would try to secure the active participation of statesmen in all future conferences.

Within the Association of Scientific Workers, a social relations committee had been formed early in 1941 at the suggestion of the Central London Branch, which had argued that the

Association should move on from the concept of "frustration" to the propagation of the possibilities of science for the people.[98] The Committee began to plan a conference for 13-14 September 1941 on the social relations of science with three sessions on food and agriculture, air raid precautions and housing, and production and efficiency.[99] However, at a meeting of the Association's executive committee on 20 July, Bernal suggested that as the British Association Division's conference on science and world order would be a very important one, the Association should try to participate in it and not hold a competitive conference.[100] The social relations committee agreed that they should strive for the maximum collaboration with the British Association.[101] Their contribution to the Division's conference was the subject of an editorial in the *Scientific Worker*:

> In March [1941] we urged that scientific workers should stand firmly for the best use of their services 'here and now.' In the present critical period this needs to be said more emphatically than ever before. We sent our representatives to the British Association conference on World Order to say it. Our realism—as the *Daily Mail* characterized Mr. Swann's speech—was resoundingly applauded. We welcomed the conference as an important contribution to international collaboration. Indeed we have put forward this policy from the beginning of the war. Mr. Riley, our other representative, made a number of recommendations for implementing the international aims of the conference, particularly in the urgent matter of Anglo-Soviet collaboration. In the interest of a better 'world order' we demanded that the dangers of the day be faced up to squarely.[102]

The executive committee of the Association of Scientific Workers stressed the necessity of the Association participating in the work of the Division,[103] which was to invite contributions from the Association on at least two future occasions.[104]

Suggestions made at the conference on science and world order were the seeds of the next two Division conferences.[105] The first of these—"European Agriculture: Scientific Problems in Post-War Reconstruction"—was held in London in March 1942. Its origins are of interest as they provide evidence that the Division's educational strategy was meeting with some success. Of all the subjects discussed at the conference on science and world order, "food and its distribution, and standards of nutrition in the post-War world, occupied perhaps the first place."[106] Among the papers delivered in this area was one by Sir John Russell, Director of the Rothamsted Experimental Station and of the Imperial Bureau of Soil Science, entitled "Reestablishment of the 'Scorched Earth' in Europe." Following the conference, the Allied Post-War Requirements Bureau discussed the subject of his address with Russell, and subsequently an Allied Technical Advisory Committee on Agriculture was set up in December 1941 with Russell as chairman. The Committee was to prepare plans for the postwar reconstruction of agriculture in the devastated areas of Europe. At the same time, it was suggested that the Division might arrange a conference on the subject. The Division's Committee agreed, and in arranging the program and in seeking the cooperation of foreign experts it received generous assistance from government departments.[107] Thus, both British and foreign experts read papers in sessions on measures for reconstruction, economic and kindred problems, the future improvement of European farming, and the problems of peasant farming.[108]

Events then were proceeding as the Division had hoped. At the conclusion of the conference, a resolution was addressed to the council of the British Association requesting the appointment of a standing committee under Russell to advance the application of scientific methods in the reconstitution and intensification of European agriculture "with a view to the restoration and maintenance of the well-being of agricultural communities, and to keep such objects before all who are or may be concerned with the resettlement of devastated lands."[109] The committee was duly appointed. In April 1943, a further conference was held on cooperative systems in European agriculture. Its full proceedings were published, significantly, in the *Year Book of Agricultural Co-operation.* This was further proof that attention was being paid to the Division's work.[110]

Like the 1942 conference on agriculture, the following one—on "Mineral Resources and the Atlantic Charter," held in London during July 1942—sprang from a suggestion made at the conference on science and world order. One purpose of that conference had been to consider from an international perspective the natural resources of raw materials, their national distributions, and the opportunities for using them. These subjects had been decided upon by the Division prior to the signing of the Atlantic Charter by Churchill and President Roosevelt in August 1941. As a happy coincidence, the fourth clause of the Charter stated that Britain and the United States would "endeavour, with due respect for their existing obligations, to further enjoyment by all states, great or small, victor or vanquished, of access, on equal terms, to the trade and to the raw materials of the world which are necessary for their economic prosperity."[111] Following this development, a suggestion was made that a Division conference be held on raw materials. It was considered, however, that the subject of the world's natural resources was much too large for a single conference, and that instead, a survey of the current state of minerals of industrial importance, together with suggestions for further investigation of their geographical distributions and for research into the production of substitutes, would suffice to show the close contact that existed between science and fundamental national and international problems.[112] At the close of the conference, a resolution presented by, significantly, the Lord Privy Seal in Churchill's War Cabinet, Sir Stafford Cripps, was adopted. It began with a preamble referring to the Atlantic Charter and continued:

> This Conference, having specifically dealt with mineral resources, submits that, as a first step, the [BA] Council should initiate forthwith consultations with appropriate scientific and technical organisations, to secure an understanding on the principles involved. The Conference would further urge that a scientific review of mineral resources, using and supplementing all existing data, should be among the first tasks of any international organisation for the social application of science, such as had been envisioned at the recent Conference on Science and World Order.[113] To this end, the Conference recommends that the Council should consider how it might help to promote the establishment of an International Resources Organisation, as a fact-finding and advisory body for Governments, as a contribution to world stability, and in the spirit of the Atlantic Charter.[114]

The resolution was approved by the council, and a committee was appointed to inquire into and report upon—from the points of view of the fourth clause of the Atlantic Charter—the state of knowledge of mineral resources, except coal and oil, essential for modern industrial development.[115]

In arranging its next conference, "Science and the Citizen: the Public Understanding of Science," held in March 1943, the Division continued to pursue a policy of dealing with subjects of especial importance to postwar reconstruction touched upon in the seminal conference on science and world order. Following that conference, the council of the Association had appointed a standing committee to consider and give effect to means of extending the public understanding of the benefits of science.[116] The conference now arranged was the first meeting in Britain to be devoted entirely to the consideration of means of increasing the public understanding and appreciation of science as well as of methods of improving those means.[117] Four sessions were held on the exposition of science, science and the radio and cinema, science as a humanity, and science and the press.[118] Again, the purpose was educational. It might well be hoped, said *Nature*, that the conference would stimulate journalists and professional associations to consider the matters further.[119] As an outcome of the proceedings, Gregory led a delegation of British Association members to the BBC. They were sympathetically received by the Director General and members of his staff, with whom they had a full discussion on future arrangements for broadcasts on scientific subjects and the best means for securing such arrangements.[120]

The Division's final wartime conference, on "The Place of Science in Industry," was held in January 1945.[121] Its purpose was to further the task of public education upon which the adequate support of research was considered to depend. Unless there was general understanding of the achievements and possibilities of scientific research, explained *Nature*, it could not be expected that there would be the necessary public support, either of finance or of men, on which the expansion of scientific effort to meet postwar demands ultimately depended.[122] The specific aims of the conference were to give the public a detailed view of the past contributions of science to industry and at the same time a clear idea of the country's future strategy and tactics in this area. Once again, governmental and industrial leaders were deliberately brought into the proceedings. Ernest Bevan, Minister of Labour in the War Cabinet, presided over the first session on "What Industry Owes to Science"; Lord McGowan, Chairman of Imperial Chemical Industries, over the second on "Fundamental Research in Relation to Industry"; Sir John Greenly, Chairman of Babcock and Wilcox, and a member of the Advisory Council to the Committee of the Privy Council for Scientific and Industrial Research, over the third devoted to industrial research and development; and Lord Woolton, Minister for Reconstruction in the War Cabinet, over the final session on "The Future—What Science Might Accomplish."[123]

Woolton's remarks, in particular, provided much encouragement to the Division.[124] He spoke of the lessons for peacetime to be derived from wartime experiences. He believed that after the war the public would demand more government involvement in certain areas of science and that there was indeed much scientific material on which to reconstruct a healthier and happier society. In explaining these beliefs, he drew on his own wartime experience. A few months after he had become Minister of Food in April 1940, Britain had been faced with a fifty percent drop in food imports. The country, however, had been saved from starvation by the application of scientific knowledge to the problem of securing the proper foods, not to satisfy appetites, but to meet nutritional needs. In the future, he said, steps taken during the war "to increase the consumption of milk, to encourage the eating of selected vegetables, to provide certain classes with orange juice, cod-liver oil, vitamins and calcium tablets, to develop communal feeding and meals in factories, and to expand the scheme for meals in schools, should form a permanent part of our health program."[125] He encouraged the British Association to continue its educational work in the field of nutrition that had been initiated at the noteworthy 1934 meeting. Then, addressing his remarks to the state and to industry, he said that these must see that medical, agricultural, industrial, and fundamental research be adequately endowed. Thus, government attitudes had clearly changed, and Gregory, for one, must have derived great satisfaction from this.

The Association's journal, the *Advancement of Science*, in which the full conference proceedings were generally published, commanded a considerable sale to nonmembers of the Association. For example, more than one thousand each of the numbers on the conferences on science and world order and on European agriculture (volume 2, numbers 5 and 6, respectively) had been sold. This was without any advertisement beyond that provided by the reviewing press and a few announcements from the Association's office. Even more successful was the volume *Science and World Order* (1942), based on the proceedings of the conference of that name. This had been prepared with the consent of the council by J. G. Crowther, the science journalist, O. J. R. Howarth, secretary of the Association, and D. P. Riley, a chemist, for publication by Penguin Books, which reported that in the approximately two-year period down to 30 June 1944 the surprising number of 58,186 copies had been sold.[126] There was also considerable public interest in reports produced by committees set up in relation to the Division's activities. For example, the final report of the Committee on Post-War University Education, which had been created following the conference on science and world order, was widely read.[127]

For half a century, the British Association had largely ignored the social relations of science. Then in 1934, it encouraged its members to include papers on the social impact of science at that year's annual meeting. Succeeding annual meetings saw increasing interest in questions concerning the social relations of science, including that of the scientist's responsibility for the use of science in the creation of weapons. By 1938, the members of the British Association were agreed that all members of society, and not just scientists, bore responsibility for the uses of science. At the same time, they enthusiastically created, in an action unprecedented in the history of the Association, the Division for the Social and International Relations of Science, declaring now that the study of these relations was just as important as the study of the natural world, the scientist's traditional concern.

The Division's objective soon changed, however, from the study of the social relations of science to that of educating the public about those relations in selected areas through the holding of public conferences. As Sir Richard Gregory, chairman of the Division's committee and president of the Association, explained, all of the Division's conferences were concerned with aspects of science in relation to the outlook and service of scientists and their contacts with problems of progressive human development. This was the cause to which Gregory had devoted his chief thoughts and works for fifty years.[128] By mid-1941, the Division had refined its objective to that of educating statesmen about the powers of science in an endeavor to foster the relationships between science and government. It proceeded not by confronting the government, or even by sending resolutions to it, but rather by inviting government leaders to participate in its conferences that ably and constructively addressed social matters of immediate national importance involving science. The Association continued to adhere to its practice of not becoming directly involved in national political affairs.

It is impossible to say just how effective the Division's educational efforts were. Unquestionably, the use of science in the war did much more to persuade statesmen and the public of the powers of science. But at least the Division's work reinforced what the war experience taught.

Neal Wood has written that with the formation of the Division the social relations of science movement had at last been "formally recognized by the peers of British science."[129] But only someone who has overlooked the development of interest in the social relations of science within the British Association from 1931, and who also considers the social relations of science movement as an intellectual one among radical scientists, could make such a statement. The formation of the Division, in which some radical scientists had a part, as members of the British Association, was one realization of what Wood says was never accomplished, namely, a formal organization and institutionalization of the movement.[130]

    1. J. B. S. Haldane spoke on "Facts and Theories concerning Human Race Differences," A. L. Bacharach on "Nutrition and Society," Hyman Levy on "The Socialization of Mathematics," P. M. S. Blackett on "Physics," and Lancelot Hogben on "The Social Background of Science." Association of Scientific Workers, "Report of the Executive Committee to Council at the Half-yearly Meeting, October 30, 1937," p. 1, ASW/1/2/18/19.

    2. *Times*, 9 September 1937, p. 13.

    3. W. H. G. Armytage, *Sir Richard Gregory: His Life and Work* (London: Macmillan & Co., 1957), p. 159. Sir Henry Tizard was much more critical of the potential use of the reports: "The forthcoming reports . . . will probably be of great interest, but of very little practical value. They will doubtless contain a mass of data which may eventually be used

as a basis for a first-class work by a first-class historian, but they are most unlikely to form a trustworthy guide for national or international policy, or to help to an understanding of the future effects of science on society" (*Nature* 141 [1938]: 736). Also, because of the composition of the International Council, which did not include medicine, engineering, agriculture, sociology, and economics, problems relating to these disciplines were not being considered (*Nature* 142 [1938]: 279).

4. Gregory seems to be overpraising the contributions of these two organizations. The Engineers' Study Group was formed in 1933 by a representative group of engineers and scientists dissatisfied that the community was not enjoying a standard of living commensurate with what they believed science and technology could provide and determined to examine this paradox and how it might be resolved. An interim report on "Food and the Family Budget" was presented at a public meeting jointly organized with the Association of Scientific Workers on 31 March 1936. It seems, however, that the Group accomplished little. See *Nature* 132 (1933): 635; The British Science Guild, *Engineers' Study Group on Economics: First Interim Report on Schemes and Proposals for Economic and Social Reforms* (London, 1935), pp. 3–4; Association of Scientific Workers, "Report of the Executive Committee to the Council for the year 1936," 6 March 1937, ASW/1/2/18/6.

Political and Economic Planning, perhaps better known as PEP, was founded as a nongovernmental planning organization in 1931. It presented its well-researched views on various subjects in its broadsheet, *Planning*, and in published reports. For example, in the area of agriculture, problems of agricultural research organization were dealt with in *Planning* no. 57 (10 September 1935), which gave a general view of the situation, and in *Planning* no. 132 (18 October 1938), which described means for disseminating scientific knowledge about agriculture. The PEP *Report on Agricultural Research in Great Britain* (November 1938) provided a comprehensive survey of the field.

5. *Nature* 141 (1938): 723–24.

6 Ibid., pp. 725–42. Those replying included: Sir William Bragg, Sir F. Gowland Hopkins, H. G. Wells, Harold Laski, A. V. Hill, J. B. S. Haldane, J. S. Huxley, Joseph Needham, Sir Daniel Hall, Sir John Russell, Sir Henry Tizard, J. D. Bernal, P. M. S. Blackett, F. A. Lindemann, Hyman Levy, A. C. G. Egerton, and C. H. Desch.

7. R. A. Gregory to Allan Ferguson, 7 April 1938, Ferguson Papers, Department for the History and Social Studies of Science, Sussex University.

8. R. A. Gregory to Allan Ferguson, 12 April 1938.

9. *Times*, 23 April 1938, p. 7.

10. *Nature* 141 (1938): 763–64.

11. The British Association's *Annual Report* was soon to be superseded by *Advancement of Science* (1939–1971).

12. "A Narrative written by P. G. H. Boswell for his wife," (typescript, 1942–ca. 1948), p. 265; Association of Scientific Workers, Exective Committee minutes, 27 May 1938, ASW/1/2/19/22.

13. *Nature* 141 (1938): 879; Peter M. D. Collins, "The British Association for the Advancement of Science and Public Attitudes to Science, 1919–1945" (Ph.D. diss., University of Leeds, September 1978), p. 189.

14. Association of Scientific Workers, Executive Committee minutes, 27 May 1938, ASW/1/2/19/22.

15. British Association, General Committee minutes, 19 August 1938, Appendix; Association of Scientific Workers, "Secretary's Report to the Executive Committee," 24 June 1938, ASW/1/2/19/24/i.

16. British Association, General Officers' Report to Council, 4 June 1938—appended to British Association, Council minutes, 4 June 1938.

17. British Association, Council minutes, 4 June 1938.

18. British Association, General Committee minutes, 19 August 1938, Appendix.

19. Ibid., 17 August 1938.

20. J. G. Crowther, O. J. R. Howarth, D. P. Riley, *Science and World Order* (Harmondsworth, England: Penguin books, 1942), p. 11.

21. Association of Scientific Workers, "Report of Executive Committee to Council," 19 November 1938, ASW/1/2/19/50.

22. *Scientific Worker* 11 (1939): 19. This was one of the two principal aims of the Association ("Policy of the Association of Scientific Workers," ibid., 9 [1935]: 4).

23. *New York Times*, 17 August 1938, p. 22; 19 August 1938, p. 7.

24. Gregory addressed the American Association in late 1938, but the outbreak of World War II prevented a reciprocal lecture being given by an American scientist at the British Association's 1939 meeting. No lectures were arranged during the war; although they were begun just after the war, they were soon discontinued.

25. *Science* 88 (1938): 277. In the following year, the Council of the British Asociation adopted a proposal from L'Association Française pour l'Avancement des Sciences whereby the president, vice-president, secretary, vice-secretary, treasurer, and secretary of council of the French Association should be honorary members of the British Association, and the president, general treasurer, general secretaries, and secretary of the British Association should be honorary members of the French Association (*Advancement of Science* 1 [1939]: 131).

26. *Science* 88 (1938): 277.

27. *New York Times*, 19 August 1938, p. 7.

28. *Science* 87 (1938): 368.

29. *New Statesman and Nation* 16 (1938): 339.

30. *Manchester Guardian*, 25 August 1938, p. 13.

31. *Nature* 142 (1938): 409.

32. British Association Council, minutes of extraordinary meeting, 9 October 1936.

33. Peter M. D. Collins, "The British Association for the Advancement of Science and Public Attitudes to Science, 1919-1945," p. 162.

34. Gabriel Jackson, *The Spanish Republic and the Civil War, 1931-39* (Princeton, N. J.: Princeton University Press, 1965), pp. 320, 327, 330, 385, 408.

35. For example, the "horrifying" air raids on Barcelona that Bernard Lovell spoke of—*Science and Civilization* (London: Thomas Nelson & Sons, 1939), p. 82—were reported by *Nature* 141 (1938): 546-47.

36. J. B. S. Haldane, *A. R. P.* (London: Victor Gollancz, 1938), p. 9.

37. Ibid., p. 11

38. This lecture series was also a reflection of the times. As MacDonald explained, Mather, who founded the series, "has been impressed by the importance of the work of the scientist in the ordinary everyday life of our people, especially at this moment; and, after a long life enlivened by scientific and social interest, he feels keenly that a recognition of that work is not only owing to the scientific worker himself but also will be helpful in inducing the public to use the advantages which the scientist has put at its disposal" (*Nature* 140 [1937]: p. 756).

39. Ibid., p. 758.

The British Association's Division · 151

40. *Scientific Worker* 10 (1938): 97; John Lewis, *The Left Book Club: An Historical Record* (London: Victor Gollancz, 1970), p. 83.

41. Blackett's handwritten notes, "The Frustration of Science Exhibition. May 16, 1938," are in the Blackett Papers, H 4.

42. H. Levy to J. D. Bernal, 19 August 1938, Bernal Papers, Box 84, Folder J 122.

43. John Lewis, *The Left Book Club*, p. 83.

44. See above, p. 110.

45. *Nature* 142 (1938): 338.

46. Ibid.

47. Ibid., pp. 310-11.

48. In regard to Snow's well-known later writings on the two cultures, it may be of interest to note here that *Discovery* first appeared in 1920 with the stated objective of giving "readers an interest both in the Sciences and the Humanities by making the work of the specialists in both as plain as possible. . . . In the past, unfortunately, there has been considerable opposition between the representatives of science and those of the humane studies. Now, since the war, it is becoming abundantly recognized that the interests of these two somewhat artificial divisions of knowledge are not hostile, but complementary; and that the welfare of everybody depends upon advances being made in both of these, and on our recognizing the necessity of both."

49. *Discovery* 1 (n.s.) (1938): 318-19.

50. Ibid., p. 319.

51. *New York Times*, 2 October 1938, section 7, pp. 4, 16.

52. British Association, General Committee minutes, Third meeting, Cambridge, 19 August 1938. The members of the nuclear group were: the president and general officers, Professor S. Chapman, Professor H. J. Fleure, Sir Richard Gregory, Sir Daniel Hall, Professor A. V. Hill, Dr. J. S. Huxley, Dr. C. S. Meyers, Professor J. G. Smith, Lord Stamp, Sir Henry Tizard, Professor F. E. Weiss, H. G. Wells, and J. S. Wilson.

53. *New Statesman and Nation*, 3 September 1938, p. 340.

54. Gregory had earlier relinquished the editorship of *Nature*. *Nature* 144 (1939): 472.

55. British Association, Division Nucleus Committee minutes, 20 October 1938.

56. British Association, Council minutes, 4 November 1938. The members of the full Committee were Gregory (chairman); Sir Daniel Hall, Sir Frederick Gowland Hopkins, Sir John Russell, and Lord Stamp (vice-chairmen); Professor F. C. Bartlett, Professor J. D. Bernal, Professor P. M. S. Blackett, Ritchie Calder, A. M. Carr-Saunders, Professor S. Chapman, Dr. C. H. Desch, Professor A. C. G. Egerton, Professor H. J. Fleure, E. W. Gilbert, Professor N. F. Hall, R. F. Harrod, Professor A. V. Hill, Sir Clement Hindley, Professor L. Hogben, Dr. L. E. C. Hughes, Dr. J. S. Huxley, Mr. D. Caradog Jones, Professor H. Levy, Dr. C. S. Myers, Mrs. Nicholson, Sir John Orr, Professor J. C. Philip, Professor J. G. Smith, Professor R. G. Stapledon, Professor F. J. M. Stratton, Professor F. E. Weiss, H. G. Wells, J. S. Wilson, Dr. S. Zuckerman. In addition, the president and general officers of the Association were *ex officio* members.

57. British Association, Divisional Committee minutes, 28 November 1938.

58. British Association, Council and Committee of Division for Social and International Relations of Science, minutes of joint meeting, 14 May 1943, Appendix (President Gregory's memorandum).

59. Sir Richard Gregory to the council of the British Association [late January/early February], 1944, *Advancement of Science* 3 (1944-46): 223.

60. Ibid., 4 (1946-48): 4-5.

61. As the general committee of the British Association considered whether or not to continue the meeting, the city of Dundee—the place of the meeting—was evacuating its women and children (*New York Times*, 2 September 1939, p. 32).

62. *Advancement of Science* 3 (1944–46): 81.

63. *Nature* 143 (1939): 550.

64. British Association Division, "Report for the Year 1938–39," *Advancement of Science* 1 (1940–41): 137.

65. British Association, Council minutes, 3 June 1939, p. 3.

66. *Advancement of Science* 4 (1946–48): 4.

67. Ibid., p. 89.

68. British Association Division, "Report for the Year 1938–39," *Advancement of Science* 1 (1940–41): 137.

69. *Nature* 143 (1939): 814.

70. *Times*, 26 May 1939, p. 11.

71. For this and the remaining information in the paragraph, I am indebted to Peter M. D. Collins, "The British Association for the Advancement of Science and Public Attitudes to Science, 1919–1945," p. 208.

72. R. Gregory to O. J. R. Howarth, 26 February 1939, Papers of the British Association, Bodleian Library, Oxford, box labeled "Correspondence concerning the 1939 annual meeting" (hereinafter referred to as BA Box 1939).

73. O. J. R. Howarth to Sir Daniel Hall, 7 March 1939, BA Box 1939.

74. R. Gregory to Allan Ferguson, 1 June 1939.

75. British Association Division, "Report for the Year 1938–39," *Advancement of Science* 1 (1940–41): 134.

76. *Nature* 143 (1939): 151.

77. *Nature* 144 (1939): 1, 2.

78. *Nature* 144 (1939): 3.

79. In addition to the report on research, the session heard a paper by A. M. Carr-Saunders on population problems—*Nature* 144 (1939): 387; *New York Times*, 1 September 1939, p. 9. At the second session, Gregory was to have given an evening discourse on "Contacts of Religion and Science"—later published in *Advancement of Science* 1 (1940–41): 163–75. There was to have been a discussion of social aspects of human nutrition at the third session.

80. Following the outbreak of war PEP continued the inquiry and published a brief report. See *Planning* no. 156 (5 December 1939), pp. 1–14.

81. *Nature* 145 (1940): 619, 892.

82. *Nature* 148 (1941): 331.

83. Ibid.

84. J. G. Crowther, O. J. R. Howarth, D. P. Riley, *Science and World Order*, p. 11.

85. Sir Richard Gregory, "Science in Britain: The British Association," *Advancement of Science* 2 (1942–43): 182–84, 183.

86. Armytage, *Sir Richard Gregory*, p. 182.

87. *Nature* 148 (1941): 331.

88. The creation of the Scientific Advisory Committee is described in the following chapter.

89. *Nature* 148 (1941): 379–80, 403.

90. *Advancement of Science* 2 (1942–43): 31–33.

91. Ibid., p. 4.

92. For a brief account of the program, see *Nature* 148 (1941): 338; for a full account, see *Advancement of Science* 2 (1942–43): 3–116. J. G. Crowther, O. J. R. Howarth and D. P. Riley's *Science and World Order* is based mainly on the papers given at the conference.

93. *Advancement of Science* 2 (1942–43): 4.

94. *Nature* 148 (1941): 678. Professor A. V. Hill attended the conference and later wrote that although the meeting itself was remarkable, even more remarkable was the widespread interest taken in it. "The press devoted considerable space to it: the BBC provided a large number of special broadcasts for British and foreign listeners; and the National Broadcasting Corporation of America arranged a special party between five of us in London and four in New York to discuss for half an hour for the benefit of American listeners some of the points brought up by the conference." To Hill, the wide interest shown in the conference was "a clear, indeed dramatic, demonstration of two things: *first* the strong public conviction that science has a great deal to say in world affairs; and *second* an eager interest in anticipating the human and material problems which will arise when the war is over" ("Memories and Reflections," manuscript deposited in Royal Society's library, 2 vols., 2: 409–10).

95. *Nature* 148 (1941): 392.

96. Ibid., p. 635.

97. *Nature* 149 (1942): 253.

98. *Scientific Worker* 13 (1941): 110.

99. Association of Scientific Workers, Social Relations of Science Committee minutes, 5 July 1941, ASW/1/2/22/15.

100. Association of Scientific Workers, Executive Committee minutes, 20 July 1941, ASW/1/2/22/19.

101. Association of Scientific Workers, Social Relations of Science Committee minutes, 25 July 1941, ASW/1/2/22/20.

102. *Scientific Worker* 13 (1941): 181.

103. Association of Scientific Workers, Executive Committee minutes, 5 October 1941, ASW/1/2/22/24.

104. Ibid., 4 January and 13 December 1942, ASW/1/2/23/1 and 14. On Julian Huxley's suggestion, the Association cooperated with the Division in requesting the BBC to give more attention to science in its programming (Ibid., 5 October 1941, ASW/1/2/22/24; British Association, Council minutes, 3 December 1941).

105. *Nature* 150 (1942): 718–20.

106. A V. Hill, "Memories and Reflections," 2: 411.

107. *Advancement of Science* 2 (1942–43): 123.

108. For a brief account of the conference, see *Nature* 149 (1942): 372–73; for a full account, see *Advancement of Science* 2 (1942–43): 123–78.

109. *Advancement of Science* 2 (1942–43): 178.

110. Ibid., p. 356. A summary of the proceedings was published in *Nature* 151 (1943): 523–25.

111. *Nature* 149 (1942): 576.

112. Ibid.

113. This refers to a proposal for an "International Resources Organisation" whose chief architect had been Bernal. See *Advancement of Science* 2 (1942–43): 188–89.

114. *Nature* 150 (1942): 171. For a brief account of the conference, see ibid., pp. 171–173, 201–3; for a full account, see *Advancement of Science* 2 (1942–43): 187–253.

115. British Association, Council minutes, 2 September 1942. Coal and oil had been excluded as requiring separate treatment. The report was published in *Advancement of Science* 2 (1942–43): pp. 339–45.

116. *Advancement of Science* 2 (1942–43): 184.

117. *Nature* 151 (1943): 595.

118. For a brief account of the conference, see ibid., pp. 382–85; for a full account, see *Advancement of Science* 2 (1942–43): 283–337.

119. *Nature* 151 (1943): 597.

120. *Advancement of Science* 3 (1944–46): 222.

121. Through the Division, a conference was arranged in London on 10 November 1944 for an Indian scientific delegation visiting Britain. However, the so-called conference was in reality a brief meeting at which Gregory and six Indian scientists spoke. See "Conference with the Indian Scientific Delegation," ibid., pp. 99–105.

122. *Nature* 155 (1945): 96.

123. For a brief account of the conference, see ibid., pp. 96–99; for a full account, see *Advancement of Science* 3 (1944–46): 106–56.

124. *Nature* 155 (1945): 91–92.

125. Ibid., p. 98.

126. *Advancement of Science* 3 (1944–46): 81, 223.

127. Ibid., 4 (1946–48): 4. For the report itself, see ibid., 3 (1944–46): 1–52.

128. Sir Richard Gregory, "Civilization and the Pursuit of Knowledge," *Advancement of Science* 4 (1946–48): 7.

129. Neal Wood, *Communism and British Intellectuals* (New York: Columbia University Press, 1959), p. 130.

130. Ibid., p. 137.

7

Scientists and War

Although the account in the previous chapter of the British Association's Division was carried down to the end of the war, through the period covered by the third phase of the social relations of science movement, the Division's activities may properly be regarded as belonging to the second phase of the movement. The Division was created during the summer of 1938; the third phase, related to the war, began with the Munich crisis in September 1938 and continued down through 1943. This phase is characterized by the greatest direct political activity on the part of scientists, which is another reason for not considering the Division's activities as part of it. A further distinguishing feature of the new phase is the Royal Society's decisive involvement in it, seeking goals also sought by the less prestigious but equally ambitious Association of Scientific Workers. A hitherto reluctant Royal Society was swept into the social relations of science movement by the war. The principal focus of this phase of the movement, dealt with in these next two chapters, is the central organization of science in the war.

The present chapter deals principally with the independent, yet related, activities of the Association of Scientific Workers and the Royal Society from 1938, as Europe slid toward and then into the Second World War. Prior to 1938, under the influence of its left-wing members, the practice of the Association had been to condemn the use, or misuse, of science in warfare. However, with the growing menace of Nazi Germany, the Association reversed its position after the Munich crisis in calling for the fullest use of science to defeat fascism. The Association's members nevertheless rejected the view of the Association's leaders that the government should create a ministry of science. Also, the Royal Society discouraged the Association's resolution to form a national science council to draw up proposals for the organization of science in the war. The leaders of the Royal Society were nonetheless convinced that some form of central coordination of science in the war was necessary, and during 1939 and 1940 they persistently made representations to the government. Among scientific organizations, the Royal Society commanded the greatest respect from the government. Nevertheless, Churchill finally authorized the creation of the Scientific Advisory Committee to the War Cabinet in September 1940 only after criticism was publicly expressed that the government was not making the fullest use of science in the war effort. The leading engineering institutions, dissatisfied that they had no representation on the Scientific Advisory Committee, soon succeeded in persuading the government to form the Engineering Advisory Committee to the War Cabinet in April 1941. Meanwhile, the Scientific Advisory Committee was busy proving its worth.

From 1935, the general secretary of the Association of Scientific Workers was W. A. Wooster, a lecturer in the department of mineralogy and petrology at Cambridge and one of the university's scientists who had joined the Association in 1932 and then brought it under the sway of their left-wing views. Many of these same people had also in 1932 formed the Cambridge Scientists' Anti-War Group, which aimed to function as a technical and advisory body to national and international peace movements and of which Wooster was the first chairman.[1] After studying the causes of international tension, the Group turned its attention to civil defense and in November 1936 began to test the efficacy of the Home Office's gas-proofing procedures,[2] publishing its controversial findings early in 1937,[3] the year in which Bernal was its chairman.[4] In 1938, the Group went on to test the efficacy of government-issued gas masks and examine the necessity for a large-scale, state-financed scheme of civilian shelters.[5]

By that year, the international situation was much more disturbing than it had been in 1932, with the growing power of fascism in Europe being seen as an evil even greater than that of war. That view was intensified by the Munich crisis. In September 1938, with the crisis at its height and war appearing imminent, Wooster wrote, apparently on his own initiative, to all members of the Association of Scientific Workers asking those who were prepared to do war work to send their names to the national office.[6] Wooster wanted to ensure that in the event of war the special abilities of scientists would be properly used. Similar steps had been taken by Oxford University, Cambridge University, the University of London, the Institute of Physics, the Institute of Metals, the Institute of Chemical Engineers, and the British Medical Association. Solly Zuckerman, an Oxford anatomist, was first pulled out of his "academic backwater" by the Munich agreement.[7] He went to Bernal, and together they drafted a memorandum on science and national defense that Bernal used as the basis of an editorial published in *Nature*.[8]

By 20 October, the national office of the Association of Scientific Workers had received 235 replies to Wooster's letter.[9] Ninety-one of the respondents were already on other lists or had been told by employers not to volunteer for work outside their firms; 103 gave permission, without comment, for their names to be placed on the Association's list; eight would not do war work of any nature; and twenty-two wrote either giving conditional permission for their names to be listed (they would be willing to do war work of a scientific nature if the government were behaving in a reasonably anti-fascist manner) or saying that the Association should not use its time and energy collecting lists of names, but should act up to its stated policy of opposing the use of science for destructive purposes.

This last view found little sympathy among the members of the Association's executive committee, who solicited the advice of the Association's several vice-presidents as to whether the Association should call a conference of the nation's scientific organizations to discuss plans for the utilization of scientists in wartime. At this time the committee sensed an unwillingness on the part of the Services to have anything to do with scientists. Sir William Bragg, a vice-president of the Association, replied that a scheme, which he did not explain, was already being worked out.[10] He may have been thinking of the card register of scientists being compiled at the Royal Society by one of its two secretaries, Professor A. V. Hill, Nobel Laureate in medicine in 1922. Hill had the assistance of "volunteers roped in from his family" and they were "at it till 10 p.m. or later."[11] Bragg did not think that it would be of value for the Association to organize the proposed conference at the present time.

Following the Munich agreement, and in view of the varied opinions expressed in response to Wooster's letter, the Association's executive committee decided to take no further action until the membership had had an opportunity to thoroughly discuss the course open to the Association in a time of national emergency. One respondent had asked for more than a verbal guarantee that his services would not be employed "in alliance with the dictators against the elected Government of Spain, against the Soviet Union or against the United States of America."[12] A second wanted the Association to maintain its opposition to the misuse of science:

> In almost every number of *The Scientific Worker* appear articles or letters containing passages in condemnation of scientists who voluntarily pursue war research. This attitude, which is supported by all those scientific workers who value their social responsibilities, is a consequence of the expressed aims of the Association, and the strengthening of the opinion of scientists against those of their colleagues who misuse their skill and calling by taking part in military preparations has surely always been a part of the Association's policy. I do not see that this is in any way modified by such circumstances (now largely removed) which had prompted your letter; on the contrary, any such retraction at such a time would be a renunciation of the Association's stand against the misapplication of science.[13]

A third respondent was quite blunt:

> I did not join the A.S.W. in the belief that it would become a recruiting station for the war preparations of this Government. Lord Rayleigh has recently attempted to show [at the 1938 British

Association meeting] that war is not due to scientists and that they only play a minor part in the execution of this activity of mankind: but this does not . . . absolve scientists from the task of discussing the purpose for which their special assistance is required.

If the A.S.W. is to fulfill its function of making the voice of the scientists heard in society then surely its first task in a national emergency is to discover the opinion of its members and to present their view (or views) to the statesmen who are responsible for this state of affairs.

As the 'crisis' is now over and whilst the victors are digesting their conquest, might I suggest that you divert your energies from collecting names for the Government to arranging for the Association's members to discuss the question of the position of the A.S.W. when war again becomes imminent.[14]

The suggestion was followed, and the Association's position was discussed at open meetings of the Cambridge and London branches and at business meetings of other branches. At the Cambridge meeting on 2 November 1938, W. L. Bragg, Cavendish Professor of Experimental Physics and Nobel Laureate in Physics in 1915, opened the discussion from the chair saying that the question of the use of scientists in the event of war was already under discussion "by those responsible," but he welcomed the meeting in the hope that it would contribute new ideas. The Association's plan was to have the various meetings submit resolutions which would then be circulated to council members who would meet together on 19 November. After receiving the resolutions of other branches, and prior to the council meeting, the Cambridge branch drew up and submitted a second resolution incorporating the principal points of the other resolutions. Then at the council meeting, Bernal presented the omnibus resolution that was adopted as Association policy.[15] An improved version, retaining the full sense of the original and entitled "Science and National Defence," was issued by the executive committee as follows:

> While we regard war as the supreme perversion of science, we regard anti-democratic movements as a threat to the very existence of science. Hence we are prepared to organise for defence in a democratic cause and would actively resist any attempt to introduce fascism or any other anti-democratic system into this country either from inside or outside. We draw attention to the fact that the most

efficient utilisation of science in time of emergency necessitates in time of peace a much wider application of science to all productive forces and social services.

We stress the necessity for the immediate formation of an organisation for the utilisation of scientific knowledge in defence and we put forward the following points:

1. It is in the best interests of the country that scientists should be utilised in scientific work, and that the organisation and control of scientific work should be in the hands of scientists.

2. The professional organisation of scientists should play an important part in drawing up and operating any scheme of organisation and also be represented on scientific advisory committees to the defence departments.

3. In any scheme of registration, scientists should be grouped on a laboratory basis, no matter whether university, special research or industrial. It is particularly important that scientific workers in industry should be included in any such scheme and not left at the disposal of individual employers. Further, scientists should be consulted *now* with regard to such organisation, firstly on the grounds of efficiency and secondly to secure its democratic working.

4. Some machinery should be provided whereby scientists, including those in service departments, may exercise a right of criticism on purely scientific and technical matters.

5. It is important that scientific and technical as well as medical education should be maintained in time of war.[16]

Wooster incorporated the statement in a letter that he sent to leading newspapers and journals.[17] He explained that the items numbered one through four were directed toward ensuring that the mistakes of World War I would not be repeated. "Our best brains must not be wasted in the trenches, and scientists should be better occupied than in having to explain the operation of technical devices to military chiefs who are ignorant of the underlying elementary scientific principles."[18] When two weeks later, on 24 January 1939, the Minister of Labour announced the creation of the Advisory Council of the Central Bureau of Scientific, Professional, Technical and Administrative Staffs available for national work in an emergency, the Association sent a copy of its statement to each member of the Council.[19] It would seem that the Council was the outcome of the discussions mentioned by W. L. Bragg at the Cambridge branch meeting in the previous November. His father, Sir William Bragg, had supported the efforts of the two secretaries of the Royal Society, A. C. G. Egerton, Professor of Chemical Technology at Imperial College of Science, and A. V. Hill, to ensure that scientific and technical manpower was properly employed in the war that loomed ahead. Hill's register of scientists has already been mentioned. He and Egerton had urged the Ministry of Labour to take the entire matter seriously.[20] The Ministry subsequently invited the Royal Society to nominate a representative to the Advisory Council. Hill became chairman of the Council's Scientific Research Committee charged with compiling the section of the Central Register (of persons willing to serve in their professional capacities in some appropriate way in the event of war) dealing with scientific research.[21]

The February 1939 issue of the *Scientific Worker* dealt with various aspects of the subject of science and war including the need for national preparation and organization. In an article on "Scientific Workers and National Service," Bernal argued that "the most effective way in which scientists as a body can assist in the defence of the country is by securing the setting up of a comprehensive organisation of science, scientifically controlled and democratically administered."[22] Such a scheme had recently been implemented in France, and Bernal thought that it could be adapted to British needs and traditions.[23] The details of its constitution and functions could be agreed upon, Bernal suggested, after full consultations among such organizations as the Royal Society, British Association, and Association of Scientific Workers.

Deciding to approach the Royal Society on his own, Bernal sent proposals to Egerton, who responded on 23 March 1939, saying: "You and I probably see things from rather a different standpoint, but we wish to achieve much the same ends!"[24] In regard to Egerton's comments and suggestions, Bernal replied: "I am particularly attracted by the idea that much of the work which I think needs to be done could be done by the Royal Society. If this were really possible I think it would be the best solution."[25]

Six weeks passed with no sign that the Royal Society was undertaking the desired work, so on 13 May, Bernal had a motion passed at a conference of the Association of Scientific Workers instructing the executive committee "to proceed, if possible in conjunction with other bodies, to arrange a meeting of the Royal Society for the purpose of discussing present needs for the further organisation of science."[26] By that time, the executive committee considered that two of the points in the Association's statement "Science and National Defence" had to some degree been implemented.[27] First, with the creation of the Advisory Council in the Ministry of Labour, the control of scientists was in scientists' hands. Second, laboratories were being organized as units—the Department of Scientific and Industrial Research was surveying the situation and collecting information on scientific and technical teams in industrial and other laboratories. Other items of "utmost importance" had not, however, been implemented. Science was not being fully used, and there was no coherent plan for its use in defense.

Arrangements were made for a delegation from the Association to meet with officials of the Royal Society, but in June the Association's executive committee decided to postpone the meeting until October.[28] P. M. S. Blackett convinced the committee that the Royal Society was so busy with the actual compilation of the Central Register that it was not a good time to approach the Society concerning broader issues. Someone else pointed out that, in addition, summer vacations were approaching! Upon learning of the Association's decision, Hill replied that he would expect to hear from them later.[29]

When Hill and Egerton finally met with the Association's delegation, the conflict they had all anticipated had begun. Upon Britain's declaration of war on 3 September 1939, the Association's new general secretary, Reinet Fremlin, who was also a member of the Cambridge Scientists' Anti-War Group, drew up a memorandum on the utilization of scientists to be considered at the September 16 meeting of the executive committee.[30] The committee decided that a statement on the Association's position in the war, based on Fremlin's memorandum, should be sent to all Association members.[31] The subcommittee charged with preparing the document included Professor F. G. Donnan, the Association's president and until recently Professor of Chemistry, University College, London, Fremlin, Bernal, Levy, Holman, and Leonard Klatzow,[32] an industrial scientist.

As science would play an important part in the war, it was

necessary, the subcommittee's controversial statement argued, that Britain's scientific resources should be fully used to support the country's defenses.[33] At the same time, social and scientific progress should be maintained and even expanded and the foundations laid for further advances when peace returned. To work toward these ends, it was of fundamental importance that there should be central coordination of the country's scientific resources. The most effective coordination and the maintenance of a suitable balance between long-term research, short-term research, development, and immediate production would be, however, a major undertaking that no scientific body had the power to effect. This task could only be carried out satisfactorily by the "setting up of an organisation on the lines of a Ministry of Science which must be constituted with the express object of carrying out the above steps and with adequate powers to do so."[34] It was pointed out once again that such an organization was already functioning in France.

The idea of a ministry of science had been discussed at the September 16 meeting of the executive committee.[35] Fremlin had previously written to the *Daily Telegraph* supporting a proposal made by other correspondents for such a ministry. Recognizing that the Association now had a great opportunity to press the government to form a ministry of science, the executive committee decided that it should put all of its energies into doing so. It resolved to seek the support of the nation's scientific organizations in a campaign for the national coordination of science. The subcommittee previously mentioned was also asked to prepare a memorandum presenting the Association's views on a ministry of science to be sent to all bodies affiliated with the Parliamentary Science Committee and to other organizations.

By 14 October, the memorandum, "The National Coordination of Science," having gone through at least four drafts and obtained the executive committee's approval, was ready for distribution.[36] Copies were sent to twenty-five selected organizations and their comments solicited. The memorandum focused on the need for "a body, armed with powers and resources resembling those of the State Departments organising Labour and Supply, to ensure the pooling of knowledge and the rational allocation of scientific work."[37] The Association hoped to win the support of other organizations so that together they could request the government to set up an ad hoc committee to draft a constitution for the proposed body. As a basis for discussion, the memorandum offered a tentative five-point scheme:

1. The head of the proposed new department would rank as His Majesty's Secretary of State for the Co-ordination of Science.

2. The permanent service head of the Department would be the Director-General of Scientific Services. He would rank in status and salary with the permanent heads of other major Departments of State. He would be selected for his scientific eminence and breadth of outlook and for his special knowledge and experience of the requirements of Government departments and of industry.

3. The Secretary of State and the Director-General would be assisted by an Executive Council. This Council would consist of three parts: (a) Members appointed after consultation with the various State scientific departments, with the Universities, and with representatives of industry; (b) Members appointed by the professional organisations of scientists; (c) Permanent members whose position would be that of senior officers of the Department, responsible to the Director-General.

4. The functions of the Executive Council would be in the first place to make a review of the scientific research and development carried out in the whole country; then to maintain constant touch with the expansion of this work, to plan the extension and reallocation of scientific resources to meet the present and future needs of the Services, of the State departments responsible for health, education, etc., of industry, and of the social welfare of the people; and finally to advise the Secretary of State on the measures necessary for carrying out these plans.

5. The Department would take over the administration of those parts of the Central Register dealing with scientists and engineers, and would be responsible for the allocation of personnel on the register.[38]

The executive committee of the Association of Scientific Workers met on November 18 to consider the responses to its memorandum, which was to be debated at the half-yearly meeting of council the following week. Although eleven organizations, including the Royal Society and British Association, had responded, none was definitely in favor of the Association's proposals.[39] Four were decidedly against, and two suggested extensions of the DSIR or nongovernmental action (in

Scientists and War · 165

this connection the Royal Society mentioned the British Association's Division). Some organizations believed that the time was not ripe, or that the creation of the proposed department was unnecessary because of the existence of the DSIR and Medical and Agricultural Research Councils. To the further disappointment of the executive committee, considerable opposition to the memorandum was reported from Association branches. Two resolutions opposing its proposals had been tabled for consideration at the forthcoming meeting of council. Also, Julian Huxley had written saying that it would be impossible to have a ministry of science formed at present.

After much animated discussion, during which R. W. Western complained that no help could be expected from the "learned societies" in the nationalization of science, the executive committee reluctantly decided for the present not to press for the creation of a ministry of science. Klatzow dissented. Fremlin suggested as an alternative that the Association should organize a national council, independent of government and composed of representative scientific and technical personnel, that would draw up proposals for the organization of science. The executive committee agreed to propose this to council.

At a lively meeting of council on 25 November, Dr. Wingfield of the Aberdeen branch reported that although his colleagues agreed that the Association should work to secure the mobilization and coordination of scientific resources, they did not believe that the establishment of a ministry of science would bring these about.[40] Furthermore, such a ministry would not be controlled by scientists but by the government, which would be likely to direct science solely to the prosecution of the war, especially in its offensive aspects, thus conflicting with the Association's objective that science be used to benefit the community. Other opponents emphasized that although the proper organization of science could only be carried out by a democratic body, scientists were not yet in a sufficiently organized position to ensure that a government department for the coordination of science would be democratic. There was, therefore, the danger of a bureaucratic organization being created.

Proponents of a ministry of science felt that fear of this danger was exaggerated. Klatzow argued that even if there was a possibility of the government organizing science in a manner contrary to the ideas of the Association of Scientific Workers, that should not deter the Association from demanding proper state coordination of science for it would maintain its right of criticism. Nevertheless, the council resolved, by a vote of twenty-eight to twelve, to abandon the quest for a ministry of science.

In connection with the executive committee's alternative proposal, Bernal reviewed both the difficulties that would be encountered by, and the potentialities of, an unofficial national council formed by the Association of Scientific Workers. J. G. Crowther, the science journalist and a member of the London branch, observed that in contrast to earlier times, the Royal Society was now a group of scientific specialists having no influence on government in regard to social-political questions. The resolution to create a national council passed without opposition. One observer concluded that the council members were fully agreed that the first step towards adequate and democratic organization of science was "*the organisation of scientists themselves within a body representing their own interests.*"[41]

In the *Scientific Worker*, an anonymous writer gave a further reason for opposing any attempt to have a ministry of science created: "Such a Ministry under the present conditions would inevitably be used not for the purposes of the utilisation of science but simply for the propaganda value of assuring people that science was being used for their benefit."[42] Nevertheless, he was dissatisfied at the lack of "means for directing the collective scientific knowledge of the country." The Central Register was not fulfilling this function and the scientific societies had been unable to take the place of a national organization for this purpose. "Even the Royal Society is actually unable to make its immense prestige fully felt."

The long contemplated meeting between the Royal Society and the Association of Scientific Workers finally took place at the end of 1939. The Society was represented by Hill and Egerton, and the Association by Bernal and Klatzow.[43] Unlike the president of the Society, Sir William Bragg, neither Hill nor Egerton was ever a member of the Association. To the contrary, Hill strongly disapproved of some of the Association's activities both before and in the early days of the war. Defining politics as the art and science of government, he later explained: "I was all in favour of bringing more science into politics; but all against bringing party politics (quite another matter) into science. And when party politics led people to say, *deliberately*, what was not true I had no patience with them."[44] The scientific establishment's feelings were not unknown to other scientists. The Tots and Quots was a small society of scientists, founded by the Oxford anatomist Solly Zuckerman and including Bernal, Needham, and Huxley, that met regularly in London for dinner and discussion of scientific issues.[45] The theme of its discussion of 23 November, 1939, was the disorganization of science, and "its keynote was the belief that scientists who wanted to help in the war effort were being regarded by the scientific establishment as meddlesome troublemakers. The leaders of science 'felt it their duty not to disturb the equanimity of the politicians and civil servants who were running the war.'"[46] At the following month's dinner, a guest, William Slater, argued for the creation of a scientific general council. A second guest, Egerton, "was diffidently unsympathetic. He disliked both the idea of excessive planning in science and any undue emphasis on its 'social aspects.' In his view, the general position would be sufficiently improved if the Secretaries of the three Government Research Councils operated in concert with the Council of the Royal Society, instead of separately. This was not shared by most [Tots and Quots] members, who felt that even if the Research Councils and the Royal Society did get together, they would still be remote from those responsible for the determination of national policy."[47]

The meeting of Hill, Egerton, Bernal, and Klatzow was cordial. Hill thought that the Central Register was working well when properly used. The chief difficulties were due to the administration of scientific work by nontechnical officers. In regard to national science policy, and presumably also to the Association's resolution to create a national science council, Egerton agreed that scientists should have a greater measure of control, but both he and Hill felt that the Royal Society itself would be a sufficient executive body. Here, Klatzow gained the impression that in this connection the Royal Society been "snubbed."

On this occasion, as on an earlier one, the Royal Society chose not to tell the Association of Scientific Workers all it knew. In

October 1939, when the Association was seeking support for its idea of a ministry of science, its president, Donnan, had approached Sir William Bragg, and together they met with Sir Edward Appleton, secretary of the DSIR. Although Appleton was interested in the idea of a ministry of science, he did not think it feasible. As for Bragg, Donnan reported to the executive committee of the Association of Scientific Workers that he "had vaguely indicated that something was afoot."[48] It may well be, Donnan had speculated, "that the announcement in *Nature* of 7th October concerning the plans of the Royal Society for making an emergency list of unemployed scientists was what he was referring to." But Donnan was mistaken; the Royal Society had more ambitious plans.

Sir Henry Tizard perhaps did more than anyone to contradict Stanley Baldwin's famous remark that the bomber would always get through. The story of his chairmanship from 1934 of the Air Ministry's Committee for the Scientific Survey of Air Defence—or, as it is better remembered, the Tizard Committee—is well known.[49] In this capacity he was well placed to observe the organization and effectiveness of government defense science. On 21 July 1938, Tizard wrote to Sir Thomas Inskip, Minister for the Co-ordination of Defence in Neville Chamberlain's National government.[50] It seemed to Tizard that it would be of the utmost value on general grounds to bring together at "reasonable intervals, say monthly or quarterly" the three directors of scientific research in the three Service departments (Admiralty, War Office, and Air Ministry) and to assist them with the collaboration of selected outside scientists. He wanted to ensure that the departments would receive the best available scientific advice and be kept abreast of scientific developments in the universities and industry. Tizard therefore proposed that a Research Co-ordination Committee be set up under the Minister for the Co-ordination of Defence with a membership to include the three directors of research, the secretary of the Department of Scientific and Industrial Research, and not more than six nongovernment scientists. He suggested that the terms of reference of the committee be to initiate and to advise upon proposals for the application of new scientific methods to defense problems, to advise on any scientific problems specifically referred to it by the Service departments or by the Committee of Imperial Defence, to advise on the selection of individuals to assist the Service departments by providing scientific advice on particular problems of defence, and generally to keep under review the organization of scientific research within those departments.

Copies of Tizard's proposal were circulated to the three service ministries but led to no innovations. The reason being that "while the Air Ministry were favourably disposed to the scheme, and while the War Office were willing to consider it, the Admiralty were opposed to the idea."[51] The matter was not pressed because it was thought that little was to be gained by attempting to force the scheme on an unwilling Admiralty.

So the general idea lay dormant for several months until a similar suggestion was independently made on 14 June 1939. This new proposal was submitted jointly by Egerton and Hill. As early as April 1938, Egerton, a former member of the Advisory Council of the DSIR, had publicly stated his belief that scientists should be invited to take a much more responsible position in the affairs of state and should be in closer touch with members of the government.[52] In the spring of 1939, he and Hill discussed a memorandum on scientific advice in relation to government written by a member of the council of the Royal Society, Major Greenwood, Professor of Epidemiology and Vital Statistics at the University of London.[53] Egerton told Hill of Tizard's somewhat similar, earlier scheme. They then proceeded to compose a memorandum in consultation with Sir William Bragg. The Egerton-Hill proposal specifically argued that communication between the government and the scientific community could be improved by the appointment of a small scientific subcommittee to the Committee of Imperial Defence.[54] The functions of this body would be to advise the government as to the best means of obtaining scientific advice and information on any question that the government might submit to it, and to inform the government of the opinions of responsible bodies or of the results of scientific research that seemed likely to be of importance in governmental activities or for the public welfare. The authors suggested that these types of communication were often delayed for lack of known and accepted methods of approach. The Egerton-Hill proposal, although it was submitted to the Committee of Imperial Defence, was much broader in scope than Tizard's, involving as it did the entire spectrum of government science and not merely defense science.

In informing the Minister of Co-ordination of Defence, now Lord Chatfield, of the Egerton-Hill proposal, the sympathetic Sir Edward Bridges (Permanent Secretary of the Offices of the Cabinet and of the Committee of Imperial Defence) noted that a suggestion from the Royal Society "carries great weight and must be carefully considered." "There is also the consideration," he continued forcibly, "that it is difficult to refuse an offer of help proffered by such a body as the Royal Society. What could be more damaging than the knowledge that the leading scientific men in the country offered their help to the Government and that that help had not been accepted."[55] When Chatfield met with Bridges three days later on June 30, he agreed that were the Royal Society, as distinct from two of its fellows, to offer its help in ensuring that leading scientists were in touch with developments within the defense services, "the offer was one which could scarcely be refused." He instructed Bridges to discuss the matter informally with Egerton and Hill. Bridges would explain that if the Royal Society were to write officially and offer to help along the lines indicated, Chatfield might then summon a meeting of the service ministers to meet with Bragg and other representatives of the Royal Society to discuss the proposal.[56] Chatfield understandably wanted to deal directly with the Society's president.

When Bridges subsequently met with Egerton and Hill, he also explained that there were difficulties involved with their proposed committee, covering as it would scientific matters in general and not merely those connected with the defense departments.[57] He therefore suggested that the proposal be altered so as to limit it to these departments. Egerton and Hill agreed to draw up a modified version for official submission by the Royal Society and to send a draft to Bridges. They experienced no difficulty in persuading Bragg to write to Chatfield a week later, on July 13, forwarding the new proposal.[58]

Bragg's letter advocated the creation of a small committee of not more than six members of the Royal Society with the following duties: first, they would familiarize themselves with the various governmental scientific organizations and whenever possible would draw the attention of the Committee of Imperial Defence either to knowledge that might be used by these organizations or to men possessing special abilities to deal with particular matters of concern; second, the committee would advise the Committee of Imperial Defence, when requested, as to the best means of obtaining scientific advice or information on any matter and would report in general on any deficiencies of which it became aware; third, on its own initiative, the committee would bring to the attention of the Committee of Imperial Defence any opinion, suggestion, or scientific result that might seem to be of importance to its purposes.[59] Thus, the committee would provide a link between the government and the scientific community concerned with quickening and improving the application of science to defense. This was a startling proposal. Unlike Tizard's, it excluded government scientists, and unlike the previous Egerton-Hill proposal, it specified that all members of the committee be fellows of the Royal Society. Furthermore, they would oversee all government scientific establishments.

Despite these features, Chatfield sent a detailed account of the proposal to his Service colleagues—Earl Stanhope, First Lord of the Admiralty; Leslie Hore-Belisha, Secretary of State for War; and Sir Kingsley Wood, Secretary of State for Air—and to the Minister of Supply, Leslie Burgin. In suggesting a meeting to discuss the proposals, Chatfield disclosed: "I am rather favourably disposed to the suggestion in Sir William Bragg's letter. It seems to me that a body on the lines proposed cannot but have beneficial results in bringing scientists in Government Research Establishments into contact with persons of the highest scientific attainment in the country. In any case I think you will agree that a suggestion on these lines, coming from the President of the Royal Society, is one which cannot lightly be put on one side, and which must be given very serious consideration."[60] A meeting was set for July 31, and each service minister and the Minister of Supply was invited to bring with him the member of his board or council responsible for research.[61] Bragg was unable to attend as he was already committed to being the principal guest at a dinner at Winchester College, but he was to be represented by Hill.[62]

While these arrangements were being made, however, the

Royal Society's proposal became the subject of an unanticipated and crippling attack. On July 24, the Man Power (Technical) Committee of the Committee of Imperial Defence held a meeting at which Bragg's letter to Chatfield came under discussion because the general subject to which it was related was before the Man Power Committee.[63] To this Committee's chairman, W. S. Morrison, Chancellor of the Duchy of Lancaster, Bragg's letter seemed to suggest that the DSIR should be superseded by the proposed committee, and that the committee should be charged with eradicating the departmentalism and isolationism that Bragg evidently considered to be rampant in government establishments. These views were supported by vigorous protests from other members of the Man Power Committee. An Admiralty spokesman heatedly "denied absolutely and unconditionally all the charges made by Sir William Bragg." It was clear, he added, that Sir William must be ignorant not only of the high degree of cooperation sustained between government establishments, but also of the scope of exchanges made between them and the scientific world. Some thirty to forty fellows of the Royal Society were already actually engaged by departments, and the best available scientific brains were habitually applied for the benefit of departmental research. With these remarks the secretary of the DSIR, Sir Edward Appleton, also a member of the Man Power Committee, agreed without qualification. He was thoroughly satisfied with existing arrangements and thought it easy to rebut Bragg's charges. Appleton did not see how the proposed committee would serve any useful purpose. He was personally in close and constant touch with the departmental directors of scientific research, and he was perfectly confident that any six fellows of the Royal Society whom Bragg would wish to name for the proposed committee would be found to be already in consultation, as advisers, with government establishments. It was ironic that the helpful but unfortunate Bragg was thus cast as the villain, while the real architects of the proposal, Egerton, Hill, and Bridges, went unmentioned.

The result was that the meeting called by Chatfield to discuss the Royal Society proposal was postponed to "some future date." In informing Bragg of this decision, Chatfield did not mention the reaction of the Man Power Committee but gave the excuse that the date set was inconvenient to some of the ministers concerned.[64] It is true that, as Bridges had informed Chatfield, the Secretary of State for War would have been unable to attend, but Bridges had also informed Chatfield of the response of the Man Power Committee and had suggested a postponement of the July 31 meeting "as one of the Departments concerned was evidently viewing the proposed establishment of [the] Committee with dislike and was regarding it in some sense as an affront to their own personal competence."[65] Chatfield viewed the issue as involving a difference of opinion among scientists—with Bragg and his colleagues desiring increased contact between scientists outside and those within the services and Appleton and the Admiralty opposing this. Chatfield did not think it desirable to hold a meeting at which this would be presented as the point at issue but at which Bragg would not be present.

Even before Chatfield had written to Bragg, Hill knew what had happened. He had discussed the outcome with Bridges and Wing Commander William Elliot, an Assistant Secretary to the Committee of Imperial Defence and as sympathetic to the Royal Society's aims as Bridges, and had agreed to the postponement. Indeed, on the day that Chatfield wrote to Bragg, Hill wrote to Elliot expressing the hope that he and Bridges would "not feel that the temporary misadventure of the R[oyal] S[ociety] proposal, and the decision to defer its consideration, are due only to its handling at your end."[66] Hill explained that he and Egerton should have foreseen the reaction that had occurred and should have gotten Bragg to word his letter differently. Hill had had a suspicion that all might not be well and he ought, he said, to have had Elliot and Bridges wait another week while he and Egerton planned the letter more carefully. Bragg had telephoned him, and Hill had told Bragg that the meeting was off and "hinted that this [was] to smooth over any misunderstanding that might arise."

Ironically, that day before telephoning Hill, Bragg had sent him a letter concerning the scheduled meeting. Bragg acknowledged that the idea for the proposed committee had come from Hill, Egerton, and Bridges. He observed that the persons before whom Hill would be making his case were so representative that Chatfield was obviously attaching much importance to the proposal. He continued:

> I do not think that there can be much contradiction of the facts in my letter. If objections were made by the representatives of research in the various services they will I think relate to the manner of working of the scheme. It *may* be said in the first instance that this new body would be inquisitorial, and that their powers would be practically those of censors, and would be resented. There would be much force in this objection, and it is not easy to see how it might be met. We discussed this point.... I did actually put into my original draft that this body was to go into the scientific relations of any research body when directed to do so, by CID. I saw that you left this out, and perhaps you are quite right: it might prevent the new body being useful. And I thought that the right tactics would be worked out by Bridges who has had so much experience. On our part we can insist that we are out to help, not to criticize. It would be very difficult to lay down rules: the success of the thing must depend wholly on personality. If the new body can be looked on as friends and helpers, and will inspire confidence that they are such, the result will be excellent and may be of tremendous importance. If not, the move will come to no good end. The members of the new body will have to be selected with great care. They must not only have a wide knowledge of science and scientists, but must also be very tactful, not aggressive, willing to let others have the credit of any suggestions they make. Of course they must be discreet men.[67]

As always, discretion was important. The ministers involved in the scheduled 31 July meeting were discreetly told that, due to the inability of the Secretary of State for War and Sir William Bragg to attend, the meeting had been postponed.[68]

Just over two weeks later, Hill and Elliot met and agreed that the next step should be for Bragg to write a second letter (after Hill had seen him and fully explained what had happened) in which he would refer to his previous letter and say that, lest their scope and purpose be misunderstood, he would like to put his proposals in slightly different form! In fact, as Bridges had suggested in a minute of August 4, Bragg should "re-state the object of the proposed Committee so as to prevent the Defence Departments from thinking that the establishment of this Committee in any way reflects on their competence or on their ability to obtain adequate outside information where necessary."[69] Hill agreed that such a letter would reopen the matter "this time rather more cautiously" and "without official cognizance of the 'objections' (which need never reappear if the matter is handled with tact)."[70]

Bragg again obliged and sent the new proposal to Chatfield on 8 September 1939, by which time Britain had officially been in a state of war with Germany for five days. Somewhat earlier, in late August, Elliot had seen Tizard and, as he told Hill, they had agreed that if any good were to come out of the coming war it would be "the fillip which it will give to the proposal for the Central Scientific Committee which we have at heart!"[71] On 9 September, Tizard was at the headquaters of the Committee of Imperial Defence where he told the authorities that Bragg's new proposal was excellent. The next day he wrote to Hill: "From what I heard there is a good chance of it going through."[72]

In his letter of 8 September, Bragg had written that it seemed almost certain that new problems requiring scientific attention would arise from time to time and that it would be very important to realize their urgency as early as possible and to apply the appropriate scientific methods immediately to their solution.[73] Since the Royal Society had a great reserve of knowledge and experience that was desirable to use as soon as a need was discovered, he was making a simple proposal that would help ensure that that knowledge and experience would be applied with as little delay as possible. The suggestion was that the two secretaries of the Royal Society, Egerton and Hill, who between them had contact with most branches of science, should be attached in some appropriate way to the establishment of the War Cabinet, which would then have at its disposal a direct connection with the scientific community. Such an arrangement would help ensure that coordination of scientific effort in governmental and other organizations was maintained. From a beginning of this kind, the letter continued, the fuller plan suggested previously in July might come to reality after trial and experience. But the times required urgent action, and the beginning suggested would probably satisfy immediate needs. As these changed, the plan would mature. In all of this, the boldness of Hill and Egerton is remarkable.

As for Bridges, this time he proceeded much more cautiously.

On 11 September, he suggested to Elliot that before they took any action on the new proposal Elliot should confer with Harold Parker, a principal assistant secretary at the Treasury, who was "very knowledgeable on these scientific affairs."[74] Indeed, Parker knew quite a bit, including what had transpired on the Man Power Committee, and he offered his impressions.[75] At the beginning of World War I, there was very little in the way of what might be called scientific research establishments in government departments, so there was much more for the Royal Society to do. At that time, for example, the National Physical Laboratory was not a government department, as it was in 1939, but was under the Society's control. In the current situation, Parker could not help feeling that there was a fear on the part of departments, however unjustified, that the Royal Society wanted to get back into a position of general executive authority. He therefore thought it important to make it crystal clear that the Royal Society was "out to help and not to dictate."

A memorandum based on the Royal Society's proposition was subsequently drawn up, and on 28 September Chatfield sent it, on Bridges' suggestion, to only the three service ministers and the Minister of Supply and not to the full War Cabinet.[76] The intention was to avoid any suspicion on the recipients' part that Chatfield was trying to rush them or that he was bringing a proposal to the War Cabinet before it had been adequately discussed with them. The office-holders concerned were the same as in July, with the exception that Winston Churchill had replaced Stanhope at the Admiralty.

Having outlined the new proposal, Chatfield went on to acknowledge in the memorandum that various objections had been made against the previous one.[77] These, he disingenuously suggested, were mainly on the grounds that, first, establishment of a new committee would mean more work and would put a strain on officials who were already very hard pressed, and second, that the necessary contacts between scientists within and those outside the Services already existed. Chatfield had no "wish to enter into a discussion of controversial matters," but so far as he could see the new proposal was "not open to objection on either of these grounds," although its adoption might be of incalculable value. He therefore strongly favored accepting Bragg's offer, "at any rate for an experimental period." If his colleagues, he suggested, affecting a casual air, were willing to support his conclusion in principle, it might perhaps be left to him to settle the details with the departments concerned.

The War Office and the Ministry of Supply were indeed generally favorable to the Royal Society proposal, but once again the Admiralty balked.[78] Churchill replied: "So far as I know there are no grounds for assuming that the Government research departments are not in close touch or that better liaison between them is needed. I am inclined to think therefore that the suggestion is unnecessary and should be dropped."[79]

In July, opposition from the Admiralty and Appleton had led to hesitation and tactful change, but this time Chatfield took the proposal directly to the War Cabinet, which was apprised of it on 29 September and discussed it on 3 October. The Prime Minister, Neville Chamberlain, was in the chair. Chatfield explained how the Royal Society's initial suggestion had not found great favor with the service ministers. They had been afraid, he repeated, that it might involve either interference with the existing satisfactory arrangements, or at least a waste of valuable time in acquainting outside scientists with devices and plans with which they would be unfamiliar. Many of these devices and plans were matters of the highest secrecy, and it was thought undesirable to widen unnecessarily the circle of those who had access to them.

In turning to the Society's more recent proposition, Chatfield said that with the exception of the First Lord of the Admiralty the view of the Service ministers, broadly speaking, "had been that if the Royal Society were anxious (as they evidently were) to offer their collaboration, it would be impossible for the Government to refuse that offer."[80] Chatfield, however, went on to say unexpectedly that he himself had somewhat revised the views that he had expressed in his memorandum to the Cabinet. In doing so, he displayed his independence of Bridges. He would now go no further than to say that Bragg's proposal had very useful possibilities. He had come to the conclusion that the offer should be accepted in principle but not in the actual form in which it had been made. It would be preferable if there were an understanding with the Royal Society that if any department, whether defense or civil, were confronted with a new problem that its existing staff had not the time to investigate, it should call in the help of outside scientists. This might be done by referring the problem to one or other of a number of advisory committees or panels dealing with engineering, physics, chemistry, food, and medicine, that the Royal Society might be asked to set up. It seemed hardly necessary to attach the two secretaries of the Royal Society to the establishment of the War Cabinet. They should, however, be available to act as a link between the office of the Minister for the Co-ordination of Defence and the suggested advisory committees or panels. Chatfield's new attitude had resulted from discussing the proposal with Sir Frank Smith, director of instrument production at the Ministry of Supply, and Appleton's predecessor as secretary of the DSIR, who had favored accepting the Royal Society's offer in the modified form that Chatfield now advocated.

Following general discussion of the matter, the Cabinet agreed that a reply should be sent to Bragg along lines proposed by the Prime Minister. In this, Chatfield explained that the proposal had been given a very thorough and sympathetic hearing by the Cabinet.[81] However, a wealth of scientific knowledge and experience was already at the government's disposal in the existing technical staffs of its departments and in various advisory committees and panels on which fellows of the Royal Society and other scientists were already serving. He reminded Bragg that the Royal Society had cooperated with the Ministry of Labour and National Service in compiling the scientific section of the Central Register. The DSIR had in addition drawn up a register of scientific teams that would be used as required. With these considerations in mind, said Chatfield, the Cabinet had concluded that the adoption of Bragg's proposal in full might involve some danger of overlapping with existing machinery. It had also concluded that the objectives that Bragg had in mind could be realized without attaching the two secretaries of the Royal Society to the establishment of the War Cabinet.

The letter, however, went on to say that the government was most deeply impressed with the importance of seeing that the fullest possible use was made of the country's scientific knowledge and that it would most certainly avail itself of the Royal

Society's offer to place at its disposal the advanced knowledge of the Society's fellows. Steps were accordingly "being taken to ensure that this offer is known to Departments, and the suggestion is being made to them that, as and when new problems arise, they should consult with the Royal Society before deciding how best to arrange for their investigation." Given the protests against the first proposal, it is surprising that it was not foreseen that such steps would elicit a further storm of protest.

Bridges and Elliot set about informing departments. At the War Cabinet meeting, one of the points to emerge in discussion had been that the medical profession would likely "experience difficulties" if the Royal Society were brought into consultation on medical questions. It was emphasized that physicians would resent any advice from the Royal Society. With this in mind, Bridges and Elliot agreed that in writing to the secretary of the Medical Research Council, Sir Edward Mellanby, they should, in addition to sending a copy of the letter being sent to other heads of departments, enclose a covering note to the effect that the Council was different from other government research organizations and that the letter was being sent only for Mellanby's information.[82]

Despite this precaution, Mellanby took alarm. He replied to Bridges that he found the letter very disquieting as it seemed to indicate a reversal of the policy, developed during the First World War and followed ever since with growing success, that the government should seek scientific advice in the first instance through its own research departments.[83] If now the various government departments began to consult directly with the Royal Society on matters of scientific research at the expense of their own officially appointed research bodies, there would be a regular mix-up that would make the end results less effective. The Royal Society did not have the machinery, the official personnel, nor—at least in the biological area—the knowledge and experience for dealing with matters of this kind. Mellanby could not see that the consequences of the proposed action would be other than detrimental, both from the point of view of practical discovery and from the undesirable clash of personalities that would most certainly ensue.

The secretary of the DSIR, Appleton, replied in similar vein to Bridges' letter which, he said, had disclosed a situation that was "extremely embarrassing" to him. He explained:

> The procedure now envisaged in your letter cuts across a fruitful practice of consultation between other Departments requiring scientific advice and the D.S.I.R., and substitutes for an orderly scientific advisory service, regulated through a single channel, a haphazard approach by individual Departments to an outside organisation not possessing, and unable rapidly to acquire, full knowledge of the whole field of Government scientific endeavour. Thus co-ordination goes by the board and the effectiveness of the assistance which the Royal Society can offer is automatically diminished at the same time as the usefulness of this Department is in large measure destroyed.
>
> Frankly, I think that in the long run it would be disastrous from the national point of view if this Department, which was born in the last war, should become one of the first casualties of this one.[84]

The secretary of the other research council, the Agricultural Research council, was Sir Edwin Butler. He, too, had been disturbed by Bridges' letter, but chose to take his complaints to the Minister of Agriculture, Earl Stanhope, formerly First Lord of the Admiralty. Stanhope spoke with Bridges who was much distressed that the secretaries had all felt that, through the circular letter, their respective organizations had received what Stanhope had described as "a slap in the face." Stanhope subsequently informed Butler that Bridges was going to ask the secretaries to be good enough to come and see him and, significantly, "provided that it can be done without upsetting Sir William Bragg and the Royal Society, to send out a further letter saying that Government Departments should approach that august body through one of your respective Councils, each of which is so largely comprised of Fellows of the Royal Society."[85] Bridges wrote to Appleton saying he was sorry to learn that Appleton had been so much disturbed by the circular letter.[86] Although Bridges had merely carried out the wishes of the Cabinet, he nevertheless assumed the blame in agreeing that the terms of the circular letter were certainly ill-judged. The last thing that was intended, he said, was to take any action that would cut across the existing practice of consultation between the DSIR and other departments requiring scientific advice. Bridges also sent a copy of this letter to Mellanby asking him "to accept the apologies in my letter to Appleton as offered to you also."[87] As Stanhope had indicated, a meeting of Bridges and the three secretaries was arranged "to settle this matter by discussion rather than by correspondence." There the entire matter rested for the time being, the Royal Society, or rather Hill, Egerton, and Bridges, having been foiled a second time.

The Society's third attempt, however, was the proverbially lucky one. It was launched after the Society learned of certain questions being asked within the government. These were initiated, surprisingly, by Churchill, who on the afternoon of 3 April 1940 spoke to Sir Horace Wilson, head of the civil service, about the need to coordinate the various research activities of the defense departments.[88] Churchill, however, had had a recent conversation with Tizard that had confirmed his view that the country was not getting the best from its research organizations. Churchill therefore suggested to Wilson that the responsibility for supervising research be linked with the Ministerial Military Co-ordination Committee of which Churchill, as First Lord of the Admiralty, was chairman. In discussing this suggestion, the question was raised whether it might not be an equally good arrangement to ask Lord Hankey, minister without portfolio, to deal with research organization, with the understanding that Hankey would keep in close touch with Churchill as chairman of the Co-ordination Committee. Hankey had for a long time— from 1912 to 1938—been secretary of the Committee of Imperial Defence.[89] Wilson undertook to think about the matter, and Churchill agreed that there was no need to reach a decision immediately.

Nevertheless, Wilson mentioned the matter to Chamberlain that evening, telling him that he would begin inquiries with a view to making a recommendation. Wilson then sought the advice of Sir Edward Bridges, who subsequently discussed the matter in turn with Wilson, with Churchill, and finally— accompanied by Wing Commander Elliot—with Sir Alan Barlow, Under-Secretary to the Treasury.[90] Barlow undertook to put on paper a scheme that he, Bridges, and Elliot had discussed and to supply a draft of it to Bridges and Elliot. This was forwarded at the end of April. It was seen shortly thereafter by Wilson and Hankey, both of whom agreed to it.

This memorandum, entitled "Research Work for the Services," began by mentioning that Churchill had raised the question of increased coordination of the research work of the Admiralty, Air Ministry, and Ministry of Supply. The vital subject of radar had been mentioned specifically in this connection within the Cabinet. Furthermore, Tizard had also submitted certain related suggestions to Chamberlain, namely, that there was need of a central authoritative review of technical progress, that there ought to be some means of deciding on the relative priorities of current researches and of new proposals, and that the chief persons concerned with technical development in each department might be asked to agree on a short list of new requirements or inventions that they considered to be war-winners. After reviewing the existing situation, the memorandum concluded that it was not likely that any new coordinating machinery of the nature of a committee was needed or would be helpful. What appeared to be needed was:

> a) Some central arrangement for considering the relative priority of researches for war purposes.
> b) Some central watching of the progress of those researches which might have an important effect on the war.
> c) Periodic reports to the Cabinet.
> d) Possibly a rule that some central approval should be obtained before new research is begun for which any priority is desired.
> e) The allocation of responsibility to one Cabinet Minister.[91]

In regard to item (e), it was suggested that Hankey be the minister responsible. A list of the most important and urgent items of research in progress in each department would be submitted to him. He would then consult with other ministers before drawing up an agreed priority list to be circulated to all departments concerned. Hankey should be assisted, said the memorandum, by a liaison officer whose business it would be to observe the development of the work and to obtain or prepare progress reports for Hankey. It was suggested that Lord Falmouth—a qualified electrical and mechanical engineer, and a member both of the executive committee of the National

Physical Laboratory and of the advisory council of the DSIR—should be asked to serve in this capacity.

Wilson was to have taken the memorandum to Chamberlain, but apparently before this was done, or at most before any action was taken on it, Chamberlain's government fell. From 10 May 1940, Churchill was prime minister. Chamberlain, however, became Churchill's Lord President of the Council—the minister responsible for government science. As such, he sent a note to Churchill on June 2 concerning the application of scientific research to defense. This referred to requirements similar to those listed in Barlow's earlier memorandum, "It seems to me that the best way of meeting these objectives is to ask Lord Hankey, who has already done a considerable amount of work in this field, to take the matter up. I suggest that he be assisted by Lord Falmouth who is well qualified for such work."[92] Chamberlain noted that these proposals were similar to ones that Churchill had considered "a little while ago." Churchill had thought then that the matter was not urgent, but Chamberlain suggested "that the importance of the objectives I have described is such as to make the adoption of the proposals desirable now."

On 3 June, Churchill remitted the matter to Hankey requesting a report, which Hankey sent to Chamberlain on 27 June.[93] Its subjects were the current state of scientific research and the part that this research could play in home defense.[94] On the very day that the matter had been referred to Hankey, the last troops were being evacuated from Dunkirk, while at home Britain was preparing for the worst. In the ensuing dire national situation, Hankey had held meetings with certain prominent physicists who were also fellows of the Royal Society: Tizard, P. M. S. Blackett, Lawrence Bragg, C. G. Darwin, and T. R. Merton, all of whom had been recommended to Hankey by Sir William Bragg. Hankey also met with various directors of government research, including Appleton of the DSIR. After having submitted his report, Hankey, as had been recommended by Chamberlain, was given responsibility for coordinating research in the service departments, a task in which he had the assistance of Lord Falmouth.[95] Meanwhile, in involving Bragg and the Royal Society in his discussions, Hankey had revealed to Bragg an opportunity for pressing the Royal Society case a third time.

In spite of the earlier failures, Egerton remained convinced that it was essential that something be done.[96] Consequently, he had had conversations with a person named Beresford, who was knowledgeable about government organization because of his Treasury connections. Beresford thought that something could be done through the scheme set on foot in the late 1920s of a council composed of the secretaries of the DSIR and Medical and Agricultural Research Councils. Although the secretaries continued to meet now and again unofficially, there seemed good reason to estabish official meetings. Egerton thought that it might be possible to incorporate officials of the Royal Society into the Council and thereby create the focus point that was so badly needed in the organization of British science. It was essential, however, that a third proposal should not fail. So Egerton and W. W. C. Topley, Professor of Bacteriology and Immunology at the University of London and a member of the Medical and Agricultural Research Councils, who had assumed Hill's duties at the Royal Society while Hill was in the United States,[97] arranged to discuss the idea at a lunch at the Royal Society to which the secretaries of the research councils were invited. After a second luncheon meeting with them, Egerton drew up a memorandum that Topley made "much better."[98] Bragg agreed "to go forward but was very anxious lest there should be another setback."

On 10 June, Bragg wrote a long letter to the prime minister and began by employing a new lever.[99] He had recently been made chairman of a committee appointed by the Lord Privy Seal, Clement Attlee, to advise the government on national food requirements and production.[100] Bragg claimed that the appointment of this committee afforded clear evidence of the government's determination to utilize all the resources of science in solving problems created by the war. But if these resources were to be applied quickly and effectively, there was urgent need for coordinating machinery that did not at present exist. He explained that although each research council was fully competent to give advice within its own sphere of activity, nevertheless there were many problems that concerned more than one of these bodies, and so to be fully effective, the councils needed "to be in constant touch, not only with one another, but with the various Government Departments, and with the Cabinet." What was needed was a coordinating body that, if it was to work quickly and effectively, must be small and have direct access to the Cabinet. If possible, it should have a Cabinet minister as chairman. As for the other members, Bragg suggested that they be the president and two secretaries of the Royal Society and the secretaries of the three research councils. This was a significant and diplomatic modification of the earlier, second proposal. The three principal government scientists were now included, and the president of the Royal Society was brought in to create a balance between them and the Society's representatives. With this arrangement, any proposal of the latter would immediately be known to, and if necessary could be opposed by, the government scientists. The proposed committee would not be the independent body that Hill and Egerton had wanted, but it would be more acceptable to the opposition and, if created, would ensure coordination of government science and some measure of critical assessment. Its suggested duties were, in general, to make recommendations on scientific matters referred to it by the Cabinet and to transmit to the Cabinet any suggestion as to the better utilization of existing scientific knowledge in the prosecution of the war. In all of its activities, its function was to be limited to that of achieving effective coordination. Most importantly, it was not in any way to usurp the functions of the research councils or any other official body.

The government paid no attention to this letter until more than a month later. During the interval, the much discussed resignation of Tizard, as Scientific Adviser to the Chief of Air Staff and chairman of the Committee for the Scientific Survey of Air Warfare, occurred on 21 June. This was accompanied and, indeed, caused by the emergence of his famous rival, F. A. Lindemann, Churchill's personal scientific adviser and later Lord Cherwell, as the chief scientific adviser in the government. Five days later, Hill also submitted his resignation from the air warfare committee to Sir Archibald Sinclair, Secretary of State for Air. By this time, Hill, having won a bye-election in February as an Independent Conservative, represented Cambridge University in the House of Commons.[101] In his election address, he had urged the need of bringing science and learning to bear upon national affairs, declaring that science had still too little influence on higher government policy.[102] In writing to

Sinclair from his politically enhanced position, he warned: "In the difficulties and confusion now confronting the Government it would do no good to pursue the matter [of Tizard's resignation] further at present; later on, however, it may be necessary, on some suitable occasion to bring the matter up in the House. I know that practically the whole Scientific Community would be of the same opinion about it."[103] In a second letter to Sinclair written soon thereafter, Hill elaborated: "In the ordeal which the country may soon have to face—and in preparing for it now—it will be disastrous if the best advice is not available and if action is not quickly taken on it. It cannot be, unless the present unhappy situation is drastically altered. This is not a matter only of Tizard, but of the effective use of our scientific and technical assets and resources in waging war—and not squandering them in wild goose chases."[104] This was a reference to various suggestions that had been made by Lindemann and that were soon listed and damned in a secret memorandum—"On the making of technical decisions by H. M. Government"—drawn up by Hill and approved by a number of scientists including Tizard.[105] In this, Hill wrote:

> The time is too grave to allow personal considerations to hinder one from making a frank statement, though I make it with regret. It is unfortunate that Professor Lindemann, whose advice appears to be taken by the Cabinet in such matters, is completely out of touch with his scientific colleagues. He does not consult with them, he refuses to co-operate or to discuss matters with them, and it is the considered opinion, based on long experience, of a number of the most responsible and experienced among them that his judgment is too often unsound. They feel indeed that his methods and his influence are dangerous. He has no special knowledge of many of the matters in which he takes a hand. He is gifted in explanation to the non-technical person; the expert, however, realises that his judgment is often gravely at fault. Most serious of all is the fact that he is unable to take criticism or to discuss matters frankly and easily with those who are intellectually and technically at least his equals.
>
> I realise that Professor Lindemann's presence may be indispensable to the Prime Minister, and the prestige and influence of the Prime Minister are now so important to the nation that some compromise may be necessary. It is impossible, however, for the present situation to continue as it is. I possess detailed evidence in various instances of the waste of time, money and effort which his

advice has caused, and of the improper methods by which he has worked. A large number of my scientific colleagues, inside and outside the Government service, will confirm these statements and they cannot be seriously disputed. I realise that the whole trouble is not due to one person, but that a system has grown up of taking sudden technical decisions of high importance without, or against, technical advice. These decisions, as in the ill-considered case of the pellets designed to destroy enemy crops, may involve questions of high policy. They may involve grave loss of human life and of national supplies, as in the failure to provide protection for merchant ships. They may lead to ill-advised adventures which slow down the production of tried weapons. Unless the system is altered and a better one evolved the situation will become highly dangerous.[106]

The urgency of the times was illustrated when in July the Ministry of Labour and National Service made successive orders for the compulsory registration on the Central Register of, first, all professional engineers, and soon thereafter of chemists, physicists, and quantity surveyors. These registrations supplemented those already made voluntarily.[107] With a certain satisfaction, *Nature* accurately observed that since the disasters earlier that year in Norway, Flanders, and France, the circumstances for making full use of science had grown more favorable than ever before in Britain's history.[108]

Meanwhile, Churchill had directed that Bragg's letter be carefully examined. It went to Bridges at the War Office and then to Barlow, who arranged a meeting with Bragg for 24 July. At this, Barlow asked Bragg for examples of cases in which the body he proposed would have been of use to the government. Afterwards, Bragg wrote to Barlow describing more clearly what he had in mind.[109] On 27 July, Barlow promised him to do all he could to get a reply.[110]

On the very same day "a tract for the times, which every scientific worker should read," was both reviewed in *Nature* and mentioned in its editorial.[111] This slim volume, *Science in War*, had been conceived, written by twenty-five scientists, printed, and published as a paperback Penguin special, all in less than a month. It appeared anonymously, but was the work of the Tots and Quots led by Solly Zuckerman. Following the outbreak of hostilities, many of their dinner discussions had concerned the use of science in war. As a guest at one dinner, Allen Lane, the publisher of Penguin books, heard of the complaints, criticisms, and constructive suggestions of the members, many of whom were in government service, and offered to publish their views. This was at the time of the fall of France, when Hitler's next initiative was expected to be a large-scale air attack on Britain. Zuckerman seized the opportunity and acted as editor.[112] The book's principal thrust was for "the effective utilisation of scientific thought, scientific advice, and scientific personnel" in the war. It described earlier successful applications of science to war and went on to suggest how science could be better used in the current war in such areas as the treatment of the wounded, the production of food, and weapons research. Demand for the book was such as to warrant a second printing four months later in November; in all, some 20,000 copies were sold.[113] According to *Nature*, the volume was successful in creating "the impression of vast potential forces insufficiently coordinated or inadequately marshalled."[114]

However, during the difficult weeks of the Battle of Britain following its publication, when Britain 'heard' much from Germany, the Royal Society heard nothing from the government. Hill and Egerton became impatient. On 10 September, Hill wrote to Wing Commander Elliot enclosing a memorandum entitled "Scientific Research and Technical Development in Government Departments," which he said he was going to send to Chamberlain asking to discuss it with him, and/or send to various ministers asking to discuss it with them, and/or publish in slightly modified form in some journal or newspaper. Hill explained that "Egerton strongly feels that something must be done about the business, or there may be an explosion: anyhow, I propose to make myself a nuisance until something is done or I am squashed."[115] Eleven days later, he decided on the first of the above courses and sent the document with a letter to Chamberlain.[116]

The memorandum began by stating that two conditions had to be met if proper use was to be made of scientific knowledge and research within a service or other ministry.[117] First, the ministry must have its own establishment and staff for research and development, and second, a place must be found in the research organization for independent scientific advice. It was with this second requirement that the memorandum, as the earlier proposals, was concerned. The very great importance in modern government, particularly in modern war, of scientific research and technical development, and the urgency with which scientific problems might arise, required that science and technology should be able to exert a direct and sufficient influence on policy. This could only be assured, said Hill, by introducing at a high level some means by which a minister could obtain, and be expected to consider, independent critical advice on scientific and technical subjects. He therefore argued for the creation of scientific advisory councils in the Admiralty and the Ministries of Aircraft Production, Agriculture, Food, and Health. Their model would be the Advisory Council on Scientific Research and Technical Development set up in the Ministry of Supply in early 1940 and whose members included himself and Tizard.[118] The various committees of this Council had both the duty and the right to observe and advise upon the work of all of the establishments within the Ministry. They could also offer advice, through the Council and its chairman, to the minister.

To complete the organization of these various proposed scientific advisory councils, continued Hill, "*a small central authoritative body*" would be required. In this connection, he referred to and endorsed the proposal in Bragg's June letter in which half the personnel was from within and the other half outside government service—that is, the secretaries of the three research councils and the president and secretaries of the Royal Society. Hill suggested that reference could be made to this body on such matters as major scientific appointments, general scientific policy, urgent new scientific problems suddenly emerging, and the coordination of, and prevention of overlap in, the work of the various government and other agencies. In the "present emergency," this body could assist in taking quick initiative to meet sudden and unexpected difficulties.

About this time, Churchill was asked in the House of Commons whether he would consider forming a full-time scientific committee, composed of fellows of the Royal Society, to make a comprehensive investigation of the country's entire war effort, to propose new ways of waging war, and to have direct contact with the Cabinet to promulgate its views and emphasize its recommendations. In replying for Churchill, the

Lord Privy Seal, Clement Attlee, said that representations that had been received from the Royal Society were under active consideration. Hill learned of this from the *Times* and wrote immediately to Attlee mentioning Bragg's letter of almost three-and-one-half months earlier and asking what exactly Attlee had meant by "active consideration." He enclosed a copy of the memorandum he had sent to Chamberlain earlier that day and explained that he had asked to see Chamberlain to discuss it with him. He then continued in a vein similar to that of his letter to Chamberlain:

> Some people might think that present events are too urgent for such discussions. The urgency in fact is just what prompts me to press the matter now. The country is having to face a number of acute and novel difficulties, to many of which science can provide essential help in guiding the way to a solution. For example, the present air raids at night provide urgent scientific problems on the military, the instrumental, the epidemiological, the engineering, and the psychological sides. This is the very time that the full resources of science are needed. Reputable scientific people usually do not enjoy pushing their own wares: perhaps that is why in competition with others they have failed to make any impression so far. [B]ut I feel now that we should be failing in our duty if we were to allow our representations to be put aside any longer by formal acknowledgments.[119]

Clearly, Hill was determined.

He also wrote to Sir Richard Gregory, former editor of *Nature* and now president of the British Association for the Advancement of Science, telling him of his letter to Attlee. Gregory responded that he too had seen Attlee's reply to the question in the House and had "naturally supposed that [Hill] had prompted the question." He was glad that Hill had asked for details on what was being done on the Royal Society proposal; and he continued:

> You are indeed working hard on behalf of science in connection with the war effort and I believe that you will be successful in getting a move on. I put my trust in you for this purpose with every confidence that you will do whatever any one man could do, and with greater authority and knowledge than any other man of science in the kingdom. All I need say now is that if you think additional support would be given to your efforts by members of the Council of the B.A., as well as the General Officers and myself, I will take

action whenever you ask me to do so. I know the difficulties in obtaining support from members of any Council of a scientific society, yet I still think it would be worth while to get the names of Members of Council of the R[oyal] S[ociety] and the B.A. attached to any memorandum which may be prepared for presentation to the Prime Minister and afterwards submitted to the Press.[120]

But the British Association's support could not enhance the Royal Society's effort, and so Hill ignored Gregory's offer.

Hill soon heard from Attlee and Chamberlain. Attlee agreed that the harnessing of the country's best scientific brains to the war effort was urgently desirable and he assured Hill that he had really meant what he had said in the House. The Lord President was in the process of making recommendations to the Prime Minister, the nature of which Chamberlain hoped would go far toward meeting the views of Hill and the Royal Society.[121] Chamberlain himself replied:

> No one is more anxious than I am of the importance of harnessing Science to our chariot and making her one of the team that will ultimately win this war. Indeed, you are pushing at an open door so far as I am concerned and ever since Sir William Bragg's letter reached the Prime Minister we have been examining his proposal and endeavouring to meet the Royal Society, whose co-operation is most welcome to the Government.
>
> For your private and personal information I may add that I have just made a recommendation to the Prime Minister on this subject which I hope he will see his way to accept and which will, I trust, go a long way towards meeting your views.[122]

Under these circumstances Chamberlain did not think it necessary to meet with Hill, as he needed no convincing.

It would appear that Hill's letter to Elliot of 10 September at least accelerated the government's action. Eight days later, Elliot had written to Barlow asking how he should like the letter answered.[123] On the same day, Barlow sent to Hankey a memorandum that, in a modified form, was that later sent by Chamberlain to Churchill for his consideration. This specified certain troubles, or dissatisfactions, of the Royal Society and commented on them.

First was a general malaise among scientists, especially younger ones, who felt that scientists were not being fully utilized and that ministers and departments tended to ignore the scientific aspects of the problems of government. In this regard, said the memorandum, it would probably be worthwhile to appoint a committee "if only to allay distrust." Unless something were done, there would be more overt agitation, which would evoke a good deal of support. The committee suggested by Bragg would be "as good as any."

A second dissatisfaction was that the Royal Society felt that the right persons were not always chosen for particular scientific positions and were hurt that the Society was not consulted on the selection of such individuals or of members for government science committees. It might therefore be a further advantage, the memorandum suggested, to have the proposed committee available for consultation. Likewise, it would also be advantageous if the creation of the committee would lead to the removal of a third dissatisfaction, namely, the fear that the government might not know of, or might neglect, promising new scientific developments.

The final dissatisfaction was that, as the Royal Society had only a casual and imperfect knowledge of the multifarious research activities of governmental departments, it feared that these researches were not properly coordinated. Unlike the preceding three, this dissatisfaction was not used to justify the appointment of the committee. The memorandum explained that "the arrangements for co-ordination of research and for ensuring that enquiries likely to give immediate results are given precedence were recently overhauled, and Lord Falmouth was attached to Lord Hankey as a liaison officer." Falmouth had, significantly, found very little to do; his experience was generally that all individuals concerned knew what their colleagues were doing and that first things were being put first.

Regarding the duties of the proposed committee, the memorandum stated that it "ought not to be given a roving commission or be encouraged to think that it will have a voice in general Government policy or on non-scientific subjects." Accordingly, the terms of reference were specifically proposed as being:

a) *To advise* the Government on any scientific problems *referred* to it.

b) *To advise* Government Departments, *when required*, on the selection of individuals for particular lines of scientific enquiry or for membership of committees on which scientists are required.

c) To bring to the notice of the Government promising new scientific or technical developments which may be of importance to the war effort.[124] (Italics added)

The sense of these had been suggested by both Bragg and Hill between them, but the independent coordinating function that both had primarily proposed had been suppressed. The committee would concern itself with government scientific affairs only when invited to do so, and then solely in an advisory capacity.

In sending the memorandum to Hankey, Barlow enclosed a letter of his own and also Hill's letter to Elliot, of which he said that it indicated "the necessity of doing something fairly soon." He continued: "I think it is clear that the Government will have to set up some committee of the kind [described in the memorandum], if only to keep the scientific people quiet, and it is conceivable that it might be of some use."[125]

Prior to writing to Hankey, Barlow had discussed the matter with Appelton who this time thought that the proposed committee, which would include himself, Butler, and Mellanby, might do some good and would probably do no harm.[126] Mellanby had also taken the same view, and Butler too had agreed with the proposal. These three had in fact discussed it among themselves and, though not enthusiastic, had collectively decided to cooperate. Indeed, the proposal was for the appointment of a committee of the same composition as that suggested in Bragg's letter except that a Cabinet minister would not be chairman. In his discussion with Barlow, Bragg had agreed that the original idea of working on the Cabinet level, rather than on the ministerial or departmental level, was misconceived. And so in his letter to Hankey, Barlow said of the proposed committee: "It seems to me also that it must be nominally attached to the Lord President of the Council as he is the Minister responsible for the Government's relations to scientific activities in general. Mr. Chamberlain, however, has no time to spare for giving any personal attention to the thing, and when some of us discussed the question some weeks ago, we felt that if you could take effective charge nominally on his behalf, that would be a happy solution. . . . Do you agree that the proposed Committee ought to be set up and would you be disposed to become Chairman?"[127]

194 · Scientists and War

Hankey replied promptly to Barlow on September 20. His revealing and prescient letter is worth quoting at length:

> If I were to answer your letter in a purely realist spirit I should have to say that in my opinion the proposed Committee was redundant. It is clear from Professor Bragg's letters that the Government is making use of scientists to a degree that is not only unprecedented but probably not far from the maximum desirable. There are the three great governmental scientific bodies, which cover scientific and industrial, medical, and agricultural research with ramifications covering an immense field: to these have to be added the scientific work of the three Service Departments and the Ministry of Supply covering practically the whole of our war effort.
>
> In the same realist spirit I should have to say that in my opinion, if we admit the need of this Committee, we should have no reply to a demand from the engineering profession for a committee to concern itself with the engineering activities of the Government.[128] Of course engineering is so wide a term and covers so large a field that any Committee of the kind would have to be so large as to have little value and to do all its work through sub-committees. But science is every bit as wide in its application as engineering.[129]

However, there was another side to the coin, and Hankey continued: "So much for the realistic attitude. But in everyday life one cannot deal with matters solely in the spirit of realism. I agree with what you say in your second paragraph—'that the Government will have to set up some committee of the kind, if only to keep scientific people quiet.'"

Having heard from Hankey, Barlow forwarded a copy of the memorandum to Sir Horace Wilson asking him to look at it and saying it was intended for Chamberlain.[130] He reiterated his conviction that some committee had to be set up and referred Wilson to Hill's letter to Elliot. Two days later, on September 23, Wilson wrote to Chamberlain:

> I do not think they [the Royal Society] have any real ground of complaint, but the setting up of a Committee, such as is recommended in the attached Minute . . . would I think satisfy the Society and at the same time strengthen the confidence of the public in that it would be a standing proof that Science is being given its due place in the scheme of National Defence, and would meet some of the ill informed, if not ill-natured, criticism contained in the Penguin Special ("Science in War") which has had a large circulation. I

therefore strongly recommend your approval of the proposed Committee under Lord Hankey's chairmanship. It should be made plain in any terms of reference that Lord Hankey's Committee should report in the first instance to yourself, as Lord President of the Council, for the information of His Majesty's Government.

If you approve . . . we would send an intimation to that effect to the Prime Minister. There seems no point in making a Cabinet matter of the Royal Society point.[131]

The last remark shows how times had changed. Chamberlain responded: "I entirely approve."

In the document soon forwarded to Churchill, Chamberlain suggested terms of reference for the proposed committee which were slightly modified versions of those in Barlow's memorandum. The first and third terms were made more specific in that the committee was to deal directly with the "Lord President" rather than with the "Government."[132] Churchill replied that he had no objection provided it was understood in addition:

(a) That the secrets upon which the various Departments are now working, particularly those concerned with A.I., G.L., A.S.V., P.E., etc., shall not be imparted to a new large circle.[133]

(b) That the time of the scientists and Committees who are at present engaged in working for the Government shall not be unduly consumed.[134]

"As I understand it," stressed Churchill, "we are to have an additional support from the outside, rather than an incursion into our interior."

The Royal Society triumvirate of Bragg, Egerton, and Hill was well aware that the remarks were directed at them and not at their former rivals, and now new colleagues on the Committee, Appleton, Butler, and Mellanby. However, their satisfaction with their accomplishment was in no way diminished. Concerning the Committee, Egerton noted: "It is a new departure: the vertical departmental organisation with its various scientific and technical Departments is now bridged, and there is a direct channel from scientific organisations outside the government organisation *via* the Royal Society to the Cabinet, *via* the chairman of the Committee and the Lord President of Council. There is a focus point in the whole scientific organisation. If it is allowed to work . . . it should be a very important new departure, for science can at last have some say in the forming of policy."[135]

When the creation of the Committee was publicly announced, it was enthusiastically greeted as "a kind of scientific power house" from which great things might be expected.[136] In scientific circles, it was remarked that although a worldwide catastrophe had been required for it to happen, there was now with the Committee a means of putting science into direct contact with the innermost councils of the Empire.[137] The *Scientific Worker* noted that:

> It has always been an aim of the [Association of Scientific Workers] that a fully representative advisory body of scientific workers should exist which might influence the utilisation and development of the sciences for the benefit of the community as a whole. It is perhaps a cynical reflection that it has needed two world wars before the embryo of such a body should come into being. It is to be hoped that the present committee will be able to prove its worth by the advice it offers, that this advice will be taken, not only about immediate problems of direct military importance, but also about the many long-range problems of social reconstruction which are arising and will arise even more acutely later.
>
> Possibly this embryo may be able to develop into something in the nature of a Ministry of Science or some such executive body, amongst whose primary tasks will be assistance in the reconstruction which daily becomes more obviously necessary.[138]

The Scientific Advisory Committee first convened on 10 October 1940.[139] Hankey was in the chair and Bragg, Hill, Egerton, Appleton, Mellanby, and Butler were present. Also in attendance were Lord Falmouth and the joint secretaries— Group Captain Elliot and W. W. C. Topley. In an opening statement, Hankey said that the formation of the Committee represented a wholehearted desire on the part of the government to obtain the utmost cooperation in the prosecution of the war from the vast storehouse of brains, scientific knowledge, research and enterprise at the command of the Royal Society. In addition, it would strengthen the public's confidence by providing a strong proof that science had been given its due place in the national war effort. Hankey went on to mention the two provisos that Churchill had laid down in approving the establishment of the Committee, and his admonition about "additional support from the outside rather than an incursion into our interior." In spite of this, as Egerton noted, Hankey was to make it "his first duty to review the arrangements in all Departments" and this "set the S.A.C. on its feet."[140] Hankey summarized the extent to which science was already being used in the war. In concluding, he expressed the hope that the Committee might make a real contribution to the war effort and also that it might possibly produce results of lasting importance in strengthening the links that already existed between science on the one hand and government and administration on the other.

The second point implied a larger role for science, and so too did the remarks of Sir John Anderson, Lord President and the Cabinet minister responsible for the Committee, when he attended its next meeting to acquaint himself with its members. The "outstanding administrator of his generation," Anderson succeeded Chamberlain as Lord President and "gradually became the supreme figure in civil administration and ran the war on the civil side so far as any one man could be said to do so."[141] He was interested to know what the Committee's general policy would be and suggested that although the Committee would probably be occupied mainly with war problems, it should not ignore wider questions that would arise when peace returned. Bragg said it was of fundamental importance that the Committee should consider the ways in which science could be related to the state; Egerton expressed the hope that the Committee would see that scientists were used to the full. Anderson agreed. Hankey informed Anderson of a suggestion, received from Sir Richard Gregory and considered by the Committee at its previous meeting, that a representative of the British Association be appointed to the Committee. The Committee's members had acted by inviting Hill to tell Gregory that his suggestion could not be accepted—not only because the Committee did not consider any increase in its membership desirable, but also because it was not within the Committee's terms of reference to increase the membership.[142] Hankey added that Falmouth, who was again present, had more recently been approached by the Institute of Electrical Engineers with a similar suggestion. In response, Anderson said that there was no room for "the representation of sectional interests," and that already "all branches of science were adequately represented."

Falmouth, however, "thought it likely that considerable pressure would be exerted by engineers and perhaps by other interests." Anderson agreed that this was highly probable, but thought that such pressure should be resisted. However, he also thought that it should be made clear to those concerned that this decision was not due to any lack of appreciation of their services, but to the obvious fact that if the small advisory body that had been created were enlarged to make it directly representative of all sectional scientific interests, it could not possibly fulfill the important functions for which it had been designed. With this, Bragg, Appleton, Mellanby, Hill, and Butler all agreed.[143]

In what he said, Anderson was undoubtedly complying with the views of Churchill who, on 8 October, five days after the creation of the Scientific Advisory Committee had been publicly announced, was asked in the House of Commons whether he was aware that a letter addressed to him by the presidents of the institutes of mechanical, electrical, and civil engineers offering the services of their members had gone unanswered in all but form. The Prime Minister was also asked what steps he proposed to take to harness the scientific brains of the country for the purpose of winning the war. Churchill replied by referring to the Scientific Advisory Committee, but in response it was pointed out that the Committee had no applied scientists as members. The brief exchange was terminated by Churchill saying that he did not propose to alter the constitution of the Scientific Advisory Committee.[144]

The day after Anderson met with the Committee, Bragg wrote to Hankey:

> I hope we shall be able to resist the request of the engineers and others for places on our Committee. We shall surely be the better for having no tied representatives. I think that the Committee that sat for a few years after the last war failed in part because it was a collection of representatives, including the Crown forces. But it would be excellent if the Presidents of the various Engineering bodies would come and tell us of all they have in mind as to how the applications of science could be made in their respective spheres. Perhaps they would be content with that. They might be promised a sort of continuous hearing.[145]

Bragg, however, misjudged the situation. The engineers would not be content with the role of advisers to scientific advisers, and the circumstances that had favored the scientists would also favor the engineers.

Shortly thereafter, the pressures that Falmouth had spoken of manifested themselves. They were generated by, among other groups, the Parliamentary and Scientific Committee which had emerged early in the war as the successor to the Parliamentary Science Committee. Following the rejection by the government in early 1938 of the Parliamentary Science Committee's memorandum on the finance of scientific research (the "Bernal Memorandum"), the Committee seems to have done little for the remainder of 1938 and for most of 1939. Then on 13 September 1939, in the first days of war, the Committee's executive committee voted to dissolve the Committee pending the conclusion of hositilities. The vote was four to three, with the chairman, S. F. Markham, M.P., being one of those opposed.[146]

The Association of Scientific Workers, which together with the now defunct British Science Guild had originally founded the Parliamentary Science Committee, received the news of dissolution with dismay. At a hastily convened meeting on 16 September, the Association's executive committee agreed that letters should be sent to the president and chairman of the Parliamentary Science Committee saying that there was important work for the Committee to do, that the Association urged it to resume its activities, and that the Association was willing to continue its subscription to the Committee.[147] In less conciliatory terms, the letters also argued that the action of the Committee's executive committee—taken, as it was, without consulting the various organizations affiliated with the Committee—was invalid, and called for a meeting of the Committee's general committee on which these organizations were represented. At this same meeting, the executive committee of the Association of Scientific Workers also decided that all of the organizations should be asked to make similar protests.

Several of the organizations supported the Association's proposal that a meeting of the general committee of the Parliamentary Science Committee be held, and the Association requested Markham to call such a meeting.[148] Markham agreed and set the date of 8 November 1939, at which time he was to lay the facts of the situation before the constituent organizations and propose that the Parliamentary Science Committee be reestablished. In advance of this meeting, the Association's representatives on the Committee were instructed to say, in the event of the possibility of Henry W. J. Stone continuing as secretary of the Committee, that in view of the manner in which the Committee had been suspended, the Association had no confidence in him as secretary. Although others were involved, Stone must nevertheless have had the principal role in the suspension action. He soon resigned.[149]

Also in advance of the meeting, and this indicates the importance that the Association attached to the Committee, the Association discussed what should be done if the members of Parliament associated with the Committee were in favor of suspension. One concern was that if that were so the Committee would become ineffective, as it functioned through its members of Parliament. The Association agreed that if the Committee's current members of Parliament favored suspension the support of other members should be sought, and that even if no such support could be found the Association's representatives on the Committee should nevertheless press for its reestablishment. It was also agreed that the secretary of the Association, Reinet Fremlin, should attend the meeting of the general committee of the Parliamentary Science Committee and place before it a set of questions designed to indicate the importance of the Committee and to arouse interest in its work.[150]

There was a large attendance at the meeting of the general committee, and a resolution was unanimously adopted creating an emergency committee and inviting members of the current executive committee and all former officers to serve on it. A meeting of this committee was held at the House of Commons on 6 December. The Parliamentary Science Committee was reformed on a new basis and was to be known henceforth as the Parliamentary and Scientific Committee. Markham resigned as chairman, as he was to leave the country to serve abroad, and Captain L. F. Plugge, M.P., who had had a scientific education and considerable experience in aeronautics and radio, was elected as his successor. Professor B. W. Holman and Alan E. L.

Chorlton, M.P., remained as vice-chairman and deputy-chairman, respectively, and Christopher Powell became the new secretary. It was reported at the meeting that fourteen organizations that had previously supported the Parliamentary Science Committee had pledged their support to the new Committee, ensuring it sufficient funds to operate. At this time it had the support of six members of Parliament.[151] By the end of January 1940, four additional members of Parliament and three additional associations were supporting the Parliamentary and Scientific Committee.[152] Soon these numbers had risen to eighteen members and some twenty organizations.[153] By early 1943, they were seventy-four and thirty-three, respectively.[154] The publication of "Science and Parliament," discontinued when the Parliamentary Science Committee was suspended, was resumed.[155]

Although the Parliamentary and Scientific Committee would prove to be a longer-lived body than its predecessor and a much more influential one in regard to governmental policies and practices concerning science, it seems to have gotten off to a slow start. As late as December 1941, the Association of Scientific Workers was sending a delegate to a meeting of the Committee to "attempt to liven things up."[156] But whether because of the Association, or for some other reason, the Committee became quite active from 1942.[157]

On 28 October 1940, the secretary of the Parliamentary and Scientific Committee wrote to the Scientific Advisory Committee saying that although the former welcomed the recent creation of the latter it nevertheless regretted that no representatives of applied science had been appointed to the Advisory Committee. He inquired whether the intention was to make use of such persons in connection with the various subcommittees that the Scientific Advisory Committee might set up.[158] An earlier letter from James R. Beard, President of the Institute of Electrical Engineers, was much more forceful. The Council of the Institute, he wrote, was strongly of the opinion that the Scientific Advisory Committee would be greatly strengthened by the inclusion of a representative of applied science and research who would be able to give advice on "(a) the special problems which arise in the practical application and development of scientific discoveries on a large manufacturing scale, and (b) on the assistance which the engineering industry can give in scientific research directed to the solution of urgent problems."[159]

In a carefully considered reply to Beard on 29 October, Hankey said that the composition of the Scientific Advisory Committee had been given very close consideration by the government, and at this early stage, when he had had only a very brief experience of the working of the Committee, it would perhaps be premature for him to express a final opinion on Beard's proposal. However, he did say that although he admired the engineering industry he had doubts as to whether it should have permanent representation on the Committee. Applied science included many other important institutions in addition to the Institute of Electrical Engineers, and it would be difficult to select a representative whose appointment would satisfy all critics. This particular difficulty had not arisen with pure science as the Royal Society embraced all of its branches. However, Hankey concluded, it was too early to form a final judgment on the matter.[160] A similar letter was sent to the Parliamentary and Scientific Committee with the addition that their suggestion that representatives of applied science might be asked to serve on any subcommittee was provisionally accepted.[161]

Hankey forwarded the original letter from the Parliamentary and Scientific Committee to Anderson saying: "I gather that this body has not yet been very important, but that it is making a great effort to extend its influence."[162] Anderson may also have heard directly from Beard, for in mid-December he suggested to the Scientific Advisory Committee that it might be desirable to invite Beard to discuss with the Committee ways in which the knowledge and experience of the members of the Institute of Electrical Engineers might be more fully utilized. Falmouth reported that he understood that the Joint-Engineering Council was taking steps in this general area, but he nevertheless thought that to invite Beard would be useful. The members of the Scientific Advisory Committee agreed.[163]

Once again Falmouth's information proved to be correct, for on 23 December, the presidents of the country's eight engineering societies wrote to Churchill. They strongly urged the establishment of a new body on lines parallel to the Scientific

Advisory Committee to be concerned with the utilization in the war effort of the resources of all branches of engineering. The matter was discussed within government departments, and some three months later Anderson and Hankey agreed that a new body should be created.[164] Anderson also discussed the matter with Lindemann, and he too agreed.[165] On 26 March 1941, Anderson sought Churchill's approval to negotiate the establishment of the new body. He explained that the internal organization of the engineering profession was far from satisfactory and that he and Hankey thought that the proposed new body, which would bring together the interests now focused in a number of independent engineering institutions, would serve a useful purpose not only in promoting the national war effort but possibly also in bringing about a lasting improvement in organization. Were such a body to be created it would be desirable that Hankey should be chairman, "at any rate in the initial stages when the risk of clashing with the Scientific Advisory Committee would be greatest."[166]

It is noteworthy that Churchill, who is said to have relied heavily on Lindmann's advice in scientific matters, should have sought Sir Edward Bridges' advice in this matter. Bridges replied that he could not speak with knowledge on the subject, but the creation of the Scientific Advisory Committee under Hankey, he believed, had undoubtedly done a lot to satisfy scientists that proper use was being made of the country's scientific talent for war purposes. He expected that the proposed committee would have equally beneficial results with the engineers.[167]

With Churchill's approval, Anderson appointed on 30 April 1941 the Engineering Advisory Committee under Hankey as Chairman to advise the government on engineering questions connected with the war. Specifically, the new Committee was:

> To consider how the resources of the engineering profession can best be utilized in connexion with the war work of government departments, and to nominate engineers who might suitably be invited to undertake particular tasks.
> To suggest means of improving or supplementing the methods adopted by government departments for utilizing engineering science for war purposes.
> To advise on problems referred to them in connexion with the

development of new engineering devices, and to advise on methods of bringing such devices speedily into production, and on means of meeting new war requirements in the engineering field.

To examine new ideas or devices in engineering likely to assist the war effort, to test their technical validity, and to bring to the notice of the government those which appear to merit further consideration by the departments concerned.[168]

In addition to Hankey, the members of the Committee were Falmouth, Tizard, and the following who were selected after consultations with the presidents of the institutes of civil, mechanical, and electrical engineers: Beard; Dr. A. P. M. Fleming, director of the Research and Education Departments, Metropolitan-Vickers Electrical Company; W. T. Halcrow, consulting engineer; Professor B. W. Holman; Dr. C. C. Paterson, director of the Research Laboratories, General Electric Company; H. R. Ricardo, technical director of Ricardo and Company, consulting engineers; and Dr. A. Robertson, principal of the Merchant-Venturers' Technical College, Bristol.[169] The engineers were now, some seven months later, on an equal footing with the scientists. For the first time in British history, both officially watched over government activities in their respective fields.

Meanwhile, the Scientific Advisory Committee had been hard at work under Hankey's energetic and competent guidance. By March 1941, when Anderson was seeking approval to create the Engineering Advisory Committee, he could assure Churchill that the Scientific Advisory Committee had "already fully justified itself."[170] The reason for his saying so was doubtless the thirty-five page report that he had recently received from the Committee. At its first meeting, the Committee had agreed to prepare, as a background to its future work, a survey of the various scientific activities of government departments. In addition to providing written reports, the directors of scientific research of each of the departments engaged mainly in war work were invited to tell the Committee about their activities. The results were included in the report forwarded to Anderson in early March. The principal conclusion drawn by the Committee from the survey was that the scientific activities of the various government departments, and particularly of the defense services, were far more extensive and more effective than was commonly realized. The Committee was convinced that much of the criticism that had been offered on this score had been due to a lack of knowledge of the facts, itself resulting to some degree from the often necessary secrecy that obscured much of what was being done.[171]

At the third meeting of the Committee, Hankey gave his views on ways in which it could most quickly and effectively fulfill its terms of reference. In referring to interviews he had recently had with ministers in charge of service and related departments, he noted their willingness to afford the Committee every facility and help in its work. With regard to their particular problems, which clearly involved secrecy, and in view of Churchill's wish that precautions should be taken to preserve secrecy, Hankey thought it desirable to form a panel from among the members of the Committee to deal with secret matters. It was agreed that the panel, called the Defence Services Panel, should consist of Hankey, Appleton, Hill, Egerton, and Mellanby.[172] This was an imaginative and significant move on Hankey's part, for in creating the Panel from the members of the Committee who were already engaged in secret work he was able to extend the Committee's scope.

One interesting part of the Panel's work was its involvement in Britain's decision to push ahead with the atomic bomb project. Investigation of the possibilities of releasing atomic energy and of creating a bomb was not initiated by the Panel, but rather by the so-called M.A.U.D. Committee, which was attached to the Ministry of Aircraft Production and under the direction of Nobel Laureate G. P. Thomson, Professor of Physics at the Imperial College of Science in London.[173] Until almost a year after its formation, the Panel had officially no knowledge of the workings of the M.A.U.D. Committee—at the direction, no doubt, of Churchill. However, on 19 June 1941 the Minister of Aircraft Production, Colonel Moore-Brabazon, consulted Hankey about the M.A.U.D. Committee inquiry that was approaching a stage when its findings ought to be considered by the highest scientific authorities. Hankey and Moore-Brabazon agreed that the Defence Services Panel was the right body to make this study. Subsequently the Lord President, Anderson, also agreed.[174]

The Panel studied the matter thoroughly and with a sense of

urgency, and Hankey submitted its report to Anderson on 24 September. Yet considerably before then, Churchill had decided to proceed with the bomb project. At the end of August, Lindemann, now Lord Cherwell, had submitted recommendations to Churchill in regard to the M.A.U.D. Committee Report. Churchill had forwarded these to the Chiefs of Staff who, in meeting with Churchill on 3 September, "urged that no time, labour, materials or money should be spared in pushing forward the development of this project."[175] As Cherwell had recommended, the Cabinet Minister responsible for the undertaking was Sir John Anderson. Thus, even before receiving the report of the Defence Services Panel Anderson knew what lay ahead. Nevertheless, in the view of the historian of atomic energy, the Panel's "report was of great value in setting forth the scientific points that had still to be settled, the developments that must now go ahead, and certain questions of policy, for example the control of the power project and the location of the bomb project. Aided by the report Sir John Anderson had to formulate high policy and decide upon the right organisation."[176]

When Cherwell had met with the Defence Services Panel on 17 September to discuss the scientific aspects of atomic energy, he knew, although the Panel did not, that he had bypassed it. Just when the Panel realized this is not clear, but it would be one of the reasons for the anger with Cherwell that would soon be expressed by members of the Scientific Advisory Committee and other scientists, as will be seen below. For the time being, however, Hankey and the other members of the Committee were apparently pleased with the way the Committee was functioning and with the use being made of it by the government.

The Association of Scientific Workers and the Royal Society had alike wanted the fullest use to be made of the country's scientific resources in the war effort. Of the two, the Royal Society, and A.V. Hill in particular, acted more decisively in starting first the central register of scientists and then pressing the government to create a central body to coordinate the scientific effort in the war. In September 1940, Churchill reluctantly approved the creation of the Scientific Advisory Committee to the War Cabinet. The significance of the Committee was threefold. It raised the status of scientists within government. It brought together the heads of the three national research councils, who had previously not met officially, and the leaders of civil science as represented by the officers of the Royal Society. Finally, it had the power to oversee almost all government science. It was a major innovation in the relationships between science and government, and for scientists its creation was an important victory that had not been easily won. A prestigious nongovernment scientific elite had exerted pressure to be admitted into the jealously preserved realms of government and government-scientific elites. Both of the latter were initially reluctant to grant that the former had any right or need of access to their domains, but eventually in the adverse war situation of 1940 political expediency dictated otherwise. Meanwhile, because of the initial reluctance, a few leading members of the Royal Society found themselves engaging in unaccustomed and unprecedented political actions at the beginning of a period now commonly recognized as a watershed in the relationships between science and government. That it was so, was in part due to their bold innovative efforts in which they were motivated by the assumption, publicly adopted by scientists at the 1938 British Association meeting, that the scientist has a responsibility to see that the nation makes the very best use of its scientific resources, especially in a time of great national danger. The Scientific Advisory Committee quickly showed that it fulfilled a useful function. In its first major report of March 1941, it stated its satisfaction with the use being made of the country's scientific resources in war.

1. Gary Werskey, *The Visible College: The Collective Biography of British Scientific Socialists of the 1930s* (New York: Holt, Rinehart, and Winston, 1978), pp. 223–34; Cambridge Scientists' Anti-War Group, *The Protection of the Public from Aerial Attack* (London: Victor Gollancz, 1937), p. 9.

2. Werskey, *The Visible College*, p. 228.

3. Cambridge Scientists' Anti-War Group, *The Protection of the Public from Aerial Attack*.

4. Maurice Goldsmith, *Sage: A Life of J. D. Bernal* (London: Hutchinson, 1980), p. 78.

5. Werskey, *The Visible College*, p. 231.

6. Association of Scientific Workers, Executive Committee minutes, 30 September 1938, ASW/1/2/19/38.

7. Solly Zuckerman, *From Apes to Warlords* (New York: Harper & Row, 1978), p. 101.

8. *Nature* 142 (1938): 685-87.

9. Association of Scientific Workers, Executive Committee minutes, 20 October 1938, ASW/1/2/19/43.

10. Ibid., 30 September 1938, ASW/1/2/19/38; ibid., 20 October 1938, ASW/1/2/19/43.

11. A. C. G. Egerton, "Personal Notes relating to Meetings and Business of the Royal Society, 1938- ," p. 6, Egerton Papers, Royal Society, Box 8.

12. *Scientific Worker* 11 (1939): 23.

13. Ibid.

14. Ibid.

15. Ibid., pp. 14, 26.

16. Association of Scientific Workers, *The A. S. W. and the War* (three-page pamphlet issued with the *Scientific Worker* vol. 11, no. 3, [Autumn 1939]), p. 3.

17. Association of Scientific Workers, Executive Committee minutes, 5 January 1939, ASW/1/2/20/6.

18. *Manchester Guardian*, 12 January 1939, p. 18.

19. Association of Scientific Workers, Executive Committee minutes, 9 February 1939, ASW/1/2/20/8.

20. A. V. Hill, "Memories and Reflections," (typescript, 2 vols.) 1:96-97. (This work is in the library of the Royal Society.)

21. The Royal Society, Council minutes, 12 January 1939; *Nature* 143 (1939): 592.

22. *Scientific Worker* 11 (1939): 4.

23. S. A. Gregory and Reinet Fremlin sketched "The Organisation of Scientific Research in France," in ibid., pp. 37-40.

24. A. C. Egerton to J. D. Bernal, 23 March 1939, Bernal Papers, Cambridge University Library, Box 81, Folder J 36.

25. J. D. Bernal to A. C. Egerton, 28 March 1939, Bernal Papers, Box 81, Folder J 36.

26. Association of Scientific Workers, "Report of Second Part of Conference held on May 13, 1939," P. M. S. Blackett Papers, The Royal Society, E 3.

27. Association of Scientific Workers, "Report of the Executive Committee to the half-yearly Council Meeting, May 6, 1939," Blackett Papers, E 3.

28. Association of Scientific Workers, Council preliminary minutes, 6 May 1939, ASW/1/2/20/27/i; Executive Committee minutes, 22 June 1939, ASW/1/2/20/33.

29. Association of Scientific Workers, Executive Committee minutes, 20 July 1939, ASW/1/2/20/36.

30. Reinet Fremlin, "Executive Committee," September 1939, ASW/1/2/20/42.

31. Association of Scientific Workers, Executive Committee minutes, 16 September 1939, ASW/1/2/20/43.

32. Klatzow came to Britain as a Rhodes Scholar from South Africa in 1930. He received his D. Phil. after three years' research work at the Electrical Laboratories at Oxford. From 1934, he was employed at the research laboratories of Electrical and Musical Industries. He died on 22 September 1942, aged 35.

33. Association of Scientific Workers, *The A. S. W. and the War* (three-page pamphlet issued with the *Scientific Worker* vol. 11, no. 3, [Autumn 1939]).

34. Ibid., p. 2.

35. Association of Scientific Workers, Executive Committee minutes, 16 September 1939, ASW/1/2/20/43. The idea for a ministry of science seems to have been Holman's. Two years earlier, he had composed a memorandum and reported to the Association's executive committee on the possibility of the formation of a ministry that could be taken up by the Parliamentary Science Committee. "The position at present is that Sir [Frank] Smith [Secretary of the Advisory Council to the Committee of the Privy Council concerned with scientific research] is now ranked as a Secretary of State, and that the duties of the Lord President of the Council are to represent the DSIR, Medical Research Council, Agricultural Commission, etc., i.e., that he represents science. Hence the Ministry of Science is already coming about." The executive committee decided to send a recommendation to the government, but the matter seems to have died (Executive Committee minutes, 23 June 1937, ASW/1/2/18/10/i).

36. Ibid., 7 and 14 October 1939, ASW/1/2/20/45. For three different drafts of the memorandum, see ASW/1/2/20/46/i–iii.

37. Association of Scientific Workers, Executive Committee minutes, 4 November 1939, ASW/1/2/20/47.

38. "The National Co-ordination of Science," *Scientific Worker* 12 (1940): 4.

39. Association of Scientific Workers, Executive Committee minutes, 18 November 1939, ASW/1/2/20/48; *Scientific Worker* 12 (1940): 4–5.

40. *Scientific Worker* 12 (1940): 2.

41. Ibid., p. 5.

42. Ibid., p. 11.

43. Association of Scientific Workers, Executive Committee minutes, 4 January 1940, ASW/1/2/21/i.

44. A. V. Hill, "Memories and Reflections," 2:352.

45. The society took its name from its motto "Tot homines quot sententiae"—"As many opinions as people" (C. H. Waddington, *The Scientific Attitude*, 2d ed. rev. [West Drayton, England: Penguin Books, 1948], p. x).

46. Solly Zuckerman, *From Apes to Warlords*, p. 110. Zuckerman writes about the Tots and Quots on pp. 60–61, 111–12, 393–404.

47. Ibid., pp. 396–97.

48. Association of Scientific Workers, Executive Committee minutes, 7 and 14 October 1939, ASW/1/2/20/45.

49. See C. P. Snow, *Science and Government* (Cambridge, Mass.: Harvard University Press, 1961); Ronald W. Clark, *Tizard* (Cambridge, Mass.: M.I.T. Press, 1965).

50. Cab. 21/711.

51. Sir Edward Bridges, Memorandum, Committee of Imperial Defence, 27 June 1939. Cab. 21/711.

52. *Nature* 141 (1938): 740.

53. A. C. G. Egerton to Sir Henry Dale, 21 August 1943, Dale Papers, Royal Society, Box 4, Folder 2.

54. A. C. G. Egerton, to Sir Edward Bridges, Committee of Imperial Defence, 14 June 1939. Cab. 21/711.

55. Sir Edward Bridges, Memorandum, 30 June 1939. Cab. 21/711.

56. Ibid.

57. Sir Edward Bridges, note dictated for purposes of record, 6 July 1939. Cab. 21/711.

58. Cab. 21/711.

59. Note by the Minister for the Co-ordination of Defence, Lord Chatfield, 25 September 1939. Cab. 66/2.

60. Cab. 21/712.

61. Lord Chatfield to Sir William Bragg, 26 July 1939. Cab. 21/712.

62. Sir William Bragg to Lord Chatfield, 25 July 1939. Cab. 21/712.

63. Committee of Imperial Defence, Man Power (Technical) Committee minutes, 24 July 1939. Cab. 21/712.

64. Lord Chatfield to Sir Williams Bragg, 28 July 1939. Cab. 21/712.

65. Sir Edward Bridges, note dictated for purposes of record, 4 August 1939. Cab. 21/712.

66. A. V. Hill to Wing Commander Elliot, 28 July 1939. Cab. 21/712.

67. Sir William Bragg to A. V. Hill, 28 July 1939. Cab. 21/712.

68. Sir Edward Bridges, note dictated for purposes of record, 4 August 1939. Cab. 21/712.

69. Wing Commander William Elliot to Sir Edward Bridges, 16 August 1939. Cab. 21/712.

70. A. V. Hill to Wing Commander Elliot, 17 August 1939. Cab. 21/712.

71. Wing Commander Elliot to A. V. Hill, 29 August 1939. Cab. 21/712.

72. Sir Henry Tizard to A. V. Hill, 10 September 1939. A. V. Hill Papers, Churchill College, Cambridge (hereinafter AVHP), 2/5.

73. Sir William Bragg to Lord Chatfield, 8 September 1939. Cab. 21/712.

74. Sir Edward Bridges to Wing Commander Elliot, 11 September 1939. Cab. 21/712.

75. H. Parker to Wing Commander Elliot, 15 September 1939. Cab. 21/712.

76. Sir Edward Bridges to Lord Chatfield, 24 September 1939. Cab. 21/712.

77. Cab. 21/712.

78. Leslie Burgin to Lord Chatfield, 28 September 1939; G. Roseway, War Office, to Paymaster-Captain R. C. Jerram, 29 September 1939. Cab. 21/712.

79. Winston Churchill to Lord Chatfield, 28 September 1939. Cab. 21/712.

80. War Cabinet Minutes, 3 October 1939. Cab. 65/1.

81. Lord Chatfield to Sir William Bragg, 5 October 1939. Cab. 21/1163.

82. Sir Edward Bridges to Wing Commander Elliot, 20 October 1934; to Sir Edward Mellanby, 21 October 1934. Cab. 21/1163.

83. Sir Edward Mellanby to Sir Edward Bridges, 26 October 1939. Cab. 21/1163.

84. Sir Edward Appleton to Sir Edward Bridges, 26 October 1939. Cab 21/1163.

85. Earl Stanhope to Sir Edwin Butler, 10 November 1939. Cab. 21/1163.

86. Sir Edward Bridges to Sir Edward Appleton, 12 November 1939. Cab. 21/1163.

87. Cab. 21/1163.

88. H[orace] J. W[ilson] to Sir Edward Bridges, 3 April 1940. Cab. 21/1164

89. J. R. M. Butler, ed., *Grand Strategy*, 6 vols. (London: H.M.S.O., 1956–72), 2:354.

90. Sir Alan Barlow to Sir Edward Bridges, 13 May 1940. Cab. 21/1164.

91. Cab. 21/1164.

Scientists and War · 211

92. Cab. 163/69.
93. Cab. 163/69.
94. Cab. 163/69.
95. "Royal Society," memorandum by Sir Alan Barlow, 18 September 1940. Cab. 21/829.
96. A. C. G. Egerton to Sir Henry Dale, 21 August 1943, Dale Papers, Royal Society, Box 4, Folder 2.
97. Hill went to the U.S. in February 1940 and stayed for three months (Sir Bernard Katz, "Archibald Vivian Hill," *Biographical Memoirs of Fellows of the Royal Society*, 24 [1978]: 71–149, 117).
98. Egerton to Dale, 21 August 1943.
99. Cab. 21/829.
100. *Nature* 145 (1940): 887.
101. Katz, "Archibald Vivian Hill," p. 116.
102. *Nature* 145 (1940): 341.
103. Clark, *Tizard*, p. 241.
104. Ibid.
105. Ibid., p. 243.
106. Ibid., p. 244.
107. *Nature* 146 (1940): 89, 125.
108. Ibid., p. 108.
109. Sir William Bragg to Sir Alan Barlow, 24 July 1940. Cab. 21/829.
110. Sir Alan Barlow to Sir William Bragg, 27 July 1940. Cab. 21/829.
111. *Nature* 146 (1940): 107–8, 112–13.
112. On the origins of *Science in War* (Harmondsworth, England: Allen Lane, Penguin Books, 1940), see the publisher's note in the first edition; J. G. Crowther, *Fifty Years with Science* (London: Barrie & Jenkins, 1970); Zuckerman, *From Apes to Warlords*. Zuckerman was author of the anonymous editorial in *Nature* that mentioned *Science in War* (*From Apes to Warlords*, p. 398).
113. Zuckerman, *From Apes to Warlords*, p. 399.
114. *Nature* 147 (1941): 35.
115. Cab. 21/829.
116. Cab. 21/829.
117. AVHP, 2/1.
118. See *Nature* 144 (1939): 1085.
119. A. V. Hill to Clement Attlee, 21 September 1940. AVHP 2/1.
120. Sir Richard Gregory to A. V. Hill, 26 September 1940. AVHP 2/1.
121. Clement Attlee to A. V. Hill, 25 September 1940. AVHP 2/1.
122. Neville Chamberlain to A. V. Hill, 26 September 1940. AVHP 2/1.
123. Cab. 21/829.
124. "Royal Society," 18 September 1940. Cab. 21/829. This memorandum is unsigned, but drafts of it and various letters in this file (Cab. 21/829) show that it was originally drafted by Barlow after he had met with Bragg. It was then sent to A. N. Rucker (Principal Private Secretary to the Prime Minister), Elliot, A. Bevir, Hankey, and Sir Horace Wilson, and emerged in the final form that has been described above and

was sent to Elliot on 20 September 1940 with a note saying that the Lord President had approved it and sent it to the prime minister.

125. Sir Alan Barlow to Lord Hankey, 18 September 1940. Cab. 21/829.

126. Sir Alan Barlow to C. A. C. J. Hendricks (Private Secretary to Lord President of the Council), 18 September 1940. Cab. 21/829.

127. Sir Alan Barlow to Lord Hankey, 18 September 1940. Cab. 21/829.

128. In this Hankey was correct, and the Engineering Advisory Committee to the War Cabinet was soon created, as will be seen.

129. Cab. 21/829.

130. Sir Alan Barlow to Sir Horace Wilson, 21 September 1940. Cab. 21/829.

131. Cab. 21/829.

132. The modified terms of reference read:

(a) To advise the Lord President on any scientific problem referred to them,
(b) To advise Government Departments when so requested, on the selection of individuals for particular lines of scientific inquiry or for membership of committees on which scientists are required, and
(c) To bring to the notice of the Lord President promising new scientific or technical developments which may be of importance to the war effort.

C. A. C. J. Hendricks to Anthony Bevir, 26 September 1940. Cab. 21/829.

133. A. I. stood for Air-Interception (i.e., airborne) radar for the detection of other aircraft, G. L. for Gun-Laying radar sets, A. S. V. for Air-Surface-Vessel radar for the detection from the air of a ship or surfaced submarine which would otherwise be undetected because of darkness or bad visibility, and P. E. for Photo-Electric fuse, a proximity fuse. Churchill had a particular interest in P. E.—see his *The Second World War*, 6 vols. (Boston: Houghton Mifflin Co., 1948–53), 2:395–96.

134. Winston Churchill to Neville Chamberlain, 27 September 1940. Cab. 21/829.

135. A. C. G. Egerton, "Personal Notes relating to Meetings and Business of the Royal Society," pp. 65–66.

136. *Times*, 3 October 1940, pp. 4, 5.

137. *Nature* 146 (1940): 469.

138. *Scientific Worker* 12 (1940): 86.

139. Cab. 90/1.

140. Egerton to Dale, 21 August 1943.

141. A. J. P. Taylor, *English History 1914–1945* (Harmondsworth, England: Penguin Books, 1975), pp. 531, 587.

142. Nevertheless, an addition was soon made. According to the Committee's constitution, Bragg had to resign on 30 November 1940 when his tenure as president of the Royal Society ended. Bragg submitted a letter of resignation, but Hankey—thinking that it would be very regrettable if the Committee were deprived of Bragg's help after so short a period—suggested to Anderson that Bragg should act as an additional member of the Committee until he completed a year in office. Anderson agreed. Bragg withdrew his resignation and continued to serve with his successor on the Committee and in the Royal Society, Sir Henry Dale. Scientific Advisory Committee minutes, 21 November 1940. Cab. 90/1.

143. Scientific Advisory Committee, minutes, 15 October 1940. Cab. 90/1.

144. *Parliamentary Debates* (Commons), 365 (1940), cols. 232-33.

145. Sir William Bragg to Lord Hankey, 16 October 1940. Cab. 90/1.

146. Association of Scientific Workers, Executive Committee minutes, 16 September 1939.

147. Ibid.

148. Ibid., 7 and 14 October 1939 and 4 November 1939.

149. Ibid., 18 November 1939.

150. Ibid., 4 November 1939.

151. Ibid., 18 November 1939; *Nature* 144 (1939): 1004.

152. Association of Scientific Workers, Executive Committee minutes, 1 February 1940.

153. M. Philips Price, "The Parliamentary and Scientific Committee of Great Britain," *Impact of Science on Society* 3 (1952): 258-76, 262.

154. Association of Scientific Workers, "Agenda for Twenty-sixth Annual Meeting, 8-9 May 1943," p. 16, ASW/1/1/23/1.

155. Association of Scientific Workers, Executive Committee minutes, 1 February 1940.

156. Association of Scientific Workers, Finance and Organising Committee minutes, 7 December 1941.

157. For accounts of the Committee's activities, see M. Philips Price, "The Parliamentary and Scientific Committee of Great Britain," and S. A. Walkland, "Science and Parliament: The Origins and Influence of the Parliamentary and Scientific Committee," *Parliamentary Affairs* 17 (1964): 308-20, 389-402.

158. Christopher Powell to secretary of the Scientific Advisory Committee, 28 October 1940. Cab. 90/1

159. James R. Beard to Lord Hankey, 25 October 1940. Cab. 90/1.

160. Cab. 90/1.

161. Scientific Advisory Committee, minutes, 31 October 1940. Cab. 90/1.

162. Lord Hankey to Sir John Anderson, 31 October 1940. Cab. 90/1.

163. Scientific Advisory Committee, minutes, 17 December 1940. Cab. 90/1.

164. Sir John Anderson to W. S. Churchill, 26 March 1941. Cab. 21/1166.

165. Norman Brook, Privy Council Office, to Eric Seal (Churchill's secretary), 31 March 1941. Cab. 21/1166.

166. Cab. 21/1166.

167. Sir Edward Bridges to W. S. Churchill, 1 April 1941. Cab. 21/1166.

168. *Nature* 147 (1941): 538.

169. Ibid.; Engineering Advisory Committee, minutes, 7 May 1941. Cab. 92/113.

170. Sir John Anderson to W. S. Churchill, 26 March 1941. Cab. 21/1166.

171. Scientific Advisory Committee's First Report, Paper SAC (41)11, p. 1. Cab. 92/2. On 2 April 1941, Hankey made a statement in the House of Lords based on the Committee's report. He gave general accounts of the organization of science within the country and of the Committee's work. Much of this was reproduced a week later in *Nature* 147 (1941): 432-35.

172. Scientific Advisory Committee, minutes, 21 October 1940. Cab. 90/1.

173. See Margaret Gowing, *Britain and Atomic Energy, 1935-45* (London: St. Martin's Press, 1964), chap. 2.

174. Lord Hankey to editor, *Times*, 11 August 1945, p. 5.

175. For most of the material in this paragraph, I am indebted to Gowing, *Britain and Atomic Energy, 1939-45*, pp. 106-8.

176. Ibid., pp. 107-8.

8

Scientists and the Central Direction of the War Effort

The principal characteristic of the third phase of the social relations of science movement, it is recalled, was the central organization of science in wartime. The preceding chapter has shown that the coming of war brought governmental and civil scientists and engineers increased responsibilities within government. By May 1941, the Scientific and Engineering Advisory Committees had been created with powers to oversee almost all aspects of the technical side of the war. Nevertheless, scientists and engineers were still only advisers to the political and military leaders responsible for determining the strategy of the war effort. In the firm belief that they should be placed on an equal footing with those leaders, scientists in 1942 went beyond private appeals to the government and made the matter a public issue. This occurred during what was "politically the most disturbed time of the war."[1] Certainly, mid-1942 was the most political period of the social relations of science movement.

This chapter begins by briefly considering the work of the Engineering Advisory Committee and then examines the wave of dissatisfaction with the uses being made of science and technology in the war effort that swept parts of the British scientific, engineering, and parliamentary communities beginning in mid 1941 and cresting the following summer. The central figure in this wave was A. V. Hill, who had been one of the principal leaders of the drive that had led to the creation of the Scientific Advisory Committee in 1940. Now Hill forcefully advocated, in the press and in Parliament, the creation of a scientific general staff that would determine the strategy of science and technology applied to war. Hill's suggestion was vigorously supported from outside government circles by the Association of Scientific Workers and the Parliamentary and Scientific Committee, and from inside government by members of the Scientific and Engineering Advisory Committees. The outcome was the appointment in September 1942 of three scientific advisers to the recently created Ministry of Production. Upon the appointment of the advisers, the Engineering Advisory Committee became defunct. Although the Association of Scientific Workers continued into 1943 to advocate the creation of a scientific general staff, in regard to responsibilities within government scientists had realized their greatest wartime gains, and the major protests of the social relations of science movement had ended.

Under Hankey, the Engineering Advisory Committee got off to a very active start. Between 30 April 1941, when its formation had been announced, and the following 26 November, when it issued its first major report, it met twenty-eight times. Its primary initial concern, addressed in the report, was dictated by the widespread feeling among engineers that in the war effort insufficient and inefficient use was being made of engineering capacity. The Committee undertook to examine the matter so as to obtain a comprehensive picture and identify areas in which improvements could be made. However, it did not attempt to cover all aspects of engineering because the larger part of British industry was engaged in war production, and a full examination would have entailed consideration of almost all of industry. The Committee confined its attention primarily to mechanical and electrical engineering. It interviewed representatives of the service supply departments and of the engineering industry in order to gain an understanding of their mutual relationships. The Committee specifically wished to discover whether departments had experienced any difficulties, either because of failure of industry to cooperate or because of unsatisfactory equipment supplied by manufacturers. It also wanted to know whether industry had experienced any difficulties in working with departments, either because of the methods and specifications of the departments or because of lack of cooperation or other causes.[2]

The completed report discussed, in order, the engineering organizations of the service supply departments; coordination of the supply services; the organization of the engineering industry in regard to design, development and production; the Inter-Service Products Panel; specifications and standards; and various suggestions made by industrial firms. In most of these areas, the Committee made recommendations for improvements, but its overall conclusion was that the public criticisms of government departments, insofar as they applied to the fields of engineering covered by the report, were either unfounded or greatly exaggerated: "Taking into consideration the unprecedented scale of the expansion of our war effort, the novelty of many of the requirements of the Fighting Services, and the difficulties imposed by condiditions of modern warfare, the achievement of the Government and Industry working in combination has been extraordinary."[3]

Yet, as the various recommendations indicated, certain specific improvements were considered desirable. In particular, the report drew attention to tanks, which were found to present some of the most important and urgent engineering questions of the war. Several representatives of industry had, when interviewed by the Committee, independently referred to tanks in illustrating difficulties. Consequently, Hankey forwarded a summary of the principal points raised to the Minister of Supply, Lord Beaverbrook.

While working on this major report, the Engineering Advisory Committee had also found time to compile and issue separate reports on other matters. These included the procedure for dealing with mechanical inventions submitted to the Committee, the specification and testing of internal combustion engines for driving electrical generators, underground chambers, the testing and inspection of steel, and the provision of electrical plants for the Empire. Thus, under the chairmanship of Hankey the Engineering Advisory Committee, as the Scientific Advisory Committee, seemed to be justifying its formation. On 23 October 1941, Hankey wrote to Field-Marshal Smuts: "I am working very hard, and not without success, in putting science on the war map."[4]

In 1941, Hankey became Paymaster General, but this did not affect his position on the Scientific and Engineering Advisory Committees. Early the following year, however, differences developed between Hankey and Churchill, and on 3 March, Churchill asked him to place the paymaster generalship at his disposal. Wrote Churchill: "Should you however desire to continue the Chairmanship of the group of Technical Committees over which I am assured you preside so efficiently, it would be possible to provide you with the same authority and emoluments as you enjoy under the present arrangement.[5] You will not be serving in a Ministerial capacity, but I hope you will give us your assistance none the less."[6] To this Hankey replied:

> If I have not resigned before it was mainly because I did not want to embarrass you at a difficult time. I also wished to complete the work of the Technical Committees, to which you refer in such kindly terms, and other war work.
>
> I could not usefully continue that work when deprived of the authority of a Minister of Cabinet rank and the consequential prestige which I have hitherto enjoyed.
>
> I, therefore, gladly place the Paymaster Generalship at your disposal.[7]

Hankey explained to Sir William Bragg that to continue as chairman of the Scientific Advisory Committee, without at the same time serving in a ministerial capacity, would be to lower the status of the Committee.[8] To A. V. Hill he wrote: "Anderson is terribly worried to find a Chairman. My own belief is that the Committee will never work satisfactorily without a minister of Cabinet rank. To get our Reports carried out and to secure the goodwill of Ministers and Departments needs someone who is not only pushing, but who also 'carries guns.' I do not think Winston has ever cared a rap about science, but Anderson has been splendid and he and I have worked together in perfect unity."[9]

Hankey's successor as chairman of both the Scientific and Engineering Advisory Committees was R. A. Butler, President of the Board of Education. Attending his first meeting of the Engineering Advisory Committee on 26 March, Butler stated that he would bring the Committee to the notice of his colleagues in government and inform them that it would be glad to consider technical questions that departments might desire to refer to it. Although the Committee members approved this statement, it must nevertheless have caused them some dissatisfaction, because it implied that the Committee would henceforward act only when asked to do so. At this same meeting, it was announced that Tizard "had found it necessary, in consequence of the many calls upon his time, to tender his resignation as a member of the Committee."[10] But as Hankey wrote to Sir Samuel Hoare: "Tizard has resigned from one of the Committees (not from his job, thank God) as a protest, and the Royal Society and Engineering Institutions have kicked up a frightful hullabaloo."[11]

The days of the Engineering Advisory Committee were, in fact, numbered. After chairing two further meetings in early April, Butler wrote almost a month later to the Committee members.[12] He explained that he had met with Oliver Lyttelton, who in March had become Minister of Production in the War Cabinet, an office created the previous month. Working under the Minister of Defence, Churchill, the Minister of Production coordinated the three supply departments—the Admiralty, the Ministry of Supply, and the Ministry of Aircraft Production. As Prime Minister and Minister of Defence, Churchill controlled the supreme strategic direction of the war effort through the Defence (Operations) Committee of the War Cabinet.[13] Butler's meeting with Lyttelton had been arranged because the Engineering Advisory Committee had become concerned that its work overlapped that of the Ministry of Production. For example, it had been suggested that the Committee should inquire into delays that were being experienced in the production of machine tools. At their meeting, Lyttelton told Butler that the Committee should not undertake the inquiry. He felt that the matter might reasonably be regarded as belonging to the sphere of the industrial division of his own office in collaboration with the controller-general of machine tools. In general,

Lyttelton was convinced that all of the technical and research organizations of the supply ministries should be overhauled with a view to achieving closer integration and coordination in the areas of design, production development, and operational and long-range research. He was contemplating instituting an inquiry into these matters and believed that it would reveal what portion of the field the Engineering Advisory Committee might most profitably concern itself with in the future. Butler decided, therefore, not to call the Engineering Advisory Committee together until Lyttleton's plans had become clearer. The Committee would meet only twice more, jointly with the Scientific Advisory Committee.

During the summer of 1941, British scientists and engineers began to complain, as in 1940, that the country's scientific and technical resources were not being used to the best advantage, that much valuable knowledge and experience was being wasted because insufficient responsibility was being given to scientists, and that scientists were unnecessarily being discouraged from exerting their maximum effort. Consequently, the Association of Scientific Workers decided to hold a series of regional conferences at which scientists, technicians, and laboratory assistants could discuss their problems and decide upon appropriate action.[14]

During August and September 1941, three conferences were held of Association members working in the Ministry of Supply, in agriculture, and in the aircraft, chemical, and engineering industries.[15] One of the principal questions considered was how science could be better used for the national benefit. From the proceedings and from written reports submitted by branches, the Association's executive committee concluded that in many areas technical personnel were not being used to the best advantage. Many Association members were not engaged on essential war work. Others who were, were not fully occupied, yet were tied to their jobs by the Essential Work Order. Duplication of work continued because firms wanted to protect their trade secrets and because there was no effective government machinery for overcoming this situation. At the end of September, the Association sent two representatives to report its findings at the British Association Division's conference on science and world order.[16]

The Association of Scientific Workers' conferences received much attention in the national and provincial press, and as one important consequence, the Ministry of Supply requested a report on the efficiency of its use of scientific personnel.[17] The Association soon submitted a detailed report stressing the need for full collaboration between management and technical staffs. Later, it believed that the subsequent establishment of joint production committees in Royal Ordnance Factories and the engineering industry was to some extent due to its recommendations.

In January 1942, the Association held its largest and most influential conference to date, on "Science and the War Effort," filling Caxton Hall, London, for two days. The event was arranged by the Association's new Social Relations Committee and was similar in format to British Association Divisional conferences. There were six sessions on the university training of scientists; the training of technical personnel; building, housing, and A.R.P.; food and agriculture; the utilization of scientific personnel; and the application of scientific knowledge to production and service problems.[18] A major objective of the conference was to bring together leading scientists and their younger colleagues from industrial laboratories and Service departments "to discuss their common problem—how to get maximum use made of science in order to overthrow Fascism in the quickest possible time."[19] The conference was most successful. During the month in which it was held, 648 new members joined the Association, which had reaffiliated with the Trades Union Congress in 1940.

The serious charges made during the proceedings concerning the utilization of science were summarized by the Association's president, Sir Robert Watson-Watt.[20] Many industrial scientific workers were either giving less than two hours each working day to the war effort or were retained in programs not directly aiding it. Exchange of information between individuals in different firms was generally discouraged and frequently forbidden. Some firms were not disclosing to others in the same field technical details that could be applied generally with great advantage. There was a widespread failure to give effect to that measure of planning in industry that could only be ensured by a scientific survey carried out by qualified, experienced, and enthusiastic scientific workers. In some industries, the pooling of research data was virtually unknown. Finally, there were parallel industries that did not exchange data, leading to the needless duplication of research effort. In closing the conference, J. D. Bernal warned that scientists should not be complacent about Britain's war effort:

> The scientific profession and particularly its senior members are most liable to this attitude. We can see this by contrasting this conference with that of the British Association last September [1941, on science and world order]. There the tone was that the country was using its scientists most effectively. Here we have looked into the matter more closely and critically. We realize that it is not sufficient to be told that an eminent Committee has been appointed to deal with a subject. We have had such committees from the beginning of the war and very little has come from them. The evidence put forward at this conference from dozens of different places where scientists are actually working is in itself enough to show that complacency in official circles on the full use of science in the war effort is completely unjustified.[21]

Immediately after the conference, the executive committee requested interviews with the Ministers of Labour and Production and the head of the Central Register. The last was quickly granted, and in March Bernal led a deputation of three to the Ministry of Labour.[22] There they discussed at length the use of technical manpower with three of the Ministry's representatives, including its scientific adviser, Dr. Guy.

Meanwhile, a significant new line of criticism was initiated by Sir Henry Tizard, who on 3 February 1942 addressed the Parliamentary and Scientific Committee at its first annual luncheon held in the Savoy Hotel, London. Tizard began by observing that in general no one could deny that the influence of science was now greater than it had ever been, and that the current government and Parliament valued the help and guidance of scientists as none of their predecessors had done. In reference to Lord Cherwell, he asked which previous prime minister had ever had a scientific adviser continually at his side, and he added that other ministers and even commanders-in-chief also had scientific advisers.

In considering the general results of the intrusion of scientists into military affairs, Tizard suggested that science, like war, had its tactics and strategy. Currently, the tactical strength of science in Britain was very great. There were many well-run and well-equipped research and experimental establishments under direct or indirect governmental control or under independent control. Above all, there was a large supply of able young scientists who were rendering great service to the state. Thus, with respect to the tactics of science, Tizard concluded that Britain had nothing to fear in comparison with any other nation.

Regarding the strategy of science, however, he was not so confident. Much had been done to improve it, but much remained to be done. Who, asked Tizard, was to decide the strategy of scientific war, that is, to decide what were the things that really mattered and where Britain was to devote its scientific strength to get the greatest results in the shortest possible time. Drawing on his previous experience as scientific adviser to the Air Ministry, he continued:

> Certainly not scientists alone, working in the void, however eminent. [A reference doubtless to Cherwell.] Nor, in my opinion, can it be safely left to the staffs of the Fighting Services, even though each Service Department may contain officers of high scientific ability. Nor, I say, can it be left to the War Cabinet, however fertile in ideas. The safest way of reaching the right decision is to have scientists working side by side in the closest collaboration with those who have the administrative and executive responsibility. And the first thing that the scientist learns when he has the benefit and privilege of such collaboration is that he has a lot to learn.[23]

Tizard hoped that due credit for adopting this policy in the Air Ministry would be given to Lord Swinton and the senior members of his staff when he was Minister of Air. The technical needs of the Royal Air Force, its staff plans, and even its operations, had freely been submitted to the scrutiny and criticism of scientists. Good dividends had resulted, so much so that the example had spread to other departments—although not yet enough. Tizard suggested to the Parliamentary and Scientific Committee, which was now apparently regarded as having influence, that it could perform no greater service than to ensure this kind of cooperation. The Committee, he said, should concern itself with the general strategy of science in the war rather than with its tactics.

This was a course that the Parliamentary and Scientific Committee chose to follow at a subsequent meeting on 24 March 1942.[24] Here Bernal, armed with information collected by the Association of Scientific Workers in the previously mentioned conferences, initiated a discussion on the use of science in the war effort. The meeting had been arranged after the Committee had received a report on the subject from the Association. Tizard was present and agreed that there were many criticisms of the present system of utilizing scientific personnel, but they all boiled down to the question of responsibility. Those who had the responsibility were not scientists and did not really understand the technical requirements. On the other hand, although Britain certainly did not lack scientists, Tizard could not name a single scientist possessing real executive responsibility. Furthermore, no scientists were being educated to take on responsibility. Yet, unless scientists were given real authority, claimed Tizard, and unless responsibility was made clearer, money and effort would be wasted. A weighty contemporary problem was that of deciding between various options, all of which were possible. To decide on priorities required knowledge of science and knowledge of where things could be done and which people were capable of doing them. But according to Tizard, these types of knowledge were not being coordinated because the necessary organization did not exist, and this was because the Cabinet and the House of Commons did not believe in giving authority to the scientists.

C. P. Snow, then an administrator chiefly concerned with providing personnel for radar work, likewise suggested that the reasons some of the problems mentioned by Bernal had not been settled was because there had been no decided policy on production and scientific manpower. However, the most fruitful suggestion came from the insightful A. V. Hill, namely, that a chief of technical staff should be appointed to deal with technical matters in the same way as the Chief of Staff dealt with the operational side of the war. Were the Parliamentary and Scientific Committee to stress this point, it would be doing much to help, he said. As all present were agreed that scientists should be given more power in administration, it was decided that a memorandum—to be drafted by, among others, Bernal, Hill, and Captain Leonard Plugge M. P., Chairman of the Parliamentary and Scientific Committee—should be prepared for submission to the prime minister and possibly the Minister of Production.

The idea of a scientific general staff, first suggested by Hill, was soon openly and authoritatively brought to the government's attention. In the House of Lords on 5 May, Hankey, no longer chairman of the Scientific and Engineering Advisory Committees, expressed the hope that it would be possible to add a scientific staff to the Joint General Staff. He explained that each of the service departments had already brought in scientists and had found them of great value in operational and other areas. It was obvious, he said, that there should be increased participation of scientists in Britain's war strategy, and the weight of opinion, both scientific and lay, was clearly in favor of this.[25]

That opinion soon drew support from several contemporary developments. First, in the United States on 26 April, Vannevar Bush—one of the country's most eminent scientists and engineers and head of the wartime Organization for Scientific Research and Development with direct access to President Roosevelt—argued that "At a level of strategy there needs to be a control of trends on new weapons, a determination of emphasis, and an insistence on progress on specific matters at the expense of other things, if the situation is to be complete. The planning of military strategy needs to be carried on with a full grasp of the implications of new weapons, and also of the probable future trends of development."[26] Such was Bush's influence and Roosevelt's willingness that by May 12 the Joint Committee on New Weapons and Equipment was at work. Its members—Bush (chairman), Rear Admiral W. A. Lee, Jr., and Brigadier General R. G. Moses—reported directly to the Joint Chiefs of Staff, who had directed them to coordinate the efforts of civilian research agencies and the armed services in the development and production of new weapons and equipment.[27]

Hill and others were soon urging, as an extension of Hill's

original idea, that there should be a corresponding body in Britain. British government leaders, however, would not readily implement the idea. Specifically, Hill's suggestion now was "to form a technical Chiefs of Staff Committee corresponding to the Chiefs of Staff Committee and on the same kind of level. The broad duty of such a Committee would be to determine the strategy of science and technology applied to war, just as the Chiefs of Staff Committee determine the strategy of operations."[28]

A serious Allied setback in the war was a second influential development. Beginning in late May and continuing through June, German forces under General Rommel drove the British army out of Libya to El Alamein in Egypt. "The people ... received the news of the fall of Tobruk with bewilderment and bitter disappointment. They had never felt so profound a shock since the dark days of Dunkirk." [29] Scientists, including Tizard, blamed the inferiority of British tanks, whose deficiencies had earlier been pointed out to the Minister of Supply by the Engineering Advisory Committee. About this same time, Tizard was also embroiled in the famous strategic bombing controversy, a third decisive development.[30]

Influenced by all three, Tizard sent a note in June to Sir Arthur Street, Permanent Secretary at the Air Ministry and Secretary of the Air Council. Addressing himself to the uses of scientific and technical resources in the war, Tizard said of strategy:

> This is or ought to be the main work of the Chief Scientific Advisers. The trouble is that they all feel that their advice is not sufficiently taken, or not taken in time. A good many instances of this could be given. There is, in fact, only one Scientific Adviser who is in a position to get something done quickly even if the Fighting Service staffs do not altogether agree with him, and he is Lord Cherwell. The feeling, which a good many of us share, is that this power to get some decision made quickly, and some action to follow, should not be altogether confined to Lord Cherwell.[31]

This was the crux of the matter. Although the creation of the Scientific and Engineering Advisory Committees had brought scientists and engineers to the Cabinet level of government, nevertheless, unlike Cherwell, they had neither direct access to

Churchill nor involvement in War Cabinet deliberations concerning the use of science in the conduct of the war.

Tizard knew of Hill's proposed solution, but correctly thought it would be difficult to achieve, as the United Kingdom's constitution differed from that of the United States—"Ministers in this country have far more responsibility than Ministers in the United States, and they might well claim that the existence of a body corresponding to the Joint Weapons Committee would interfere unduly with their responsibilities."[32] He also doubted very much "whether this would work unless there was in existence a man who could be appointed Chairman and who would have the full confidence of all the Service Departments, the two Production Departments, and the independent scientists and engineers. I do not think such a paragon exists; and if the wrong chairman were put in only confusion would result. He would have too great executive powers."[33] And thus the new arrangement would be similar to the current unsatisfactory one.

Tizard favored a suggestion of the experienced and knowledgeable government scientist Sir Edward Appleton, secretary of the DSIR and a member of the Scientific Advisory Committee, who during the war noted in a book of random jottings: "Lindemann [Cherwell]: I say of him that if you work alone you make mistakes. You need the checking of your colleagues to be sure."[34] Of Appleton's proposal, Tizard wrote:

> I think I should be right in interpreting his suggestion as implying the abolition of the Scientific and Engineering Advisory Committees of the Cabinet and the appointment in their places of a Board of Advisers of about six people and not more than ten, which would be attached to the Defence Committee of the Cabinet.[35] Such a Board might include the present Chief Scientific Advisers of the Prime Minister, and the other Ministries, strengthened by some really prominent representatives of engineering. Such a reconstituted Board should have, in Appleton's opinion (which I share) much greater powers than the present Scientific and Engineering Advisory Committees. The Board as a whole through its individual members would be empowered to initiate any enquiry which was thought to be of value to the war effort and to report direct to the Defence Committee. It would not be executive. I do not see how it could be executive, unless the whole present departmental machinery is changed. In fact, it would be no more executive than Lord

Cherwell; but as a Board in a corresponding position it would have equal power to get something done if the Defence Committee accepted its advice.

I must confess that when I try to think of concrete terms of reference I find myself a little at a loss. My view is that if scientists who are trying to determine technical policy do not work in the closest collaboration with the staffs of the Service Departments concerned, they will not be of great value. The trouble now, however, is that most of the highly placed officers on the staffs of the Service Departments are so pressed with urgent matters and day-to-day problems, that they have not the time to put the right amount of thought to the more distant problems, and broadly speaking people in my position can help far more on the more distant problems than they can on the immediate ones which are well looked after by other people. So we are in this difficulty, namely, the Chiefs of Staff and the highly placed officers serving under them must trust someone to form a good judgement on these more distant problems, or we shall go on throughout the war taking hurried and even panic action to get over difficulties as they arise, which not only might have been but often have been foreseen. I think that the kind of body envisaged by Appleton would really help in giving the Scientific Advisers rather more authority in fact, although not on paper. If this was so I think it would be a step in the right direction.[36]

Shortly thereafter, on 1 July, Plugge also pressed the issue on behalf of the Parliamentary and Scientific Committee by writing a lengthy but less weighty and less incisive letter to R. A. Butler. Plugge recommended that a scientific and technical general staff, closely linked to the departments of defence and production, be created. Such a staff "should insure the fullest strategic use of our scientific manpower and resouces and the proper organisation and exchange of scientific and technical information relating to the conduct of the war."[37]

A much more effective means of gaining not only governmental, but also public, attention was employed by Hill, who on the same day had a letter published in the *Times*. He argued that the defeat in Libya had been due largely to a single cause—the inferiority of British to German tanks. In characteristic forthright fashion, he alleged that this inferiority resulted from a system that had failed "to anticipate future tactical requirements in guns, projectiles, armour, and performance, failed to collect, analyse, and profit by previous operational experience, failed sometimes even to obey the elementary rule that production must follow, not precede, development."[38] Too many disasters in the war, he continued, had been due either to technical mismanagement of this kind, together with unjustified optimism, or to improper or inefficient use of available weapons, resources, or materials. But such disasters had not been unforeseen:

> Those who have urged for years that our scientific and technical set-up was faulty at the top have hoped against hope that they were wrong: each disaster in turn has shown that they were right. Scientific and technical control is entirely departmental, not central, and usually at a low level, subordinate to administration. For operations, planning has been centralized in the Chiefs of Staffs organisation; for supply, in the Minister of Production. The third member of the trinity, dealing with research, design, and development and quantitative planning of the use of technical resources, is altogether unrepresented at the highest level. There is no central technical staff to advise the Cabinet, or the Chiefs of Staffs, directly on the scientific and engineering aspects either of operations or production, or to ensure, on their behalf, that design and development are efficient and far-seeing. All such functions are left to departments.[39]

Hill then referred to the United States' Joint Committee on New Weapons and Equipment, commenting that: "What the Americans can do early in their own war we ought to be able to do now: unless we do, we shall continue to fumble on from disaster to disaster."

Hill's letter appeared at an opportune time. On 1 and 2 July, the Commons debated a motion of no confidence in the central direction of the war.[40] At the same time, the Lords debated a motion calling attention to the conduct of the war with special reference to events in North Africa and the Mediterranean. A series of letters published in the *Times* supported Hill. Plugge wrote as chairman of the Parliamentary and Scientific Committee to endorse Hill's "important" suggestion for the early establishment of some form of central technical and scientific general staff. He reported widespread agreement among the technical bodies affiliated to the Committee that Tizard's "demand" for the better strategic use of Britain's "technicians" was one that called urgently for recognition at the very top if Britain was to avoid the disasters that must inevitably follow from lack of intelligent central direction of its formidable scientific and technical resources.[41] A mechanical engineer, Loughnan Pendred, ventured to think that he spoke for mechanical engineers as a body in welcoming Hill's letter. He thought it deplorable that, quoting Hill, "scientific and technical control was entirely departmental, not central, and usually at a low level, subordinate to administration."[42] To William Cullen of the Society of Chemical Industries, these words were the crux of Hill's letter. Many, including himself, felt grateful to Hill for voicing what had been widely thought for a long time. The secretary of the Association of Scientific Workers wrote that Hill's suggestions would be welcomed by the entire scientific profession, and that the Association had approached the Minister of Production on the matter.[43] H. E. Wimperis, who had been Director of Scientific Research at the Air Ministry from 1925 to 1927, and who in 1935, as scientific adviser at the Air Ministry had created the Committee for the Scientific Survey of Air Defence and appointed Tizard its chairman,[44] explained why British tanks were inferior while at the same time air armament was good.[45] At a critical moment in the history of aviation, there chanced to be a Secretary of State for War—Lord Haldane—with a mind imbued with scientific principle. In 1909, Haldane had recommended the creation of a strong and independent scientific and technical committee to watch over and guide aeronautical development. This was the Advisory Committee for Aeronautics (later the Aeronautical Research Committee), under the presidency of Lord Rayleigh and the later chairmanship of R. T. Glazebrook and Tizard. This Committee had an aerodynamic laboratory under its own scientific direction at Teddington, and it gave advice on the work done at the Royal Air Force experimental stations under the direction (after 1925) of the Director of Scientific Research at the Air Ministry. From his experience as director, Wimperis testified to the immense value of the services rendered by that Committee of distinguished scientists and to the wise and able direction of its successive chairmen. The personnel of a fighting service was, he continued, rarely scientifically minded, and officers placed in high position for the usual limited term of years were humanly desirous of being able to see the results of what they did while still in office. This led to short-term thinking, while long-range investigations, however scientifically important, were liable to be pushed aside. It was here that Wimperis thought that an independent and forceful chairman of a research committee could be such a power for good. If the advice of the committee was ignored, he could take both his own views and those of the committee to the highest level—and Wimperis knew of scarcely any occasion when this action did not immediately produce the desired result. When, he concluded, the difficult problem of Army mechanization had had to be faced by the War Office, a committee for the scientific survey of mechanical warfare had been created, with an independent civilian chairman of the highest standing in the world of applied science having direct access to the secretary of state, the story of recent years might have been different.

In the wake of this correspondence, the *Times* observed on 9 July that fortunately it was becoming more and more accepted that strategic planning must involve both science and industry and that action must follow this acknowledgment. As yet, science had not been given its proper function and station in the war, except perhaps in relation to air defense. It had been called into consultation in a junior capacity, but had had small chance of making original contributions. This was the more unfortunate, the paper continued, since the scientist is a natural innovator, and the war would be won by innovations or not at all. The time had come, it concluded, to give scientists and technologists their due at a higher level of responsibility. On 7 July, the *Manchester Guardian* had said that Hill's proposal "should be thoroughly discussed." At the same time, its London correspondent opined that "research and design are fundamental considerations now, and it will be surprising if the Government can get through the coming debate [on production] without granting the demand of Professor A. V. Hill and others for a central direction of operations and supply."[46]

The principal figures behind this mounting criticism were, of course, fully aware of what had happened to the Engineering Advisory Committee. Hill and other members of the Scientific Advisory Committee had an informal meeting with members of the Engineering Advisory Committee, and several from both Committees then requested a joint meeting with their common chairman, Butler, which was held on 5 July. This "amusing deputation," as Butler later described it to Anderson, was led by Sir Henry Dale, who as Bragg's successor as president of the Royal Society was a member of the Scientific Advisory Committee, and included also Hill and Ricardo, Fleming, and Beard from the Engineering Advisory Committee. Prior to meeting with Butler, this group had held a meeting attended also, significantly, by Tizard and Sir Ralph Fowler (Acting Secretary of the Royal Society and Plummer Professor of Applied Mathematics at Cambridge). The group had prepared a memorandum for Butler which seemed to him to be directed almost entirely to a transformation and promotion of the Engineering Advisory Committee to form a central body with executive powers, or powers of initiative.[47] Butler was also informed about the profound disquiet among scientists and engineers concerning particular designs of new weapons and mechanisms. In the group's view, Britain was not fulfilling Lyttelton's words when he had said that "the weapons in the hands of our troops should surpass, either in power, or in mobility, or in concealment, or in surprise, or in all, the weapons in the hands of the enemy."[48]

The deputation further claimed that the current arrangements for research and design were organized on "too departmental a scale" and that there was neither sufficient initiation of new ideas nor coordination of existing plans. They therefore suggested the creation of a central body composed of representative physicists, chemists, and engineers, in addition to a technical member from each of the three services. The chairman of this body, they said, should be able to represent the views of scientists and engineers at early stages in the processes of forming national policy and operational plans. In claiming that the body should have executive powers, the group observed that the Scientific and Engineering Advisory Committees had neither the requisite powers nor were they suitably constituted to carry out the work of the proposed central body.[49] This was to be an addition to, and not a replacement of, the Scientific and Engineering Advisory Committees. To be effective, it should be under the immediate direction of a member of the War Cabinet, have a full-time chairman of outstanding energy, and be composed of highly qualified scientists and technicians working full time (in contrast to the part-time activities of the members of the Scientific and Engineering Advisory Committees).[50]

In "cross-examination," Butler elicited that the deputation did not actually intend that the proposed central body should have executive powers, but rather that it should have the right to initiate new plans and ideas. The group had in mind some body similar to the New Weapons Committee in the United States. Butler then asked about methods of coordination: about whether the current heads of the three service research departments would be the three technical representatives on the central body. Here again the deputation appeared to Butler to adapt its plan to meet his cross-questioning, saying that they would not want the research departments of the three services as currently constituted to be represented on the central body. Butler therefore countered that the proposed body's functions were appearing more and more advisory.

Perhaps it was because the group was so pliant that Butler found it amusing. The deputation was not as prepared as it might have been, but on Butler's part there was a condescension towards these scientists and engineers who could easily be made to shift their ground under courtroom tactics. Between Butler and the Committees there was a gulf that had not existed between the Committees and Hankey, with his immense knowledge of and sympathy for the nation's scientific affairs. In chairing the Committees, Hankey had shown an initiative never displayed by Butler; under Hankey the Committees had an active role, under Butler a passive one. This may not have been entirely Butler's fault, but in part merely the result of giving responsibility for the Scientific and Engineering Advisory Committees to an already busy minister.[51] In any event, it contributed to the scientists and engineers' dissatisfaction.

In spite of the group's unpreparedness and pliancy, Butler recognized that they were quite in earnest and sincere. This was assured by, for example, Ricardo and Fleming "from the wealth of their experience" drawing attention to "the weakness of the current army organisation research *vis-a-vis* the Ministry of Supply." Thus Butler concluded: "From my own experience it does appear that the present Scientific and Engineering Advisory Committees, though useful in undertaking such work as the consideration of patents and patent designs, are not suitable for the bigger task recommended by this body and that the suggestion put forward by my visitors does deserve the most serious consideration by the Government."[52]

In forwarding the minute giving this account of the meeting to Anderson, Butler enclosed a most interesting letter:

> Sir Henry Dale tells me he is seeing you on Wednesday and I think it would be a very good thing if you would have a talk with him and put him in a position to hold the other conspirators at bay.
>
> Though the plan may not work in its present form it is clear that Hill's original idea has been much revised, and that there is considerable anxiety in the world of Science and Engineering.
>
> I used all the delaying tactics, which you recommended, including the statement that we did not want to draw the Scientists from practical tasks on to Central Committees. But they are in dead earnest about the lack of confidence—as they describe it—in the manner in which our best scientific and engineering brains are being used. Even the staid President of the Royal Society warmly supported the others in a merciless attack on Lord Cherwell, which I tried to get put in its right perspective.
>
> The atmosphere of the talk was perfectly friendly, and they accepted my criticism of the executive potentialities of any such body, as well as my probing as to how it would "co-ordinate" in good part.
>
> But I should say that some British "pendant" of the American new weapons Committee is desirable. It is at any rate important to handle this distinguished posse with Tizard and others in the background with distinction.[53]

For a week nothing happened, and then the issue surfaced in the House of Commons on 14 July. In opening a debate on production, Lyttelton referred to criticisms that the country's scientific organization for war was faulty. He said that although he and Anderson were looking into the matter, he nevertheless thought that the scientific work that was being done was of a very high order. He explained that Hill, who represented Cambridge University as an Independent Conservative and who was present, was a member of the Scientific Advisory Committee which had reported to the government that the country's scientific organization had reached a high state of efficiency.

Hill replied that Lyttelton completely misunderstood what

had been urged regarding higher technical control.[54] The need was to provide for some high-level central body that would ensure that the scientific and technical resources of the departments and the country were properly and effectively used. He explained how the Scientific Advisory Committee had been instructed by Churchill not to "meddle with our innards," that is, it was not to examine the organization handling the central direction of the war. It was to deal only with the scientific work of the various departments. Yet the current complaints were not about what was happening tactically within the departments, upon which the Scientific Advisory Committee had indeed reported favorably, but about what was happening strategically in the use of the work of each department and of the work of all the departments taken together.

Earlier in the debate, it had been pointed out that Tizard was a member of the Air Council. But, responded Hill, the question was not whether Tizard was on the Council; rather it was whether he had any authority on it or on the Council of the Ministry of Supply, whether his advice was taken, and whether he had any influence. And Hill implied that Tizard was continually thwarted in his efforts. Yet, continued Hill, strategy depends on tactics, and tactics upon weapons; the man who knew something about weapons and tactics should be enabled to have some influence upon strategy because it was certain that most of those who determined strategy knew nothing whatever about weapons.

Hill then referred to a committee that had been set up some sixteen months earlier by the First Lord of the Admiralty to examine the operation of the research and development establishments within the Admiralty. The chairman of this committee was the individual in charge of the staffs of the establishments that might need to be criticized by the committee. This, said Hill, was hardly a satisfactory way of obtaining a truthful report on the performances of the establishments. The only way to get the truth was "to have some properly constituted authority, with technical knowledge, centrally placed, that could insist on getting the information it wanted."

Hill next raised the subject of missions to the United States, explaining that their large number constituted one of their problems. Again, there was no properly organized central agency for bringing together the work of the various missions, and in Hill's view there never could be so long as departments were trusted individually to send missions not in contact with one another and thus not coordinated with one another's activities. However, if there were a central technical body it would be easy to attach to it a central information bureau from which scientific and technical liaison could be conducted, not in order to prevent departments from having their own missions, but at least to keep these missions in touch with one another and to have people at the center who would know what they were all doing.

Two days later Hill joined Butler, at the latter's request, to receive what the *Times* described as an influential deputation of scientists, members of Parliament, and peers from the Parliamentary and Scientific Committee and led by its president, Lord Samuel.[55] The Association of Scientific Workers had requested the Committee to arrange the meeting.[56] It would seem that on this occasion Hill did not speak, but then the principal ideas presented were very similar to what he himself had suggested within the Parliamentary and Scientific Committee. Indeed, the degree of unanimity among scientists at this time is illustrated by *Nature*'s claim that the deputation clearly had as its objective the views expressed by the journal's joint editor, A. J. V. Gale, in the leading article of 18 July.[57] Once more, the establishment of a central body was advocated as the proper cure for the troubles in the scientific and technical fields. The War Cabinet, the deputation argued, currently received strategic advice from the Chiefs of Staff who coordinated the activities of the three services. The Minister of Production was in a position to coordinate activities in the production field. However, in the areas of scientific research and development no permanent central body existed. In proposing the creation of such an organization, the deputation said it should be under a strong chairman possessing engineering and administrative experience and be a small, full-time board of eminent scientists and engineers in close touch with industry and assisted by a large panel of expert advisers who could be called in to deal with special topics. It should balance and assist the Office of the Minister of Production in the scientific and engineering fields and should be in a position to

Scientists and Central Direction of the War Effort · 237

tender advice to the Chiefs of Staff Committee and to the War Cabinet with such authority that it could:

> a) Deal effectively with large scientific and engineering questions, in which measures taken by the Departments were inadequate.
> b) Decide on priorities of men, laboratories and apparatus between the various Departments, and the best strategic uses of man-power.
> c) Act as a clearing house for scientific troubles and ideas.
> d) Maintain contact with the Research Departments, with industry and with the Professional Institutes.
> e) Assist in the co-ordination and control of the general strategic direction of the war from the point of view of science and engineering, and provide Ministers with advice free from departmental bias.[58]

It had become a trying situation for Butler, who, having little taste for it, would have been happy to wash his hands of scientists. On the following day, he poured out his thoughts to Anderson in a letter:

> You will remember that, when you asked me to take on the Chairmanship of the Scientific and Engineering Advisory Committees, I said that I would be only too glad to help the Government and fit them in with my other activities here.
> After the short period during which I have been connected with these Committees, I have come to the conclusion that, as constituted, they are not capable of undertaking the sort of work which some of the leaders of the recent agitation would wish undertaken by a central scientific co-ordinating body. This is particularly true of the Engineers.
> I was in fact asked by the Minister of Production... not to allow the Engineers to interfere in the day to day difficulties of production, or in the commercial rivalries which very often hold up production.
> The long and short of all this is that on the Scientific Advisory Committee we have been able to do useful work in planning and maintaining the scientific mission in America and in undertaking an inquiry into the complicated and detailed question of Patents—at your request.[59] We have also acted as a lightning conductor for the recent anxiety on the scientific conduct of the war.
> With the Engineers I have confined myself to obtaining their general views and directing their attention to the Channel Tunnel, with which considerable progress has been made.[60]

You will see, therefore, that with both Committees I have conducted the work on a minor key, but we have been fairly active and it has certainly taken up a good deal of time.

Since I do not think that either of these Committees, as at present constituted, is suitable to be formed into a central co-ordinating body, which should give confidence to the world of Science and Engineering, I hope that you and Lyttelton will bear this in mind when considering the possibility of forming some such body. I do not fancy myself appearing in a false position. I am not trained as a scientist and have little knowledge of production, and cannot fulfil any more ambitious role than that which I am already taking. My training in politics has been in Imperial and Foreign diplomacy—of which I appear to have a great need at present in dealing with the characters and susceptibilities of members of these Committees.

I think you would want for any new body a Chairman quite differently equipped. If you want, therefore, to review the whole position of the Scientific and Engineering Advisory Committees, please do so, feeling that I shall be only too glad to conform to any decision which you may reach.

Meanwhile I will go on with the work I have been doing on the lines that I have been following, if that is found by the Government to be convenient and suitable.[61]

Pressure was building on the government, and on Anderson and Lyttelton in particular, to do something.

Once more Hill acted forcefully. On 18 July, he outlined his now familiar views in a biting two-page article entitled "A Practical Plan to End our Military Weakness" and published in the popular weekly magazine *Picture Post*. Hill explained that although the influence of the Scientific and Engineering Advisory Committees had been very valuable, the Committees met only at intervals, had no executive authority, and their members were very busy with many other tasks. He illustrated the type of problem whose proper handling required the attention of a full-time joint technical board:

The Admiralty, let us suppose, claims that if it had, for its Atlantic patrols, more of the long-range bombers at present employed in attacking German cities, it could make better use of them than the Air Ministry does. The decision whether to accede to the Admiralty's demand depends on a balancing of claims. The Air Ministry, by unconscious bias, multiplies the effect of its own bombing by three and divides the effect which the Navy might produce with the same aircraft by 10; the numbers of course are only for illustration. The Admiralty, not so single-hearted about the use of air power, and less experienced in propaganda, is realistic about the bombing of German cities, and multiplies the effect which the Navy might produce with the same bombers only by 2. There is a fifteenfold divergence between the two assessments of the relative values. How is higher authority to decide? An accurate assessment, based on the best available evidence by the best scientific and statistical methods, by an impartial staff attached to the [proposed joint technical board], would give a better basis for decision. Otherwise ballyhoo and propaganda may decide.[62]

Nine supportive letters, including ones from Plugge, Sir Richard Gregory, Sir Robert Robinson, Professor of Chemistry at Oxford, and Wimperis, appeared in the next two issues of *Picture Post*. Edgar Granville, a Liberal member of Parliament, wrote that for two years he and two other Liberal and two Labour members—Clem Davies, T. L. Horabin, Aneurin Bevan, and Emanuel Shinwell—had "been pleading for this mentality."[63] It is true, he said, that "we now have a War Production Minister, but as usual this Government makes alterations but not real changes."

When Butler had met with the deputation from the Scientific and Engineering Advisory Committees, it had been agreed to continue the discussion at a future meeting attended both by Anderson and Lyttelton.[64] This took place on 20 July, and beforehand Dale convened a small informal meeting of interested members of the two Committees. Those present at the subsequent meeting were: Anderson, Lyttelton, Butler, Dale, Hill, Fleming, Beard, Ricardo, Fowler, and Thomas R. Merton (treasurer of the Royal Society and formerly Professor of Spectroscopy at Oxford). Here Dale referred to the principal conclusion of the Scientific Advisory Committee's First Report, which Lyttelton had attempted to use against Hill in the House of Commons.[65] He reiterated that it should not be assumed from this that the members of the deputation, or of the Scientific Advisory Committee, now regarded all scientific activities of the government as satisfactory. The Committee's terms of reference, as interpreted, had precluded it from meddling in the central machinery of government. Furthermore, being a part-time body whose members were fully occupied in other work and seldom able to meet more than once a fortnight, the Committee was unable to embark upon a more detailed investigation of the actual working of the various scientific departments.

Regarding the Engineering Advisory Committee, Dale said that many of the technical difficulties encountered in the war had been engineering rather than purely scientific ones, and yet Lyttelton had suggested that the Committee should not investigate engineering activities that fell within the sphere of his Ministry. Furthermore, even when the Engineering or Scientific Advisory Committee was able to investigate a particular problem and its recommendations were approved by the department concerned, it often happened that no effective or early action was taken in the matter. As an example, Dale mentioned an investigation by the Scientific Advisory Committee into photography in the services, where after rapid investigation a report was made suggesting certain improvements which, although approved by the Ministry of Aircraft Production, were not implemented for more than a year and then only after pressure from another quarter.

Dale next mentioned that with few exceptions practically all government scientific research was conducted on a departmental basis. In some cases no machinery had been provided for dealing with interdepartmental matters and no provision was made for communicating central advice to the War Cabinet on scientific and engineering matters not covered by any department. As an instance of the lack of central scientific and engineering advice, he mentioned the Besa machine gun adopted for British tanks. These guns gave off carbon monoxide from the breech in such quantities that an elaborate ventilation system had to be provided to enable tank crews to remain conscious under battle conditions. Dale understood that the tanks in question had been designed purely from the engineering point of view and that in the early stages no particular attention had been directed to the effects on the crews of the plans adopted. In relation also to central advice, Dale referred to the economic section of the War Cabinet under Professor L. C. Robbins which, unlike the Scientific and Engineering Advisory Committees, reported directly to the Lord President and formed a valuable advisory team. He suggested that an analogous body of scientists and engineers might work as full-time advisers to a suitable member

Scientists and Central Direction of the War Effort · 241

of the War Cabinet. In concluding, Dale requested Anderson to permit the Scientific and Engineering Advisory Committees to consider means of strengthening the machinery of government and to submit proposals for setting up a new scientific advisory body with suggestions for its place in the structure of government, its members, and its terms of reference.

Hill spoke next. In the autumn of 1940, he had learned that certain scientific departments of the Admiralty, including that engaged on antisubmarine devices, were not working well. After consulting Hankey, he decided that the matter lay outside the powers of the Scientific Advisory Committee, so he reported it to the First Lord of the Admiralty. As a result, a panel was set up in February 1941 under the official responsible for the department in question. The panel held its first meeting in June 1941, and though its reports to the Admiralty were secret, he understood that it had so far produced little result. Fowler, a member of this panel, interjected that though it had done some good work, its effects had proved much too slow. He felt that the Admiralty suffered from the lack of a body similar to the Aeronautical Research Council at the Ministry of Aircraft Production.

Hill next mentioned that a distinguished fellow of the Royal Society had criticized the present position of the Chemical Defence Establishment at Porton and commented that no machinery existed that would enable such a statement to be investigated. He bluntly said that it would be useless to report the complaint to the Minister of Supply since the most probable result would be that the head of the Porton establishment would be asked for his views in the matter and that no further action would be taken. Merton suggested that a body should be created that could go into complaints and, when necessary, advise the service scientific establishments and report to the Minister of Production. He suggested that such a body could also do much useful work without bringing all matters to the attention of the Ministers concerned.

The members of the Engineering Advisory Committee were equally critical. Beard stated that although the terms of reference of the Committee were wider than those of the Scientific Advisory Committee, in practice this had made little difference. The engineers had found it very difficult to follow up their recommendations. He suggested that a full-time body was needed to cover the broad planning of the war. Fleming also suggested the need for a full-time engineering organization. He added that the Engineering Advisory Committee had not in practice been effective, particularly because very few problems were ever submitted to it. Ricardo considered the Ministry of Aircraft Production's practice of placing designs for engines and other parts in the hands of eminent experts from industry to be effective, and he contrasted it with the system adopted in the Ministry of Supply and certain other departments where design was in the hands of drawing office staffs under the direction of transient amateurs from the services. He also held that Britain had suffered from indecisiveness regarding the types of weapons needed, that extensive secrecy had led to overlapping by certain departments, and that some technical departments such as the tank design department were actually overstaffed, resulting in unnecessary delays on minor matters.

To these forthright criticisms and suggestions, Anderson replied that in view of the Scientific Advisory Committee's First Report, which at his direction had ranged widely over the organization of government science, he had been under the impression that the existing machinery was considered adequate. He then stressed the necessity of preserving the responsibility of an individual minister for the operation of his department and said that if responsible people had criticisms to make of existing government establishments the facts ought to be laid before the minister concerned, whose clear duty it was to see that in all proper cases such complaints received adequate investigation. He also emphasized that if ministers ceased to treat their highly placed subordinates as trusted collaborators they would not be likely to get the best work from them. Nevertheless, Anderson undertook to consider in consultation with Lyttelton, who apparently only listened at this meeting, whether, as Dale had requested, the Scientific and Engineering Advisory Committees should be invited to make concrete suggestions for modifying the government machine to meet the points to which the deputation had drawn attention.

The invitation was made, and two days later the members of the Scientific Advisory Committee, with the exception of Dale who could not be present, met to discuss the functions and authority of a proposed Joint Technical Board. A hastily-written document based on this discussion was submitted by Hill to Anderson, and copies were sent also to Lyttelton and Butler.[66] It enumerated the terms of reference for the Board, namely:

1. By contact with the Chiefs of Staff's organisation and the Ministry of Production to advise on major technical aspects of operations and production, and to advise where emphasis on research and development should be laid and how available resources used.
2. To maintain enquiry whether departmental and interdepartmental scientific organisations are
 a) functioning efficiently;
 b) in proper contact with the expert 'user' and to report to their Minister.
3. To plan for the better use of the scientific and technical uses of industry, and institutions and professional bodies, and the universities, and to plan research and development on special projects.
4. Through associate members to ensure that special knowledge or skill is fully utilized.
5. To maintain a centre for technical information, liaison and intelligence.
6. To review and co-ordinate operational research in ministries and commands.[67]

Clearly, a body with such functions and authority would be more powerful than either the Scientific or Engineering Advisory Committee. The latter had been forbidden to involve itself in the technical affairs of the Ministry of Production, but under the first point above the proposed Board would advise not only the Ministry of Production but also the Chiefs of Staff Committee. The second point would permit the Board, in contrast to the Scientific Advisory Committee, to investigate such matters as the one involving the Admiralty that Hill had mentioned in the meeting with the Lord President and Minister of Production. But it also incorporated Anderson's point of the necessity of preserving the responsibility of individual ministers for the operation of their departments. Point three would confer powers of initiative upon the Board, again in contrast to the purely advisory functions of the Scientific and Engineering Advisory Committees. Likewise, points five and six would ensure a more powerful and quite different organization.

For their part, Beard, Ricardo, and Fleming on 23 July personally submitted a five page "Memorandum on Technical Planning in the War Effort" to Butler.[68] Its principal idea was, once again, that the present central planning organization should be completed by having attached to it a group of engineers and scientists who would advise the War Cabinet free from all departmental loyalties. In agreement with Hill's suggestions, the memorandum explained that such a group would be in organized contact with the Chiefs of Staff Committee and the Production Staff—and thus in a position to keep itself informed of operational facts, needs, and plans on the one side, and of production possibilities and limitations on the other. By this means, it would be able to ensure that:

> 1. The Chiefs of Staff Committee is kept correctly and quickly informed of the technical properties and possibilities of the weapons available or planned and of the conditions of their production.
> 2. The Joint War Production Staff and appropriate Departments are kept correctly and quickly informed of operational needs and plans and of operational experience.

The War Cabinet, the memorandum continued, had the beginnings of such an organization in the Scientific and Engineering Advisory Committees. However, the latter Committee was far from fulfilling the required functions outlined above, and it was believed (correctly) that the members of the former Committee were of the same opinion about it.

Why this was so for the Engineering Advisory Committee was explained. The broad term of reference of the Committee—to advise the government upon engineering questions connected with the war effort—was a definite obligation, but it was impossible of fulfillment except by close association with the Chiefs of Staff Committee and the Minister of Production, whereas there had been no such association. The first, second, and fourth of the specific terms of reference could not be implemented without the willing collaboration of the departments, and that had not yet been forthcoming.[69] Any approaches that had been made had been met by "a defensive departmental attitude, by an indication to 'keep off the grass' and by emphasis on the necessity for secrecy." The third of the specific terms of reference assumed that problems connected with new engineering devices would be referred to the Committee. Yet none of importance ever had been.

The members of the Engineering Advisory Committee believed that the reason for the current unsatisfactory situation was that from the start the help the Committee could have given was never recognized by the service and supply departments. As an advisory committee having no departmental restriction, it could have been able, had it been in close contact with the planning of war strategy and operating experiences in the field, to advise the War Cabinet as to what engineering limitations might affect war plans and how these limitations could be minimized or removed. New ideas depended fundamentally on such contacts. Furthermore, although it was true that developments requiring merely modifications to tanks, guns, aircraft, or other existing war devices could properly be dealt with by the engineering staffs of the appropriate government establishments, nevertheless, even here situations might have arisen where the Committee's help in finding experts or in coordinating engineering activities could have been employed.

The engineers' essential idea was that the Engineering Advisory Committee should have been consulted when the War Cabinet needed advice as to whether new plans and new strategy were possible from an engineering point of view. The Committee was able to draw on the engineering personnel resources of the entire country; any expert needed would almost certainly be found in the membership of the three senior engineering institutions. Functioning in this way, the Committee could have nursed along any new engineering developments until they could be taken care of by the appropriate government establishment. However, at the present stage of the war, its members doubted whether any reorganization of the Committee's work would now go far enough to meet present requirements. A "more positive solution," such as the creation of the proposed full-time central group of experts, was required: "Such a group of full-time engineers and scientists in close and continuous contact with each other and with proper facilities for consultation with the higher controls of operations and production would not only exert a beneficial influence on the future conduct of the war but, if carefully chosen, would give confidence to the engineering and scientific world that this important aspect of war problems was receiving full consideration on a broader basis than departmental requirements."

As he informed Anderson, Butler was encouraged that the engineers had "got on to this 'group' idea." He himself had been influenced by members of the Scientific and Engineering Advisory Committees to think in this direction even before the engineers had presented their memorandum.[70]

While Anderson and Lyttelton were deciding what should be done, the Parliamentary and Scientific Committee, through Lord Strabolgi, pressed the matter in the Lords. On 29 July, Strabolgi asked the government what further measures would be taken to make the most effective use of the nation's scientific and inventive talent in the war effort. He urged that a scientific general staff be created.[71]

During the ensuing debate Hankey spoke. He described the terms of reference of the Scientific and Engineering Advisory Committees as "very limited, especially in the case of the Scientific Advisory Committee." The main term of reference of the latter was to advise the government on any scientific problems referred to it, but Hankey, a master of institutional understanding,[72] had been under no illusion that government bodies would be falling over one another to refer questions to it. From the beginning, he had realized that the Committees would have to take their own initiative. So they had begun by assessing the state of science and engineering in government, and had subsequently made a great many recommendations, most of which had been adopted. Nevertheless, many loose ends had remained, and the next step was to take care of these.

> What we wanted was to get the confidence of Departments like the Committee of Imperial Defence. Once we got their confidence then the flow would come from the other end. It is no vice in the Departments that they do not come to a body like that. It simply is that everybody is frightfully busy and overworked. That makes them rather self-centered, and they do not go to an outside body until that body has won their confidence and made its usefulness felt. Until then such a body does not reach full usefulness. That is what we were trying to build up. I had brought the two bodies [the Scientific and Engineering Advisory Committees] together. We had quite a number of inquiries touching both science and engineering, so that the opportunity came to sit together in panels; not always the

same group but different groups very often sat together, and that is where I left it. I was asked to retain the Chairmanship of both these bodies, and I considered the matter very seriously, but I know that to carry the work farther and to get right into close contact with the Departments it did require a Minister of Cabinet rank. I felt it could not be done without Cabinet rank, and so, with great reluctance, I gave it up and it was taken by Mr. Butler, who is very keen and I know has handled it with very great ability. Even so, I knew that the difficulties were going to be rather great.[73]

Hankey recognized that in recent months various setbacks, including shipping losses, and inferior tanks had all helped create a sense of frustration among scientists and engineers. Consequently, he had not been surprised to learn of the attempts of scientists and engineers "to get rather more into the picture, to get people in whole-time instead of part-time, realizing, as I think they did, that they must win the confidence of the Departments." Yet, while Hankey sympathized with these efforts, like Tizard he did not think that as it stood the American system was quite applicable to Britain. His advice to the government was to make as much use of science as it possibly could—not only in the supply departments, but yet further than had already been introduced into operations, and even into planning. On the other hand, he advised scientists not to open their mouths quite too wide at first and not to expect absolutely everything they were asking for straight away because it might be that when scientists really got "inside" they would see for themselves that that was not the way to do it. They should take a further step if the government would give it to them now and be content for the moment with that. "Once you have established the right contacts your sheer ability, used with tact and energy, will do the rest."[74]

In this debate, Lord Snell replied for the government, and for the first time hints were obtained as to what it might do in response to the scientists' representations. But Snell's reply was more important for another reason—it expressed the principles that would determine the position of scientists in government under British parliamentary democracy.

> It has been suggested that what is needed is a Scientific General Staff. The idea appears to be that this body should have direct access to the War Cabinet or to the Defence Committee, and that it

should have authority over the great Departmental organisations, or that its advice should be taken in preference to the advice of those organisations. I would ask the supporters of this idea to consider how it can be reconciled with our system of Government, and in particular with the individual responsibility of Ministers of the Crown. It is clear, I think, that it cannot be so reconciled. Decisions must be taken by Ministers who are responsible for the consequences of those decisions, and it is to the responsible Ministers that the War Cabinet in the first place must look for advice. Nor could the executive heads of the organisations for scientific research in the great Government Departments be expected to exercise their great responsibilities with enthusiasm and courage if their decisions were at any moment liable to be over-ridden by some central body of scientists. . . .

With all our admiration of scientific achievement we need to remember, of course, the equally great achievement of our nation in the development of a democratic form of government, and we have to take care that nothing undermines the strength of that.[75]

The Parliamentary and Scientific Committee's response to the debate was to have Plugge and others table, on 14 August, a motion that its parliamentary members had placed, with the support of 145 members of Parliament from all parties, on the Order Paper of the House of Commons on 28 July.[76] The motion read: "That this House is of the opinion that present circumstances require the early establishment of a whole-time Central Scientific and Technical Board to coordinate research and development in relation to the war effort and to ensure that the experience, knowledge, and creative genius of British technicians and scientists exert a more effective influence over the conduct of a highly mechanized war."[77]

However, Anderson had already written to Dale on 11 August informing him of the conclusion to which he and Lyttelton had come:

The Minister of Production and I have given careful consideration to the representations made to us by yourself and some of your colleagues on the Scientific and Engineering Advisory Committees in regard to the Government's Scientific and Technical Organisation.

The existing Organisation has, as you know, been very carefully built up over a period of years. Mr. Lyttelton and I do not believe it to be as defective as might appear from some of the criticisms which have been made. We cannot, however, ignore the feelings of uneasiness to which you and your colleagues have given expression; and we would naturally be the first to wish to introduce any innovations which gave promise of improvement in the existing machinery. It is in this spirit that we have considered the various proposals which you and your colleagues have put before us.

The conclusion to which we have come is that the wisest course will be, in the first instance, to give effect to the suggestions that have been placed before us within a limited and clearly defined field. The Minister of Production accordingly proposes to ask three men of scientific and technical eminence to join his staff on a whole time basis.[78] Their field of activity will be co-extensive with the responsibility of the Minister and their first task will be to consider and advise how they can best contribute to the successful prosecution of the war within that field. These three officers will not be expected to interfere with the work being done by the Scientific and Technical Staffs of the Service Departments. They will, however, be available to assist the Departmental organisations for Scientific Research and Technical Development and to make recommendations from time to time when they think that improvements can be made. It will be part of their standing instructions to keep closely in touch with Lord Cherwell throughout their work. They will of course, be responsible to the Minister of Production. Mr. Lyttelton is, however, burdened with many other heavy responsibilities and he is anxious to ensure that this experiment should not fail owing to his being unable to give sufficient time to its supervision. He has accordingly accepted an offer which the Lord Privy Seal [Sir Stafford Cripps, also a member of the War Cabinet] has made to co-ordinate and supervise on his behalf, the day-to-day work of the team.

The selection of the right men for these appointments is clearly a matter of great importance. We have accordingly decided to ask you, in the absence of Mr. Butler, to be good enough to arrange for the Scientific Advisory Committee, in conjunction with the Engineering Advisory Committee, to assist us by suggesting from six to ten names of persons who might be regarded as suitable if available. We would wish you and your colleagues in compiling such a list to pay no regard to the existing commitments of the persons recommended.[79]

This obviously was considerably less than the scientists and engineers had pressed for. Anderson knew this and shrewdly invited their assistance. They could hardly refuse to make the nominations, and having made them, they could hardly continue to criticize the government.

Addressing the annual luncheon of the Parliamentary and Scientific Committee some months later in February 1943, Anderson referred to the problem of creating a scientific general staff and enumerated four principal conditions that had to be fulfilled. The organization must be firmly built into the government machine, and at the same time it must not be insulated from the great body of scientists outside government. It must not cut across normal ministerial responsibility, but it must be linked up with some minister whose special business it is to see that considerations of a general scientific character receive proper attention.[80] These requirements ruled out not only Hill's proposal, but also Appleton's, favored by Tizard as more acceptable given the British constitution. However, government ministers gave that an even stricter interpretation. The appointment of the three advisers would amount to a transfer of the Engineering Advisory Committee to, and a limiting of its potential sphere of action to within, the Ministry of Production. As before, Lord Cherwell continued to occupy an unequalled position as scientific adviser to Churchill.

In 1943, Anderson could look back to the appointment of the three advisers and say that with it the government had created an organization that "promises well and is capable of considerable development." But before that had occurred Lyttelton had had, of course, to obtain Churchill's approval for the innovation. The proposal irritated Churchill, who thought that Lyttelton was trying to expand his 'empire' at the expense of Churchill's. (Churchill was his own Minister of Defence and closely attentive to the uses made of science by the services.[81]) Lyttelton got nowhere at first, and the question remained unsettled for some time. On one occasion during this period, Churchill said in the presence of Lyttelton and others: "Here's Oliver, always avid of power: now wanting to run the scientific side of the war; he's going to take it over from me: he first has a spearhead of three graces, and so we may expect to see everything in the scientific field better run."[82] On the next day, however, Lyttelton was asked to meet with Churchill, and the following exchange took place, as recalled by Lyttelton.

    C. "Good morning, my dear. I hope you are not vexed with me?"
    L. "Well, I was vexed, but I've got over it."

Scientists and Central Direction of the War Effort · 251

    C. "I know I shouldn't have chaffed you so much, but why do you want to run everything?"
    L. "I don't, really I don't but it does require people of the *métier* to compose some of the quarrels amongst the scientists. A won't talk to B, because B has invented something A thinks he ought to have invented, and so on. That's all there is to it."
    C. "Oh well, go ahead. I don't like it, but if you want it I suppose you had better do it."[83]

As with the earlier creation of the Scientific and Engineering Advisory Committees, again under external pressure Churchill reluctantly assented to an increased role for scientists in government. This time, however, the scientists and engineers were granted very much less than they considered necessary.

Soon after he had heard from Anderson, Dale chaired a joint meeting of the Scientific and Engineering Advisory Committees.[84] He suggested that they first decide which types of experience should be represented by the three appointments and then select three names to be submitted for each of the positions. There was general agreement that the appointments should include a physicist and an engineer. However, it was also agreed that the area of concern—the entire field of war production—could not be covered adequately by a staff of three. For, in regard to the third appointment, this would mean choosing between a chemist and a physiologist with physical and engineering knowledge who could advise on the production of weapons in regard to their efficient use by personnel.

Regarding the type of engineering experience that would be most valuable, there was some difference of opinion. Initially, the view was that only mechanical engineers should be considered. However, it was later urged that, because of the widespread military use of electrical signalling and remote control devices, the name of one electrical engineer be included. Thus, it was agreed that four names, including that of an electrical engineer, be submitted. As its first choice, however, the joint meeting named W. A. Stanier, past president of the Institute of Mechanical Engineers and chief mechanical engineer to the London, Midland, and Scottish Railway.

Although Anderson had specifically directed that no regard be paid to the existing commitments of persons who might be considered suitable for appointment, the joint committee nevertheless felt that this should be understood to mean that they should be guided solely by the interests of the war effort and that they should not recommend the appointment of someone if in their view it would entail a risk of serious loss to the efficiency of an existing research organization. Consequently, in the physics category they excluded from consideration such physicists as John Cockcroft, P. M. S. Blackett, and Appleton. But for this concern, Appleton, who had been obliged to leave the meeting early, would have been placed at the top of their list. As first choice, they named T. R. Merton.

Finally, having decided that with a staff of three it would not be practicable to cover adequately the whole field of war production and deeming it desirable to include not only a chemist but also someone whose training and experience would enable him to appreciate the importance of biological factors in the design of weapons, the committee suggested the appointment of both Dr. Ian M. Heilbron, F.R.S. (Professor of Organic Chemistry at the Imperial College of Science and Technology and a scientific adviser to the Minister of Supply from 1939) and Dr. Bryan Matthews, F.R.S. (Assistant Director of Research in Physiology at Cambridge and Head of the R. A. F. Physiological Research Unit). In forwarding the nominations to Anderson, Dale noted, significantly: "I should like to record also my impression of the general satisfaction of the members of both Committees at the action which it has been possible to take thus to strengthen the scientific advice directly available to the War Cabinet, and at the opportunity given to them to submit suitable names for consideration."[85]

This joint meeting was the last meeting of the Engineering Advisory Committee. Upon the appointment of the advisors to the Ministry of Production, it became superfluous.

In apprising the War Cabinet of the imminent appointments, Anderson and Lyttelton explained that they had been concerned for some time as to the best way to deal with "agitation" in Parliament and in the press concerning the nation's scientific and technical organization in the area of production.[86] They considered the agitation to some extent ill-conceived and largely based on ignorance of the existing organization and its achievements. At the same time, they did not feel they could ignore the general uneasiness and unrest in scientific circles, nor did they feel that the existing organization would enable them to answer adequately all the criticisms that had been expressed.

The appointment of three, not four, full-time scientific advisers—Stanier, Merton, and Heilbron—to the staff of the Ministry of Production was announced on 4 September. In welcoming the appointments, the *Times* commented that the part that science could and must play in interrelating strategy and production had, with the progress of the war, been increasingly recognized by all competent observers.[87] However, what had been widely urged upon the government was the association of an authoritative group of scientists not so much with the Ministry of Production—for it had to be presumed that Lyttelton had from the start enjoyed the assistance of scientific advisers—as with those responsible for the central direction of the war. The *Times* was therefore baffled to find that the three advisers would work under the Lord Privy Seal, Sir Stafford Cripps, who would be acting in this capacity not on behalf of the Prime Minister or of the War Cabinet as a whole, but on behalf of the Minister of Production. This arrangement, when pressed to its logical conclusion, appeared to place a serious limitation on the scope of the functions that they were required to fulfill. However, the paper continued, the appointments would be meaningless if the advisers were not intended to survey the whole conduct of the war from the standpoint of science and, free from direct departmental responsibility, tender their general advice on scientific problems and scientific opportunities in whatever field was required. This was the spirit in which the *Times* expected the advisers to approach their work. Their success, it concluded, would be measured by the extent to which the same spirit actuated others to avail themselves of the services that science could render.

Three days later, a meeting of the Parliamentary and Scientific Committee at the House of Commons considered the government's action and adopted, on the proposal of the Association of Scientific workers, a resolution expressing dissatisfaction with it:

> This Committee, while welcoming the appointment of three full-time scientific advisers to the staff of the Ministry of Production in so far as it establishes the nucleus of a central scientific and technical

board, regrets that their field of activity is apparently to be limited to the sphere of production and does not include the scientific and technical activities of the service Departments or the other Ministries outside the strict field of production. An extension of its functions is needed to ensure that all scientific considerations are coordinated and given full weight over the whole field of the national effort. The committee considers, therefore, that in order to cover this wider field, scientific advisers should have direct access to the War Cabinet and that accordingly the Lord Privy Seal should exercise his supervisory functions over the new body directly on behalf of the War Cabinet.[88]

Copies of the resolution were sent to the press and to appropriate ministers. Sir Stafford Cripps replied that it would be wise at present not to press for any more precise definition of the functions of the advisers, but rather to let him handle the situation as best he could with a view to their gradual establishment and the extension of the scope and value of their work.[89] Very often, he said, if one tries to make the directives on this sort of question precise, one only ends by limiting rather than extending the powers.

In addition to working with the Parliamentary and Scientific Committee, the Association of Scientific Workers had taken other steps during the summer. In late June, it sent Lyttelton a memorandum on the wastage of scientific manpower written by E. D. Swann, requesting, unsuccessfully, that Lyttelton receive a deputation.[90] In July, the Association's secretary, Reinet Fremlin, conveyed the Association's good wishes to Hill in the "campaign," but he replied that "the case of consultation on the job" was overdone.[91] These experiences contributed to a sense of frustration within the Association, described to Hill by one of its members, John Humphrey, on 10 September:

> I think that the A.Sc.W. is pleased enough that things are going in the right direction—for after all it has been pressing for them in an amateurish way for quite a long time. But it does raise a knotty general problem. How is an unofficial body, such as itself, to know what is going on at the top (without itself being "inside")—and thereby to avoid wasting its breath and putting its foot in things? Perhaps it is just a part of democracy that such difficulties should arise, and can't be helped. But I dare say that you'll agree that, however much they are a nuisance at times, it's a good idea for

unofficial bodies to exist and to take a serious interest in affairs of state—"The price of liberty is eternal vigilance" being a true saying. Anyway if you do tell the A.Sc.W. when something is likely to be worthwhile agitating for I'm sure that they'll agitate.[92]

The Association enjoyed better communication with the TUC. During the second week of July, Fremlin sent a copy of Swann's memorandum to E. P. Harries of its organization department who promised to bring it to the attention of the Production Advisory Committee.[93] In September, the Association's resolution calling for a central planning board was adopted by the TUC congress.[94] The TUC general council subsequently discussed the matter with officials of the Ministry of Production and raised it on the Production Advisory Committee.[95] After the three scientific advisers to the Ministry of Production had been appointed, Fremlin went to Harries, but he did not think that further pressure could be brought by the TUC for a central scientific and technical board.[96]

For its part, Nature was enthusiastic about the appointment of the advisers. L. J. F. Brimble, its joint editor, commented: "To say the least of this announcement, it is an excellent example of democratic government, and those who have made representations to the Government recently can be assured that their efforts have not proved altogether sterile."[97] Science had again raised its voice, and the government's response would give considerable satisfaction. Although Brimble found little to criticize, he did note that, referring to the Parliamentary and Scientific Committee, "some of us" would have preferred that the advisers were entirely free from any departmental organization such as the Ministry of Production. But in Brimble's view, this drawback was happily offset by the fact that the advisers would keep in close touch with the scientific advisers of the service and supply departments and would be available to assist the departmental organization of research and development.[98]

As had been planned by the Parliamentary and Scientific Committee, questions were asked in Parliament concerning the functions of the advisers.[99] On 11 September, Cripps was questioned about the relationships between them and the Scientific and Engineering Advisory Committees. He reiterated that the advisers' field of concern was coextensive with that of the Minister of Production. Their appointment did not in any way affect the functions of the Advisory Committees, which would continue to advise the government on matters coming within their terms of reference. (But, as previously mentioned, the Engineering Advisory Committee became defunct.) In appropriate cases, Cripps continued, these Committees would no doubt consult with the scientific advisers of the Ministry of Production, as they had been accustomed to consult with the scientific advisers of other ministries before submitting recommendations on matters connected with the departments.[100]

These answers did not satisfy all, so on 29 September, Lyttelton himself made a brief statement in the Commons on the powers and duties of the advisers. He began by making it clear, as indeed had Cripps, that he had not created a new scientific board. The advice and recommendations of the advisers would normally be presented to the Minister of Production through the Lord Privy Seal; it would be for the Minister of Production to bring their advice and recommendations to the War Cabinet as necessary. The appointments had been made with a view to completing the existing organization for research and development in the production field. The advisers would advise the Minister of Production on existing research organizations in any case in which it appeared to them that the current arrangements were deficient or inadequate. They might act for the Minister with the technical side of any service or other department engaged in scientific work that bore on the general field for which the Minister was responsible. The Minister would call upon them from time to time to make reports upon particular matters, but, in addition, they might initiate inquiries into any matters within their field. The advisers might obtain through the head of the department or establishment concerned, or, in the case of certain highly secret information, as might be approved between the Minister of Production and the minister concerned, any information relevant to the execution of their duties.[101]

Asked if it would not be better to have one minister take complete charge of scientific research rather than having two or three ministers divide the field between them, Lyttelton replied that he thought that it would be unsound to have a single minister. The closest liaison existed between the three Ministers (Lord President, Lord Privy Seal, and Minister of Production), and the ultimate coordination had to be done by the War Cabinet.[102]

The Association of Scientific Workers continued to advocate the creation of a central technical board. The motion tabled in the House of Commons, which eventually was signed by 150 members of Parliament, was the subject of much discussion at branch meetings before it lapsed at the end of the parliamentary session.[103] Statements supporting the motion were sent to the prime minister, Lord Privy Seal, other members of Parliament, the Parliamentary and Scientific Committee, the local press and other bodies.[104] The Manchester Central branch secured the support of five local members of Parliament. In January 1943, the Association held a public conference in London on the planning of science with the theme of making science serve useful social ends. More than a thousand scientists attended.[105] E. D. Swann told the first session, on "The Centralized Direction of Scientific Research and Development," that the Association still sought a central scientific and technical board.[106] In giving an account of the government's scientific committees, Sir Stafford Cripps, now Minister of Aircraft Production, avoided the subject of a central board but did say that the country was "now fully alive to the fact that our survival and our victory depend to a great extent upon the output of our scientists and our research institutions, and that everything must be done to utilize to the full that very high degree of scientific intelligence which this country undoubtedly possesses."[107] As 1943 progressed, the Association of Scientific Workers turned its attention increasingly to postwar policy for science, still believing that time would demonstrate the need for central coordination.

Unlike the Engineering Advisory Committee, the Scientific Advisory Committee continued in existence and on 3 March 1943 interviewed the three advisers as to how their function had developed in practice.[108] Stanier explained that he had been active in connection with the working of the Tank Board. He had been able to make progress with regard to the problem of carbon monoxide poisoning in gun turrets, which the Lord President and Sir Edward Mellanby had brought to his attention. Heilbron's chief interest had been in the fields of chemicals and raw materials. He had given assistance in problems relating to alginates, pyrethrum supplies, synthetic rubber processes, chemical warfare matters, nontoxic smoke, and rocket projectile investigations. He had been urging the adoption of a scheme for technical advisory services within the Ministry of Production. Merton was unable to attend, but an assistant explained his interests in operational requirements for the services. He had particularly interested himself in air requirements and had assisted in dealing with problems relating to fog disposal, antisubmarine methods, service photography, and altimeters. In general, the advisers had found after the first few weeks that they received full cooperation from all departments, and now on their own initiative the departments were seeking help from the advisers. The latter thought that, at any rate for the present, there was no reason to consider increasing their number. They had found that in practice they could always call upon experts in subjects with which they were personally unfamiliar, and there was the danger that if their organization became too big it would be unwieldy and too formal.

With the creation of the first national research council, the Department of Scientific and Industrial Research, in 1915, the highest level of the scientist, as of other experts, in British government had been established. The scientist would advise cabinet ministers; he could not deliberate with them as their political equal. The scientist's advisory role had developed significantly with the creation of the Scientific Advisory Committee in 1940. On the Committee, the heads of the three research councils now officially deliberated together not only with one another but with the leaders of civil science as represented by the officers of the Royal Society. The Committee oversaw almost all government science which was now coordinated with civil science. However, the Committee and also the Engineering Advisory Committee advised only the Lord President, who was not directly involved in determining the strategy of the war effort. As Prime Minister and Minister of Defence, Churchill controlled the supreme strategic direction of that effort through the Defence (Operations) Committee of the War Cabinet. Under him the planning of operations was centralized in the Chiefs of Staff Committee and the planning of supply in the Ministry of Production. In mid-1942, in the most political period of the social relations of science movement, scientists vigorously campaigned to be brought into the strategic planning of the war effort. Although they were denied such a role, their efforts did result in the appointment of a team of three scientific advisers to the Ministry of Production. In contrast to the weak Engineering Advisory Committee, which now became defunct, the advisers enjoyed a much more authoritative position in the field of production. Nevertheless, the Scientific Advisory Committee still represented the greatest gain within the machinery of government achieved by scientists during the war.

1. A. J. P. Taylor, *English History 1914-1945* (Harmondsworth, England: Penguin Books, 1975) p. 658.

2. Engineering Advisory Committee, "Report on Mechanical and Electrical Engineering" (26 November 1941), p. 2; Cab. 92/113.

3. Ibid., p. 12.

4. Hankey Papers, Churchill College, Cambridge, General Correspondence, 4/33.

5. In addition to the Scientific and Engineering Advisory Committees, Hankey chaired the Committee for the Coordination of Allied Supplies, the Committee on Enemy Oil Supplies, the Committee on Skilled Radio Personnel, and the Committee on Bacteriological Warfare.

6. Stephen W. Roskill, *Hankey: Man of Secrets*, 3 vols. (London: Collins, 1970-74), 3:544.

7. Ibid.

8. Lord Hankey to Sir William Bragg, 4 March 1942, Hankey Papers, 4/34.

9. Lord Hankey to A. V. Hill, 5 March 1942, Hankey Papers, 4/34.

10. Engineering Advisory Committee minutes, 26 March 1942, Cab. 92/114.

11. Lord Hankey to Sir Samuel Hoare, 12 March 1942, Hankey Papers, 4/34.

12. Paper A.C.E. (42) 18, Cab. 92/114.

13. Ronald W. Clark, *Tizard* (Cambridge, Mass.: M.I.T. Press, 1965), p. 381.

14. *Nature* 148 (1941): 624.

15. Association of Scientific Workers, "Executive Committee report to Council 1942," p. 1, ASW/1/1/21/3.

16. See above, p. 142.

17. Association of Scientific Workers, "Executive Committee report to Council 1942," p. 1, ASW/1/1/21/3.

18. Association of Scientific Workers, "A.Sc.W. Conference 'Science and the War Effort,' 1942," (typescript of all speeches given at the conference), ASW/1/10/1.

19. *Scientific Worker*, April 1942, p. 6.

20. *Scientific Worker*, May 1942, p. 10.

21. Association of Scientific Workers, "A.Sc.W. Conference 'Science and the War Effort,' 1942."

22. Association of Scientific Workers, Executive Committee minutes, 29 March 1942, ASW/1/2/23/4.

23. A. V. Hill, "Memories and Reflections" (manuscript, 2 vols) 1:119-20. This work is in the library of the Royal Society.

24. "Discussion at a meeting of the Parliamentary and Scientific Committee, House of Commons, 24/3/42." (manuscript, 5 pp.) A. V. Hill Papers, Churchill College, Cambridge, 2/2.

25. *Nature* 149 (1942): 549-50.

26. J. P. Baxter, *Scientists Against Time* (Boston: Little, Brown, & Co., 1946), p. 28.

27. Ibid., p. 29.

28. Sir Henry Tizard to Sir Arthur Street in Clark, *Tizard*, pp. 427-30, 428.

29. J. R. M. Butler, ed., *Grand Strategy*, 6 vols. (London: H.M.S.O., 1956-72), 3:part 2:609.

30. See Clark, *Tizard*, pp. 305-13.

31. Ibid., p. 428.

32. Ibid.

33. Ibid., p. 429.

34. R. W. Clark, *Sir Edward Appleton* (New York: Pergamon Press, 1971), p. 122.

35. Presided over by Churchill and including two members of the War Cabinet (Clement Attlee and Lord Beaverbrook), the Chiefs of Staff, and the service ministers. Taylor, *English History 1914-1945*, p. 586.

36. Clark, *Tizard*, pp. 429-30.

37. Cab. 21/1175.

38. *Times*, 1 July 1942, p. 5.

39. Ibid.

40. Butler, ed., *Grand Strategy*, 3:part 2, 609.

41. *Times*, 3 July 1942, p. 5.

42. *Times*, 4 July 1942, p. 5.

43. *Times*, 6 July 1942, p. 5.

44. C. P. Snow, *Science and Government* (Cambridge, Mass.: Harvard University Press, 1961), pp. 25-26.

45. *Times*, 8 July 1942, p. 5.

46. *Manchester Guardian*, 7 July 1942, p. 4.

47. H. Everett (War Cabinet Offices) to W. Gorell Barnes (Lord President's Office), 7 July 1942, Cab. 21/1175.

48. R. A. Butler to Sir John Anderson, 6 July 1942, Cab. 21/1175.

49. Ibid.

50. H. Everett to W. Gorell Barnes, 7 July 1942, Cab. 21/1175.

51. One of the critics of the organization of government science, Lord Samuel, president of the Parliamentary and Scientific Committee, said of Butler in 1942: "[He] is the head of a great Department of State, President of the Board of Education, and he is engaged not only in many tasks of great urgency arising out of war conditions but also, as we know, is planning a very wide educational policy, to be put into effect after the

Scientists and Central Direction of the War Effort · 261

war: a man of keen mind and of great constructive energy, much interested in this particular duty [chairman of the Scientific and Engineering Advisory Committees], but unable—he cannot be expected—to give to it the close attention which the urgency and the volume of the undertaking require" (*Parliamentary Debates* [Lords] 124 [1942], col. 84).

52. R. A. Butler to Sir John Anderson, 6 July 1942, Cab. 21/1173.

53. Cab. 21/1173.

54. *Parliamentary Debates* (Commons), 381 (1942), cols. 1108, 1147-51.

55. *Times*, 17 July 1942, p. 2. The other members of the deputation were: Peers—Lord Hinchingbrooke, Lord Leverhulme, and Lord Pentland; M.P.s—Captain L. F. Plugge, Hugh Linstead, and J. M. Woolton-Davies; Scientists and Engineers—J. D. Bernal, Sir Lawrence Bragg, C. S. Garland (British Association of Chemists), B. W. Holman, R. B. Pilcher (Institute of Chemistry), Gower Pim (Institute of Structural Engineers), Colonel Thomson (Institute of Mechanical Engineers), W. Wooldridge (president, National Veterinary Medical Association).

56. Association of Scientific Workers, Executive Committee minutes, August 1942, ASW/1/2/23/10.

57. *Nature* 150 (1942): 116.

58. "Note of a Meeting at the Privy Council Office on Thursday, 16th July, 1942," Cab. 21/1168.

59. For the scientific mission to the United States, see Clark, *Tizard*, Chapter 11. In April 1942, Anderson had requested the Scientific Advisory Committee to investigate the subject of patents, particularly those arising out of research carried out for the government during the war.

60. During 1942, representatives of the Engineering Advisory Committee and of the Defence Services Panel of the Scientific Advisory Committee held meetings to discuss the possibilities of the Germans driving a Channel tunnel and steps that could be taken for its detection. A report on the subject was forwarded by Butler to Anderson in December 1942. A few days later, Anderson sent it to General Ismay asking that it be considered by the Chiefs of Staff Committee. This the Committee did, "but in view of the indication given in this Report that such a tunnel could not be constructed in less than three years and could not therefore be completed before the autumn of 1943, decided to defer further consideration of the matter until next June" (Paper SAC [D.P.] [43] 1, Cab. 90/7).

61. Cab. 21/829.

62. *Picture Post* 16 (18 July 1942): 25.

63. Ibid., (1 August 1942): 2.

64. Scientific Advisory Committee minutes, 13 July 1942, Cab. 90/3; Engineering Advisory Committee minutes, 13 July 1942, Cab. 92/114.

65. "Draft note of an interview in the Lord President's Room, July 20, 1942," Cab. 21/1175.

66. H. E[verett] to Miss Goodfellow (Board of Education), 22 July 1942, Cab. 21/1175.

67. "Functions and Authority of a Joint Technical Board. Suggestions by Professor A. V. Hill," (22 July 1942), Cab. 21/1175.

68. Cab. 21/1175.

69. See above, pp. 203-4.

70. R. A. Butler to W. Gorell Barnes (Lord President's Office), 23 July 1942, Cab. 21/1175.

71. *Parliamentary Debates* (Lords), 124 (1942), col. 77.

72. Snow, *Science and Government*, p. 61.

73. *Parliamentary Debates* (Lords), 124 (1942), col. 89.

74. Ibid., cols. 90–91.

75. Ibid., cols. 95–96.

76. M. Philips Price, "The Parliamentary and Scientific Committee of Great Britain," *Impact of Science on Society* 3 (1952): 258–76; 270.

77. *Times*, 7 August 1942, p. 2.

78. Soon after becoming Minister of Production, and in order to relate production closely and continuously to strategical requirements, Lyttelton had created a general staff of war production to include his Chief Adviser on Programmes and Planning, the Assistant Chiefs of Staff of the three services, and the highest technical officers of the three production ministries. This staff served the War Cabinet Defence Committee of which Lyttelton was a member (Butler, ed., *Grand Strategy*, 3:425).

79. Annex, Paper SAC (42) 45, Cab. 90/3.

80. Parliamentary and Scientific Committee, *Annual Report* (1943), p. 3.

81. *Nature* 151 (1943): 152.

82. Oliver Lyttelton, Lord Chandos, *The Memoirs of Lord Chandos* (London: Bodley Head, 1962), p. 169.

83. Ibid., p. 170.

84. Cab. 90/3.

85. Sir Henry Dale to Sir John Anderson, 14 August 1942, Cab 21/1166.

86. Memorandum to War Cabinet from Lord President and Minister of Production, 3 September 1942, Cab. 66/28.

87. *Times*, 5 September 1942, pp. 4, 5.

88. Association of Scientific Workers, Executive Committee minutes, 13 September 1942, ASW/1/2/23/11.

89. M. Philips Price, "The Parliamentary and Scientific Committee," p. 271.

90. Association of Scientific Workers, Executive Committee minutes, 12 July 1942, ASW/1/2/23/9.

91. Ibid., August 1942, ASW/1/2/23/10. The Association took pains to maintain good relations with Hill. During 1941, Hill was one of many who were upset by an article in the U.S. press on the nonutilization of scientists by the British Government. The article made exaggerated statements and quoted the Association as an authority. Fremlin and Wooster were directed to draft a letter of clarification to the American writer. Hill was to be informed of this and also sent back copies of the *Scientific Worker* that would clarify the Association's position (Executive Committee minutes, 24 August 1941, ASW/1/2/22/21/i).

92. John Humphrey to A. V. Hill, 10 September 1942, Sir Henry Dale Papers, Royal Society Library, Box 51, Folder A.Sc.W.

93. Association of Scientific Workers, Executive Committee minutes, 14 July 1942, ASW/1/2/23/9.

94. Ibid., 13 September 1942, ASW/1/2/23/11.

95. Association of Scientific Workers, "Agenda for Twenty-Sixth Annual Council, 8-9 May, 1943," p. 17.

96. Association of Scientific Workers, Executive Committee minutes, 11 October 1942, ASW/1/2/23/12.

97. *Nature* 150 (1942): 303.

98. Nevertheless, *Nature* continued for some time to press the idea of a central scientific and technical board with executive, rather than advisory, powers. See, for example, Rainald Brightman's editorial in *Nature* 151 (1943):204.

99. Association of Scientific Workers, Executive Committee minutes, 13 September 1942, ASW/1/2/23/11.

100. *Times*, 12 September 1942, p. 2.

101. Ibid., 30 September 1942, p. 8.

102. Ibid.

103. Association of Scientific Workers, "Agenda for Twenty-Sixth Annual Council, 8-9 May, 1943," p. 17.

104. *Scientific Worker*, December 1942, p. 71.

105. For nine photographs of the conference, see *Picture Post* 18 (13 February 1943): 16-18.

106. Association of Scientific Workers, *Planning of Science* (London: A. Sc.W., 1943), p. 15.

107. Ibid., p. 11.

108. Scientific Advisory Committee minutes, 3 March 1943, Cab. 90/4. A deputation from the Association of Scientific Workers consisting of Bernal, Fremlin, Swann, and Wooster, had met with the scientific advisers at the end of October 1942. They discussed the memorandum on wastage of scientific manpower drawn up by Swann and sent to Lyttelton in June (Association of Scientific Workers, "Agenda for Twenty-Sixth Annual Council, 8-9 May, 1943," p. 17).

Toward the end of 1943, one of the advisers, Heilbron, spoke to the Parliamentary and Scientific Committee on the organization of research (Association of Scientific Workers, Executive Committee minutes, 12 December 1943, ASW/1/2/24/13).

9

Freedom and Planning in Science

A fourth phase of the social relations of science movement, overlapping the second, third, and fifth chronologically, involved a major polemic among scientists themselves concerning freedom and planning in science and the consequent formation of a new organization, the Society for Freedom in Science.[1] This chapter deals with the history of the Society from its founding in the winter of 1940–41 to the end of the war including its growth in membership, the defining of its position, its activities, the individual contributions of its members and especially of two of its founders and its leading lights, John R. Baker and Michael Polanyi, and the opposition to the Society on the part of certain prominent scientists, the Association of Scientific Workers, its journal, the *Scientific Worker*, and the influential weekly science journal, *Nature*.

The Society for Freedom in Science was created in reaction to the vigorous advocacy of the planning of science by left-wing scientists and the Association of Scientific Workers. The preeminent work on planned science was Bernal's *The Social Function of Science*, published early in 1939. Never before had proposals for changes in the organization of British science been presented so forcefully and at such length. Bernal shared with many of his contemporaries a resentment of "the inefficiency, the frustration and the diversion of scientific effort to base ends."[2] This thwarting of science was to him a very bitter thing. It showed itself as "disease, enforced stupidity, misery, thankless toil, and premature death for the great majority, and an anxious, grasping, and futile life for the remainder."[3] However, this state of affairs could be avoided provided that science would work with "those social forces which understand [science's] functions and which march to the same ends."[4] To Bernal, "the traditional piety of a pure unworldly science [seemed] at best a phantastic escape, at worst a shameful hypocrisy."[5]

The first part of the book described what science was doing, and the second what it could do. The latter called for a comprehensive reorganization of science brought about by scientists, the state, and economic organizations working together. By this means, science would be greatly enriched and made more productive, while at the same time high scientific standards would be maintained and the freedom and originality of research would be preserved.[6] The reorganization would be beneficial to science, for "science is a gift that demands to be used to the full for the material and cultural benefit of humanity, and ... if science is not so used it will itself be the first to suffer."[7]

The *Social Function of Science* became the principal document of the movement that would be characterized by its opponents as attacking the conception of science as a search for truth and as denying the rights to free research directed solely to that end. To these opponents, three contentions underlay the movement. First, science originated in attempts to satisfy the material needs of ordinary human life. Next, the legitimate purpose of science is to meet these needs on an expanding scale. Finally, scientists could not be left free to choose their own subjects of research, but must submit to central planning so that their work might be specifically devoted to the satisfaction of human material needs.[8]

In Bernal's view, such planning would be compatible with freedom of scientific research. But in his oponents' view it would not. This is where the issue was joined. His opponents regarded Bernal as paying lip service to the necessity of freedom in research and saw an irreconcilable contradiction between this and the main argument of his book.[9]

To unsympathetic but as yet silent observers of the earlier expression of such views as Bernal's, the appearance of Bernal's book had made the situation more alarming. However, it also provided a focus of attack. Indeed, the work invited attack through its use of provocative comments. For instance, it stated that it was not surprising that "many scientists, particularly of the older school ... prefer science to remain inefficient and obscure so long as it is preserved as a free playground for the fortunate few to whom accidents of birth and temperament have made it accessible."[10]

One who resented such views was John R. Baker, then thirty-nine years old and a lecturer in zoology at Oxford University. Baker believed that everyone should have a social conscience that would desire more, not less, liberty. "I think," he explained to Joseph Needham in mid-1942,

> that the man who has a real aptitude for pure science wastes his talent if his social conscience drives him to become a technologist. He has plenty of opportunity for making the world a better place in a quiet way, by his actions towards those whom he meets in everyday life. It worries me that so many people nowadays, who have seldom acted altruistically, are advocating reduction in liberty and at the same time enjoying a feeling of moral superiority just because they press for that reduction. It is curious to notice that they think that they themselves will be in a favored position in the new society, and will be in a position to dictate what they think will be a better life to others. They do not realize that unwittingly they are working to establish a regime which needs a dictator at the top, and that he will decide—not they—what sort of life the people shall lead.
>
> I greatly regret that scientists should lend their aid to the movement for the reduction of liberty. In doing so they are of course swimming with the tide just at the moment when people are required who will struggle against it.[11]

Baker began his own struggle when he reviewed Bernal's book under the eye-catching rubric "Counterblast to Bernalism." In characteristic fashion, Baker attacked what he polemically defined as "Bernalism," namely, "the doctrine of those who profess that the only proper objects of scientific research are to feed people and protect them from the elements, that research workers should be organised in gangs and told what to discover, and that the pursuit of knowledge for its own sake has the same value as the solution of crossword puzzles."[12] Baker was appalled by the possibility that if the "Bernalists" obtained power to effect their plans for science, scientific discovery would cease. He himself believed that if science were to continue to be free, then incalculable material and intellectual benefits for human life would continue to accrue. He agreed that to serve human welfare was certainly one of science's two functions, but he chose to stress the second which he was to champion over the ensuing years:

> To many who are contemplative there is nothing more worthwhile in life than the increase of knowledge for its own sake. To pretend that this is "escapism" and comparable to interest in crossword puzzles is nonsense. . . . every discovery in science separates falsehood from truth and makes an accretion to that vast body of demonstrable knowledge whose possession is the most valid criterion of distinction between cultured and savage communities. There are those who are unrewarded by the contemplation of nature, just as there are those who find nothing in music or art. Nevertheless, knowledge and music and art are among the ultimate things in life for many people who regard just keeping alive and healthy as merely means to an end.[13]

Baker regarded science primarily as a cause that man served rather than an instrument existing to serve man. In reply, Bernal insisted that science must and could be organized "without damaging the freedom of thought and activity of the individual worker on which ultimately everything depends."[14]

Michael Polanyi had never met Baker, but in reading his review of Bernal's book he recognized a kindred spirit. He had himself, and unknown to Baker, critically reviewed Bernal's book in a paper that would soon be published.[15] Against Bernal's Socialist view, Polanyi, like Baker, defended the Liberal view of science that sees science munificiently showering its gifts on mankind when allowed freely to pursue its own spiritual aims but collapsing into barren torpor if required to serve the needs of society. Polanyi, a forty-seven-year-old Hungarian, had been at the Kaiser Wilhelm Institute for Physical Chemistry from 1925 to 1933 when he resigned in protest against the Nazis. He went to England where he became professor of physical chemistry at the University of Manchester.[16] It was from there that he wrote to Baker, and thus began a unique friendship.[17] This would intensify and involve a close collaboration after Baker, more than a year later, took the step that would lead to the founding of the Society for Freedom in Science.

In early November 1940, Baker wrote to forty-nine distinguished British scientists inviting each to join a new, informal, and nonpolitical society for the protection of freedom and individualism in scientific research. To become a member, one had to agree in general with four propositions based upon Baker's perception of current and future threats to pure science in Britain. There was at present, he said, an influential group of scientists who regarded pure science with contempt and thought only of material benefits. Furthermore, he feared that after the war the modern insistence on quick economic results might be accentuated and that there would be a powerful attempt to put an end to free individual research in pure science. Baker sensed a danger that creative research workers might be swamped under a deluge of mass-produced scientists who would clamor for work in organized groups on dictated problems directly related to human affairs.

As for those of his fellow scientists who allegedly thought only of the material benefits of science, Baker considered them shortsighted for the reason expressed in the first of his four propositions, namely, "Great material benefits to mankind result from research in pure science along lines whose application to human affairs is not at the time obvious." Coupled with the insistence on science being made solely utilitarian was the idea that scientists should work in organized groups towards dictated ends. These Baker opposed in his third proposition—"So far as possible, research workers should be free to decide the subjects of their own research"—and also in the fourth—"Those scientists who find that they do their best work by themselves should not be forced to carry out research in organised groups." The remaining proposition, the second, stated the traditional view now ignored by Baker's opponents, namely, "The advancement of knowledge by scientific research has a value as an end in itself."[18]

Just over a month later, Baker had received replies from thirty-three recipients of whom twenty-seven had accepted his invitation. To the latter, the founding members of the new society, he explained that the responses had been extraordinarily interesting, the refusals no less than the acceptances. "One eminent scientist refuses because he is unaware that any movement exists in opposition to freedom and individualism in scientific research: another, equally eminent and in full agreement with the four propositions, refuses to join because he thinks that we are fighting for a lost cause!"[19]

In the belief that the truth lay between these two extremes, Baker suggested that the founding members co-opt additional distinguished scientists. However, the membership of the society should not exceed one hundred so as to avoid undue secretarial difficulties. The society also required a formal name. Baker asked the founding members to submit suggestions or at least to let him know their preferences among six suggested names.[20] The name soon adopted was the Society for Freedom in Science.

On 1 March 1941, certain members of the nascent Society met informally in Oxford. A provisional executive committee was created consisting of Arthur George Tansley, a leading plant ecologist and Sherardian Professor of Botany at Oxford until his retirement in 1937, as acting chairman, Baker as acting secretary and treasurer, Polanyi, Arthur Elijah Trueman, professor of geology at Glasgow University, and Frank Fraser Darling, a biologist and director of the West Highland Foundation and Survey.[21] Baker's idea that a small society could be effective was discarded. To become truly effective, the committee thought it necessary to increase the membership "very considerably." Toward that end it was decided to draw up a statement of the objectives of the Society in a form suitable for handing to prospective members.[22] Subsequently, a first draft by Baker and Polanyi was considerably revised by Tansley and circulated under their three names to the other members of the committee.[23] Several further drafts were completed before the final document was printed and sent out to prospective members in late May and June 1941.[24]

The content of the four-page circular, *Proposed Society for Freedom in Science*, was in several respects different from that of Baker's letter of the previous November.[25] Most apparent is the fact that it was international in outlook, doubtless reflecting Polanyi's influence. It opened by stating the belief that should totalitarian dictators be successful in the current war they would eventually put an end to the freedom of scientific research throughout the world. Defense of scientific freedom, as of other freedoms, was therefore an integral part of Britain's war effort. However, even after victory there would be in democratic countries a "less direct though perhaps as dangerous" a threat to scientific freedom from some of the adherents of the doctrine of central planning.

The circular recognized that this doctrine had a strong appeal for many of the more active-minded and socially-conscious scientists. Furthermore, to uphold, as the circular did, the conflicting view that science has a value that is independent of the practical benefits it yields to society was to appear to be socially indifferent. The circular argued to the contrary, however, that the vindication of scientific independence was a positive assertion of rights and duties. One of the principal social duties of the scientist, it explained, was the defense of scientific freedom, as that freedom is essential to scientific discovery and the origin of those practical benefits that are the natural by-products of the scientist's work. Furthermore, almost every professional scientist had duties such as teaching, consultation, or administration, in addition to pure research. "In fulfilling these duties the scientist should be guided by a realization of their wider social implications and should steadily help to make society more humane, juster, and more efficient."[26] Those, therefore, who were prepared to fight for freedom in science, claimed to be as eager as anyone to make contributions to social progress.

In order to maintain scientific freedom in countries where it still existed and to assist in its reestablishment in those where it was suppressed, the circular stressed the necessity of organizing those scientists who supported the ideal of free science. It mentioned the nucleus of members already in the Society and the desire to increase its numbers, primarily among active research scientists. All would subscribe to five propositions that, since they differ considerably from Baker's original four and because they would continue to be the fundamental motivating beliefs of the Society, are quoted here in full:

1. The increase of knowledge by scientific research of all kinds and the maintenance and spread of scientific culture have an independent and primary human value.
2. Science can only flourish and therefore can only confer the maximum cultural and practical benefits on society when research is conducted in an atmosphere of freedom.
3. Scientific life should be autonomous and not subject to outside control in the appointment of personnel or in the allocation of the funds assigned by society to science.
4. The conditions of appointment of research workers at universities should give them freedom to choose their own problems within their subjects and to work separately or in collaboration as they may prefer. Controlled team-work, essential for some problems, is out of place in others. Some people work best singly, others in teams, and provision should be made for both types.
5. Scientists in countries not under dictatorial rule should cooperate to maintain the freedom necessary for effective work and to help fellow-scientists in all parts of the world to maintain or secure this freedom.[27]

The first of these was a broader statement of Baker's second proposition; the second was a considerably modified form of Baker's first proposition. The fourth covered the concerns of Baker's third and fourth propositions. The fifth was novel as was, largely, the third.

The circular had a mixed reception among the scientists receiving it. Of the 30 who had eventually responded favorably to Baker's initial invitation, 20 did not sign and return the membership form sent with the circular. However, 38 other scientists joined the remaining 10 in signing, to give a membership of 48 by late August 1941.[28] By late November, that number had increased significantly to 89 and thereafter grew less slowly to reach 107 by October 1942 and 128 a year later.[29] Then in the early spring of 1944, the Society for Freedom in Science published the first edition of the pamphlet *The Objects of the Society for Freedom in Science*, a revised and expanded version of the 1941 circular, and used it in a new membership drive. This proved to be most successful and by March 1946 the membership had more than tripled to reach 420.[30] By June of that year, it had climbed further to 457, of whom 230 resided in Great Britain, 176 in the United States, 38 in the British Dominions and Colonies including Canada and South Africa, and 13 elsewhere. That same year a British government committee on scientific manpower estimated that there were about 55,000 qualified scientists in the United Kingdom.[31] However, the great majority of members of the Society for Freedom in Science were research scientists at universities and other institutions where research was carried on.[32] They included 64 fellows and foreign members of the Royal Society, all of whose fellows had been approached in November 1945.[33]

Throughout this period, the Society continued to function in an informal manner, as had been intended. It operated on a small budget generated solely from voluntary contributions. A committee of seven, later eight, elected members led the Society, whose affairs in practice were conducted by the executive committee consisting of Baker (as secretary/treasurer), Tansley, Polanyi, and, later, J. A. Crowther, professor of physics at Reading University.[34] The moving force was Baker. Only one general meeting was held during the war years—in London on 2 July 1942. Its purpose was to bring the executive committee into closer touch with the membership. However, only fifteen persons attended. None had any criticism of the organization or operation of the Society, and so it was agreed that its affairs would continue to be directed by the executive committee which would remain in touch with the other members of the general committee mostly by correspondence.[35] As for the general membership, the executive committee communicated with it mainly through duplicated, typewritten notices. Most of these carried the heading: "Confidential: for members of the Society for Freedom in Science only." From time to time, each member received a membership list so that he might know who his fellow members were. Unsuccessful attempts were made to get local branches of the Society started in Britain at centers where interest appeared keenest.[36] Likewise, an effort to find an American member of the Society who would attempt to organize a sister society in the United States failed.[37]

In writing to the forty-nine scientists in November 1940,

Baker stated his belief that after the war there would be a powerful attempt to put an end to free individual research in pure science. Therefore, that would be the time for the proposed society to make its influence felt.[38] In replying to those who accepted his invitation, Baker spoke of "the end of the war, when the politicians will be planning a new sort of life for the people of our country and will be apt to be led astray in scientific matters by the very vocal scientists who hold opposite views to ours."[39] On the previous day, Baker had written to Tansley: "I view with horror the prospect that after the war University research workers will be told that a certain piece of research has been allocated to their department. Against that I will fight with all the energy there is in me."[40] Somewhat later, in early 1943, the membership was informed of the two chief functions of the Society. One of these was "to provide a solid body of scientific opinion ready, in the period of reconstruction after the war, to oppose any steps that may be suggested which would limit freedom in science or magnify technology at the expense of pure science."[41] This concern for what might happen at war's end was again expressed during the membership drive of 1944.[42]

Although, as it turned out, the condition that was feared might occur at the end of the war did not happen, this does not mean that the fear was groundless. But it was an exaggerated fear in the Britain of the early 1940s and not one that was widely held. Nevertheless, for those who did hold it the opposition was very real. The fledgling Society for Freedom in Science was convinced that the threat to freedom in science had to be taken seriously because of the enthusiasm it saw evoked by the doctrine of the central planning of science in the interests of the community. The Society considered that many of the adherents of planning were unaware of the decisive limitations to the freedom and progress of science implied by their aims. Others appeared to minimize or disregard these dangers in their determination to follow the aims of general social planning whatever its consequences for science might be.[43] Tansley believed that the "propaganda" of which Bernal's *The Social Function of Science* was an able expression found many outlets, to be then eagerly assimilated by many younger scientists who felt a need for a positive and forward-looking "gospel." He looked to the future and saw that:

From among these—many of them no doubt able scientists—we may expect to find prominent leaders in the next 10, 15 or 20 years, and if the scientific advice of the Central Government, especially a Socialist Government which we are very likely to have, falls into the hands of convinced adherents of such a creed, anything like genuine freedom of research would necessarily disappear. It *may* be that the English have too much common sense to allow any such happenings, but it is by no means certain and we cannot afford to take chances.[44]

Many scientists who did not join the Society in its early days believed that the British did indeed have sufficient common sense.

A response common to many of those who received the Society's 1941 circular initiated an important discussion among the Society's leaders. The outcome was a modification and clarification of the Society's position in regard to planning. The nature of the response was that research should be free except in the case of scientists in institutes founded for the pursuit of certain subjects, mostly of direct or indirect social utility. Such scientists were engaged and paid on the tacit but quite obvious understanding that they were to work at those subjects under the control of a director.[45]

Related to this issue was that of the freedom of research workers in technology. The question was whether the Society should be concerned about this. Baker thought not, suggesting that planning was probably necessary in technological research.[46] Polanyi was brought into the discussion, the outcome of which was a confidential memorandum to the membership drawn up by Tansley, modified in minor ways, and issued in the name of the executive committee.[47] The document stated that it was now clear that the liberty of research scientists to choose and pursue their freely-chosen subjects of research without interference from any external authority could not be universal or unlimited. For in industrial and government laboratories and other institutions founded to pursue specific practical aims, the appointment and activities of the staff were obviously subject to these goals; in addition, activities had to be carried on under the control of a director. Thus, the Society concluded that it was primarily in the universities that the liberty existed to pursue completely free and independent research and that it was there in the first place that it should be "jealously" maintained.[48]

Although it thus recognized the necessity of planning applied science in such fields as agriculture, medicine, and nutrition, the Society nevertheless was determined to fight the "dictation and regimentation," or "central planning," of scientific research by a state bureaucracy. It held that such organization and planning as was necessary should be carried out by responsible scientists within the area concerned and not by some external government authority. It correctly saw that in the future most money for fundamental research would come from the government, and it advocated that a considerable share of such funds should be awarded to properly qualified investigators freely working at their own problems without any regard to practical use. At the same time, the Society recognized that much state money would also quite properly be allocated to collective research planned for the promotion of the general welfare.[49]

In regard to a program of action for the Society, Baker proposed that its first effort should be the production of a book of essays by Society members to be ready for publication within a month of the end of the war. The provisional committee endorsed the idea, and Baker was assigned to edit the projected book.[50] In May 1941, Polanyi wrote to Arthur Koestler, who was living in England and whose *Darkness at Noon* Polanyi had recently read and found deeply moving, asking if he would collaborate in "a book which a few scientists are planning to write as a statement of the case for Freedom in Science. We need a chapter on totalitarian oppression of science and scientists. It is fairly easy to get the information about Germany, but very difficult for the U.S.S.R. The fundamental facts of the position of scientists under that regime can be properly stated only by a former Party member like yourself. You cannot fail to see the tremendous human interest in the lessons of that experiment of State directed science."[51] Koestler, however, was "unable to collaborate."[52] Later, the executive committee decided not to proceed with it. The feeling was that the freedom in science movement had not reached a sufficiently developed stage for it to be profitable to lay down what could be regarded as a formal doctrine. Nevertheless, it was considered essential to stimulate as effectively as possible further work on problems relating to freedom in science and to keep the Society's views before the public by encouraging individual members to publish articles. An attempt, apparently unsuccessful, was made to reach an understanding with some editor whereby most of the publications would appear in the one journal.[53] In fact, the Society was for some years "unable to get its case made public through most of the recognized channels of publicity in Britain."[54]

Baker was doubtless disappointed by the decision to abandon the book project, but he soon produced a book of his own, *The Scientific Life*, which appeared in 1942. A second edition was published in 1944. The principal purpose of the small book was "to describe the human nature of the good research worker, and to show that it would be futile to confine him within the rigid boundaries of a central plan for the advancement of science."[55] It presented Baker's views on the free scientific life and on the development of a wider and deeper scientific culture. At the same time, it attacked the teachings of the planners. In this connection, various views expressed by Bernal in *The Social Function of Science* and by J. G. Crowther in *The Social Relations of Science* (1941) are mentioned. These works were clearly regarded as the principal volumes of the "planners' library," to which, in Baker's view, there had hitherto been only one little book published in opposition, namely, Polanyi's *The Contempt of Freedom*.[56] In preparing his study, Baker had received advice from Polanyi and many suggestions from Tansley. Thus, although no explicit connection with the Society for Freedom in Science was indicated in the book, it upheld all that the Society espoused. The continuing concern about the period of reconstruction after the war, when "the future of science will be decided," was again expressed.[57] A choice would then have to be made between the desirable scientific culture depicted in the book and a vastly different prospect, various aspects of which had been attacked throughout the book but which in the conclusion Baker depicted with blunt conciseness: "An ugly new god called the state demands worship. Nourishment, shelter, health and leisure are falsely regarded as ends in themselves. Culture is looked down on with contempt. Science is equated with technology and decay. Individualism and free inquiry both are ridiculed. Everything is planned from 'above.' A dreary uniformity descends. Each person is a cog in a vast machine, grinding towards ends lacking all higher human values."[58]

Meanwhile, by mid-1941 several members had suggested that the Society was in a weak strategical position because its attitude was negative in the sense that it was opposing a positive view and plan of action without advancing constructive counter-proposals. The defense of freedom, the finest of causes, said Tansley, could be made to appear as mere reactionary conservatism. Laissez-faire was no battle-cry that could compete with a forward-looking gospel. Therefore, he suggested that the Society should formulate a positive scheme for the reorganization of certain social scientific activities that seemed to require reform and extension. These included the allocation and administration of public moneys assigned to scientific work and science education, activities in which organization and planning would be of definite advantage. At the same time, the Society would defend the freedom of able and qualified investigators to pursue their own work in complete independence.[59]

The provisional committee subsequently decided to develop a general scheme of national scientific organization to be outlined in a pamphlet issued by early 1943. It would aim at a rational coordination of the numerous agencies that financed and directed research while paying attention both to the efficiency of the agencies in pursuit of their immediate objectives and to the maintenance of independent free research.[60] This was planning and in an area for which Bernal had offered a plan of his own, but strong emphasis was laid upon free research. The executive officers were quite aware of the direction being taken. "The Society recognizes the necessity of planning in many of the complex and increasingly important relations of Science to the life of the community, and here is its proper field. Such planning should justly delimit the appropriate spheres of freedom and of control."[61]

It was in connection with this proposed pamphlet that the formation of local branches had been suggested. Their primary objective was to discuss and clarify ideas and submit constructive suggestions to the committee for use in the pamphlet. However, the branches did not materialize, and it is perhaps for that reason that the pamphlet, although it continued to be discussed throughout 1942, never appeared either.[62]

Thus, neither of the major projects contemplated by the Society reached fruition. However, and most appropriately for a Society that valued the freedom of the individual, members singly made numerous contributions that propagated the philosophy of the Society. This was a practice encouraged by the committee, which suggested to members in October 1944 how they could promote the Society's cause by individual action.[63] One could, for example, encourage suitable people to join the Society, speak on the subject of freedom in science when it was relevant in conversation and in discussions at public and private meetings, contribute letters on the subject to *Nature*, the American journal *Science*, and other suitable journals and newspapers, and write articles or books on the subject. When these suggestions were made, it was not with the intention of starting something new, but of extending a practice already in progress from the formation of the Society and in which the members of the executive committee, especially Baker and Polanyi, were prominent.[64] But it was not always easy to have papers published: difficulty was experienced in finding editors who would publish articles opposing what had become the accepted view.[65] However, there were successes, and Baker kept the membership informed of these and other activities by means of duplicated, typewritten notices.

One of November 1943 explained that the subject of freedom in science had been clarified and kept before the public in various ways.[66] Polanyi had published in the *Manchester Guardian* on 7 November 1942 a long letter describing and attacking the planning movement, which had led to further correspondence in that newspaper. His paper on the autonomy of science had been published.[67] He had also taken the opportunity to dispute publicly with Sir Robert Watson-Watt, the radar expert and a proponent of planning, during the latter's recent visit to Manchester.[68] Professor A. E. Trueman, a member of the provisional executive committee, had published his pamphlet *Science and the Future*.[69] A. D. Ritchie, Professor of Philosophy at Manchester and a member of the Society, had delivered the Herbert Spencer Lecture, part of which was relevant to the Society's cause, at Oxford. Baker himself had addressed the Institute of Physics at the Royal Institution[70] and the International Student Service at Oxford.[71] He had also contributed an invited article to *Electronic Engineering* and subsequently disputed with Bernal in its columns.[72] In addition, Baker had "exposed" in the *Manchester Guardian* a document produced by the Association of Scientific Workers, which strongly supported the planning movement, whose ostensible meaning was that all research workers should be forced to communicate the subjects of their research to a central body. Finally, members were informed that at least three books dealing with various aspects of freedom in science were in preparation by members of the Society.

This was by no means a full description of activities. Baker somehow overlooked the fact that certain developments within the USSR, to which the planners continued to look with great admiration, were used to significant advantage both by himself and Polanyi. The latter's paper, "The Autonomy of Science," which had been first given as a lecture in Manchester in February 1943, exposed the degradation of Soviet genetics perpetrated by T. D. Lysenko. Polanyi used this subject as an illustration of the outcome of a genuine effort to organize science directly for the public good.

Before the state had intervened beginning about 1930, he explained, genetics in the Soviet Union had existed as a free science guided by the standards recognized in other countries where science was practiced. However, in 1932 the All Union Conference on the Planning of Genetics and Selection decided that genetics and plant-breeding should henceforth be conducted with a view to obtaining immediate practical results and in keeping with the official doctrine of dialectical materialism. Research would be directed by the state. No sooner had these blows against the authority of science been made, continued Polanyi, than the inevitable consequences set in. Anyone claiming a discovery in genetics or plant-breeding could henceforth appeal directly over the heads of scientists to gullible agricultural practitioners or to politically-minded officials. "Specious observations and fallacious theories advanced by dilettantes, cranks and imposters could now gain currency, unchecked by scientific criticism."[73] This was illustrated by the case of I. V.

Michurin. He had claimed the discovery of new strains of plants produced by grafting, and thereby the achievement of revolutionary improvements in agriculture and a striking confirmation of dialectical materialism. The opinion of responsible science, however, was that Michurin's observations were mere illusions, that they referred to a spurious phenomenon, "vegetative hybridisation," that had frequently been described before. Nevertheless, his work fulfilled both of the criteria that had replaced the standards of science: it appealed to the practitioner and it conformed to the philosophy imposed by the state. Hence, it was given official recognition.

As Polanyi assessed the situation, the breach thus made in the autonomy of science laid the fields of genetics and plant-breeding open to further invasion by spurious claims. The leader of this invasion was Lysenko, whose rise to power Polanyi briefly described. Against his influence, such scientists as N. I. Vavilov, internationally recognized as the most eminent geneticist in the USSR and recently elected a Foreign Member of the Royal Society, became ineffective in upholding true science. In a passage that is most important for the understanding of Polanyi's efforts in the cause of freedom in science, Polanyi uses Vavilov's case to illustrate that even eminent scientists can err. At the All Union Planning Conference mentioned above, Vavilov, "yielding at the time perhaps to pressure, or believing it wise to meet popular tendencies half way, little expecting in any case the far reaching consequences to follow from his relinquishment of principles . . . had then allowed himself to say: 'The divorce of genetics from practical selection, which characterizes the research work of the U.S.A., England and other countries, must be resolutely removed from genetics-selection research in the U.S.S.R.'"[74] The lesson that Polanyi had drawn from that, and which he would have all scientists draw, was that the scientist must always protect the ideal of scientific truth pursued and upheld for its own sake.

Soviet genetics illustrated the corruption of a science, and yet the Soviet government had striven to advance science. The large subsidies to Soviet science had been beneficial, argued Polanyi, only so long as they had flowed into channels controlled by independent scientific opinion. However, as soon as their allocation was accompanied by attempts at establishing governmental direction they exercised a violently destructive influence.[75] To Polanyi, only one thing was necessary to remedy the situation and that truly indispensable, namely, the restoration of the independence of scientific opinion.[76]

Soviet genetics was also discussed in the same month by Baker in his lecture at the Royal Institution.[77] This was obviously a subject of much importance for the Society for Freedom in Science, and Baker took it up again in an address, "Science, Culture, and Freedom," given in August 1944 at a conference called by the London Centre of the International Association of Poets, Playrights, Editors, Essayists, and Novelists—the PEN Club—to commemorate the tercentenary of the publication of Milton's *Areopagitica*.[78] His strong anti-Communist and anti-Soviet comments were attacked by all who participated in the prolonged discussion that ensued. The chief criticism was heard from J. B. S. Haldane, the professor of biometry at University College, London, and a Marxist.[79] Materials from each of these lectures were incorporated in Baker's next book, *Science and the Planned State* (1945), in which the case of Soviet genetics was again treated, this time in some detail.

The volume was devoted to an analysis and criticism of what Baker termed the totalitarian view of science. The Society for Freedom in Science was in no way a political organization; all shades of political opinion from conservative to socialist were to be found among its members. It was, however, opposed to totalitarianism in the sphere of science whatever the political complexion of the government imposing it.[80] As in *The Scientific Life*, Baker again stressed the role of chance discovery in science. He observed that most of the four kinds of freedom that he considered necessary to the scientist—freedom to become a research scientist, freedom of association, freedom of inquiry, and freedom of speech and publication—had been lost in totalitarian countries, and he feared that the movement toward central planning in Britain threatened them there.[81] He was critical of Soviet scientific work in general—it had displayed "an unduly high proportion of bad and suspect science"—and used the example of Soviet genetics to show what can befall science when the freedoms necessary for it to flourish are denied. Baker's, as Polanyi's, sole source of information on the subject was the report of the Soviet conference on genetics held under the auspices of the journal *Pod Znamenem Marxisma* in 1939. This report had been reviewed in *Nature* as early as 1941.[82] However, Baker and Polanyi had now read it much more critically. They alerted British readers to what was to become notoriously known as the Lysenko affair. Doubtless their telling exploitation of this report had much to do with the growth of the Society for Freedom in Science during 1944–45. It provided what scientist and layman alike found compelling, namely the convincing 'experimental' confirmation of 'theoretical speculation.'[83]

Within the Society, such theoretical speculation was frequently the work of Polanyi, whom Baker admired as the philosopher of the Society, and who came to be regarded also by those outside the Society as its chief spokesman.[84] Polanyi described how his generation on the continent of Europe had set out with immense hopes for the future. With science as its polestar, it had been determined to make "a clean sweep of all ancient stupidities, of all silly obstructions to human happiness, and to rearrange life in a thoroughly rational and scientific fashion." It subscribed to the scientific view of man: "That man was fundamentally an animal; that man's ideals were mere passing shadows while his appetites were firm, tangible and eternal forces."[85] Polanyi, however, had become convinced that this scientifically-minded generation, which would suffer no obstruction in advancing what appeared to it to be the necessary progress of mankind, had struck the first blow against freedom and tolerance in society.[86] Subsequently, he had come to hold that science did not have to teach a materialist view of man and society, and indeed that, in contradiction to the Continental outlook, it taught faith in ideals:

> Every discovery of science has its starting-point in a guess which is yet much more than a guess, and represents an act of faith. In fact the scientific method as a whole must be taken on faith by the scientist before he can even make a start in science. To become a scientist he must unquestionably accept the main body of scientific tradition and fully adhere to the ideals transmitted by that tradition. In this light the triumphs of science confirm rather than impair the roots of our Christian civilization. They testify to the power of traditional ideals on which our civilization rests. The new scientific outlook which I see approaching will clearly recognize that science

is only one form of truth which is of the same substance as all the other forms of truth. It will recognize that we cannot believe in science without becoming involved in the whole range of human ideals of which the ideal of science is only the youngest sister. In this light science may help in reconquering our faith in traditional ideals.[87]

Polanyi's views were developed in numerous articles some of which had begun as critical reviews of pro-planning books but which had gone on to become creative essays. His review of Bernal's book had followed such a course, as did his review of J. G. Crowther's *The Social Relations of Science*. Polanyi was asked in May 1941 to review the book for *Economica*. Its editor, F. A. Hayek, best known today perhaps as author of *The Road to Serfdom* (1944), wrote: "it clearly needs rather careful analysis. A glance at his chapter on Freedom and Science will immediately show you how big the problems are which he raises."[88] Two years earlier, Hayek had told Bernal after reading his book, *The Social Function of Science*, that he found himself "in complete agreement on ultimate ends and almost as much in disagreement on the methods of social change which you propose."[89] By July, Hayek and Polanyi were talking of a review article. Hayek now told Polanyi that he "attached very great importance to these pseudo-scientific arguments on social organisation being effectively met" and that he was becoming "more and more alarmed by the effects of the propaganda of the Haldanes, Hogbens, Needhams, etc."[90]

Crowther, science correspondent for the *Manchester Guardian* and also a member of this loosely-knit group of exponents of the Marxist view of science, was convinced that a national policy for science had to be fashioned. His book was offered as a contribution toward the creation of such a policy. In its first part, Crowther sketched a history of science. This was intended to demonstrate "the nature of science as a social product." He wrote that: "Complete freedom of thought is not the chief condition of the progress of science. There are other social conditions which can assist science more."[91] The section on science and freedom particularly incensed Hayek and Polanyi. There it was stated that: "Freedom is beneficial to science when it provides opportunity to a rising class. Control is beneficial to science when it protects a rising class. Freedom is inimical to science when it preserves the power of a declining class. . . . If the scientist wishes to enjoy freedom, he must be able to choose the progressive side."[92] Crowther went on to say that industrial research laboratories produce much fundamental science and that planned research can achieve wonders,[93] that more money for research should be made available by the government, and that the "progress of science has owed much to the desire to serve humanity, and it is probable that in the future, through better organization of the expression of this motive, science will owe more to it than to any other personal motive."[94] The distinction between pure and applied science was either disregarded or denied.

In his review, Polanyi criticized not only Crowther's book but the entire movement denying the justification of pure science. He named Crowther, Bernal, and Hogben as its leaders. Polanyi regarded their attack on pure science as part of a war against all human ideals, and their attack on the freedom of science as part of the totalitarian assault on all freedom in society. In contrasting totalitarian and liberal views of the state, he developed the idea of science as an example of a dynamic order within society; that is, an ordered arrangement resulting from the spontaneous mutual adjustment of the elements composing it. According to this view, new scientific claims were made with due consideration of all previously established ones; the results thus obtained continuously modified those previously achieved.[95] One generation handed on to the next a public intellectual heritage accessible to all; each new contributor made contact, of a more or less consultative or competitive character, with the elements that others had established in the same field before. When the new contributors proposed their own additions or reforms, they claimed publicly that these be accepted by society—to become in their turn a part of the common heritage.[96] The maintenance and growth of an autonomous dynamic order in society entailed, he argued, a freedom that was not for the sake of the individual but for the benefit of the community. This was freedom with a responsible purpose, a privilege combined with duties. To Polanyi, it was impossible to transfer the task of ordering achieved by a dynamic system to any planning authority. Nothing but chaos would result from such an attempt.[97]

Each dynamic order within society was governed by a fundamental idea that often could not be expressed precisely, but that came into play every time the standards of the day were challenged. Scientific truth was the ideal underlying the practice of science. Its guardianship, Polanyi maintained, was the most important function of scientists. The proponents of totalitarianism attacked scientists' standards and ideals when they charged that those who pursued the ideal of scientific truth were neglecting their duty to society for the sake of insubstantial values of purely formal significance.[98]

It was mainly through such individual efforts—in books, articles, speeches, and published letters—rather than through the collective efforts of the Society that its views were disseminated during the years 1941–45. To start a new journal was illegal under wartime regulations. Even the distribution of duplicated, typewritten circulars was made difficult by the restrictions on the sale of paper.[99] But as the war neared its end, the Society was able to publish its views. The series of Occasional Pamphlets, distributed to and through the members, was the first step. Several hundred copies of each pamphlet were printed. The initial pamphlet was written on the invitation of the executive committee by the historian of science, F. Sherwood Taylor, curator of the Museum of the History of Science at Oxford. He was asked to examine a recent publication of the Association of Scientific Workers in which it was argued that science was a by-product of industry.[100] The pamphlet, "Is the progress of science controlled by the material wants of man?", which answered the question in the negative, was issued in April 1945 together with one of the increasingly frequent notices to members. Members were asked to give the pamphlet wide publicity. "It would be desirable to put a copy in every laboratory and scientific library to which our members have access."[101] Two further pamphlets were issued during 1945, one a reprint of Polanyi's 1939 article "Rights and Duties of Science," which Baker described as "the first comprehensive answer to our opponents' propaganda, and . . . one of the foundation stones of our movement"[102]; the other "Free Science" by Warren Weaver, director of the division of natural sciences at the Rockefeller Foundation, which the Society considered to be one of the best short statements of the case for freedom in science. Another pamphlet by Polanyi issued early in 1946, "The Planning of

Science," was regarded by the executive committee as the best statement of the Society's case published.[103] Thus by early 1946, the Society had issued only a half-dozen official publications, including the 1941 and 1944 circulars on the objectives of the Society. In August 1946, it began to issue its *Bulletin*, usually of one or few pages, in place of the earlier notices. The Society had now grown to such a size that it was found to be cheaper to have circulars printed rather than stenciled.[104]

In the Society's early days, support for it was limited. That is not surprising since the Society was formed to combat a popular view. It attempted to recruit those research scientists whom it judged to be unsympathetic to central planning. In May 1941, Tansley approached the president and two secretaries of the Royal Society, Sir Henry Dale and A. V. Hill and A. C. G. Egerton, respectively. To Dale he wrote: "One cannot of course say how serious the threat to freedom in research may turn out to be after the war, but Bernal and his associates are very active indeed, and one feels that there is a Marxian or quasi-Marxian drive behind this movement. You will probably agree that politics in science is the Devil! No one, of course, wants to interfere with extensively organized work for practical ends, but some of us feel that it is highly desirable to establish a focus in order to concentrate the large body of opinion that is certainly in sympathy with the ideal of freedom in science."[105] Tansley heard first from Hill and Egerton, both of whom declined to join the Society. Although Hill was in close general agreement with the Society's principles, he felt, however:

(1) that they should be equally applicable to all other forms of intellectual activity, and not limited to science;
(2) that a more conscious planning of our social and therefore of our intellectual activities is inevitable, and that it is wiser to cooperate and guide rather than resist;
(3) that we must all learn to temper order with freedom and freedom with order if we are to live together;
(4) that there is great danger in mixing politics with science; and therefore
(5) that scientific people will guard their freedom better by not forming themselves into rival political groups.[106]

Hill feared that the Society might too easily be driven into a reactionary position with progressive individuals tending to side with those whom the Society opposed. The opponents would say that they agreed with the Society's principles, but that they were being used to cover political aims. "We know in fact," Hill continued, "that Bernal and company are using science themselves as a cover for political propaganda: it would be wiser not to give them an excuse for saying the same of those who disagree with their politics: otherwise the wretched cover itself will get badly torn in the resulting conflict."[107]

Hill would not join the Society for another reason, namely, a desire not to compromise the Royal Society, which "must not be associated with any brand of politics or it will lose its influence and position." He elaborated:

> If its officers were to join in a movement to oppose the application of certain political ideas, whether of the right or left, sectional divisions and differences would be bound to occur within it, instead of the very pleasant harmony which at present exists. Remember, that Haldane and Blackett, for all their queer political notions, are useful and co-operative members of Council: I am sure that Bernal and Hogben will be the same when their turn comes to serve, for they have always been most helpful whenever we have called for their advice on scientific matters. We can keep them in order better by co-operating with them in scientific affairs than by formally setting up to oppose their political ideas in the name of science.[108]

Dale had known of Hill's views for almost two months when he replied to Tansley. He found it difficult to believe that in Britain freedom in science was in any particular danger. Upon receiving Tansley's letter he had doubted the need for, and the wisdom of, raising the fear that after the war there might be an attempt to fasten a more permanent political servitude on science. Since then, the changed situation following Hitler's invasion of the USSR on 22 June, seemed to render the activities of the Society additionally open to misunderstanding:

> Since Russia has been forced by Germany to fight on the same side as this country, a suggestion that the freedom of science is imperilled here, by the admiration of some of our colleagues for the Russian political system, encounters a new difficulty. At its last meeting, the Council of the Royal Society, the members of which are certainly

Freedom and Planning in Science · 289

not predominantly admirers of Bolshevist parties, requested the Officers to send a message of fraternal sympathy to the National Academy of Sciences in Moscow. For reasons which I do not need to elaborate to you, we felt the commission to be slightly embarrassing, but the message we sent emphasized the fact that Russia is now, like this country, fighting for the freedom which has been a condition of the achievements of the great scientists of Russia, as of our own.[109]

Although Dale had the "fullest theoretical sympathy" with the Society's objectives, he had no wish to have his consistency questioned and so declined to join. He assured Tansley that "those colleagues of ours, whose activities you apparently suspect of creating an atmosphere dangerous to [the] freedom [of science] are, in fact, as eager as anybody else to help their country in its present conflict. Bernal, in particular, was eagerly at work doing an absolutely first-rate job for the Ministry of Home Security, long before there was any suggestion of Russia being involved in the war against Germany."[110]

Tansley acknowledged that the Russian alliance had made the Society's work much more difficult. Regarding Bernal, he had no doubt that he and his left-wing colleagues were as eager as any British scientists to defeat Nazi Germany. However, Tansley could not forget

> that they strive, politically, for a system which is the very opposite of that in which we believe. That system involves the entire subordination of science to the state as well as a completely "planned" economic organisation. There are those who think that it is possible to have comprehensive economic planning without the regimentation of science, and some of them have already joined us. There are few of us who think we can avoid a very great expansion of economic planning accompanied by far-reaching organisation of applied research. All that is essential in our position is the maintenance of complete freedom in "pure" and so-called "fundamental" research. That, we believe, is seriously threatened by the believers in a totalitarian state on anything like the U.S.S.R. pattern.[111]

The Society also failed to recruit the eminent atomic physicist and Nobel prizewinner, Max Born, who as a Jew had been dismissed from his position at the University of Gottingen in 1933.[112] Tansley asked Polanyi to write to Born, then at the University of Edinburgh.[113] Born replied that the 1941 circular gave the impression that planning and other changes in economic affairs were to be fought together with any restrictions on the freedom of science and thought. He begged the Society to reformulate the circular, omitting all allusions to economic questions or, preferably, emphasizing the principal importance of keeping the question of freedom of thought completely separated from the questions of expediency connected with political and economic restrictions. He said he would then be very glad to join the Society and assist with all his heart.[114] The circular was not reformulated. When some months later Polanyi published his essay review of Crowther's book, he sent copies of it to Born and others. Born agreed with Polanyi's defense of pure science and the freedom of thought, but as before he regretted that Polanyi coupled this defense with that of economic liberalism which Born would not accept. He held that scientists should refrain from mixing their ideals with "the very unideal purposes of business affairs."[115] Still Polanyi persisted, and two years later during the membership drive of 1944 he sent Born a copy of *The Objects of the Society for Freedom in Science*, again inviting him to join the Society. Born replied:

> You are quite right to assume that I agree with many of your aims, in particular the importance of "pure science" or how I should prefer to say, of the philosophical foundations of science and their consequence for the development of human thought in general. However, I am still reluctant to join your society. I cannot see an urgent necessity for its existence. You quote my name, in conjunction with that of A. V. Hill, in your Circular as being the only opponents [at the 1941 London conference on science and world order sponsored by the British Association for the Advancement of Science's Division for the Social and International Relations of Science] to the prevailing trend of considering science from its practical side alone. I cannot agree with that statement. The Meeting in London was convened for the purpose of discussing the social functions of science; I said only, that this aspect and that of "pure science" are hardly in opposition, but complementary, somewhat in the manner of the particle and wave aspects in quantum mechanics. Under the present social and political conditions I am in fact convinced that the utilitarian aspect of science has to be stressed, and I am also convinced that this involves political measures of a decisive character. But I am not in the least afraid that a reasonable nation like the British will strangle freedom of research

if they consider it necessary to restrict freedom of industrial exploitation. These are two very distinct freedoms, and to mix them up seems to me to endanger that freedom which we love, the freedom of thought and research. . . . I also cannot quite agree with the 5 points of your circular; points 1 and 5 are alright—but take e.g. point 3: Why should those whose funds are allocated not have some right to some kind of control or, at least, advice? It would be our task to convince them that what we are doing is worthwhile. I think I can do more good by keeping outside of your society and use my little influence amongst your "enemies" just as well as with your friends.[116]

Polanyi also sent a copy of his review of Crowther's book to Sir Richard Gregory, president of the British Association for the Advancement of Science and chairman of its Division for the Social and International Relations of Science. Gregory had also been invited to join the Society in 1941, but he too had declined. The British Association's Division was regarded by the Society as being a platform for the planners. Baker was particularly incensed by its conference on science and world order. He had attended the conference throughout its three days and had heard only A. V. Hill and Born advance ideas that would have been upheld by the Society, no member of which had been invited to participate in the conference.[117] The leaders of the Society had shown no sympathy with the sentiments that had led to the creation of the Division. Baker, Polanyi, and Tansley had had no part in it; Hill, and especially Gregory, had. So also had Bernal and Crowther; but it was the moderates, the Gregorys and Hills, who exercised control. The Society made the mistake of not seeing that.

In replying to Polanyi, Gregory wrote that:

A cause, whether totalitarian or the liberty of thought and expression which most of us cherish as closely as you do yourself, is greater than any man or group of men. I think, therefore, that you give too much attention to what Hogben, Bernal, Crowther and a few others say about the organisation of science on totalitarian principles. This gives a personal outlook on the subject instead of a strictly rational one. Your case for freedom in science is strong and convincing enough to stand by itself, and if all such personal views were omitted from it, I believe that the vast majority of natural philosophers would find themselves in full agreement with you.[118]

There were many others besides Gregory in British scientific circles including some foreigners like Born, who, because of their faith in the stability of British society with its basis in freedom, saw no great threat in the views of Bernal, Crowther, and other would-be planners. Although Solly Zuckerman found Bernal's book a "fascinating challenge," some of his colleagues feared that what Bernal wanted was to dragoon scientists into work that they would not necessarily wish to undertake, in the pursuit of ends of which they would not approve. "I recall," wrote Zuckerman,

> how one of them, a micro-anatomist who was an expert in the staining of tissues, remonstrated with me, perhaps not in all seriousness, about Bernal's presumed purpose. 'Why,' he exclaimed, 'I know what he is after, he is going to tell me that I dare not ever again use gentian violet as a stain for my sections, and that from now on I shall have to restrict myself to methylene blue.' No words of mine could persuade him that I was as ready as he to defend to the end our right to carry on with science as we knew it, and that no one was going to coerce me to undertake researches that I had not decided on myself; but that knowing Bernal I felt certain that he had no dictatorial ambitions, however extreme the effect of his printed words might appear to some.[119]

As Gregory remarked in an address at the Royal Institute of International Affairs: "Liberty of thought, work and expression is highly cherished in the commonwealth of science, and in Great Britain no conditions which would limit it would be tolerated."[120] Consequently, he and others could not entertain the Society's fears about what might befall British science at war's end. On the contrary, they realized that they could learn from the planners without going all the way with them. The planners and the Society's leaders occupied opposite ends of a spectrum. Between them lay the majority of British scientists, many of whom were initially, in the late 1930s and early forties, swayed away from traditional views by the planners, but who later, by war's end, were fully conscious, as Born, of the importance of freedom in science as well as of the desirability of some compatible form of general planning.

One who would not have been expected to be a member of the Society was Joseph Needham. In 1942, Needham learned of the Society through reading in the American journal *Science* the

Society's statement of principles that *Nature* refused to publish. Needham wrote to Baker inquiring about membership in the Society. His letter gave Baker "much pleasure," and not only because the Society was "relatively weak in Cambridge."[121] The executive was "particularly pleased" that Needham joined the Society.[122] In response to Needham's statement that freedom in science existed in the USSR, Baker wrote: "I can't help thinking that the present worship of the U.S.S.R. may be a real menace to the cause of liberty in every sphere. It is good to know that some liberty still exists in science, even if not in other spheres. The purges of 1937/8 made a profound effect on me, from which I cannot free myself. However one interprets the trials and mass executions, the regime under which they occurred seems self-condemned."[123] In contrast, Bernal, according to C. P. Snow, was a totally committed Marxist who "couldn't entertain the thought that men in power, even if they had arrived at power through impeccable Marxist channels, didn't necessarily behave with the sweetness that he would have shown himself.... He couldn't imagine that the Stalin purges weren't precisely what they were officially said to be."[124]

In August 1941, Tansley enumerated the most common negative reactions to the Society's circular sent out earlier that year. There was doubt or disbelief that there was any real danger in Britain of interference with scientific freedom. Consequently, the Society was regarded as being unnecessary. There was also a suspicion that there was an underlying dislike of all planning of scientific activity on the part of the founders of the Society, but it was clear to many of those invited to join the Society that extensive planning of such activity in the future would be both good and inevitable. Finally, there was a desire on the part of some eminent men of science, "including officials of the Royal Society," to cooperate as far as possible with those who avowedly believed in central planning as a general method of organization of society, including some who were professed Communists. One or another, or a combination of, these three types of reaction had led the majority of the more eminent scientists approached, including the president and two secretaries of the Royal Society—Sir Henry Dale, A. V. Hill, and A. C. G. Egerton, respectively—as well as Sir Richard Gregory to refuse adherence to the Society.[125]

In his presidential address to the Royal Society later that same year, Dale touched on the questions of planning and freedom. He had attended the British Association Division's conference on science and world order and had observed much support for the planning of science in the postwar world "for its proper purposes of enriching life and enlarging the opportunities of happiness for all men alike." However, he had noted that Hill had sounded a warning of dangers that might be entailed by such fullness of association between science and government as was being advocated with conviction and enthusiasm. Dale had found himself in agreement with the view that freedom and opportunity rather than organization provide the conditions for the highest types of research, and thus, in the end, for the greatest services that science can give to mankind. That was also the Society's view. At the same time, he also referred to the remarkable development within Britain since 1914 of state-supported research administered by the Department of Scientific and Industrial Research and the Agricultural and Medical Research Councils; this had occurred without any obvious detriment or danger to the freedom of science. He pointed also to the large amount of free research supported by the state through grants to universtities, administered without any trace of detailed government control. Dale thus found no reason to fear any threat to the freedom of science from the existing mechanisms for the support of science by the state or from any likely development on these lines.[126]

At that time, the Society did not enjoy the support of *Nature*, either. In 1941, a "careful, sober declaration" of the Society's views submitted to this influential journal for publication was "flatly rejected."[127] That was not altogether surprising since *Nature* had previously criticized some of Polanyi's ideas. It considered that his plea, in "Rights and Duties of Science," for the reestablishment of man's right to pursue truth regardless of social interest, would have had more reality had he recognized the contemporary lopsidedness of scientific development. The reorientation of scientific effort for which so many scientists were calling, argued *Nature*, involved no threat to fundamental research that it was desired to promote in neglected fields.[128] Later, in commenting in a leading article in *Nature* on an address given by Polanyi at the University of Leeds in January

1941 and entitled "The Social Message of Science," Rainald Brightman described Polanyi as "obsessed with the independence of science." His rejection of all ideas of the central planning of science and other human activities, Brightman continued, did not lead to the offer of any constructive alternative to "the anarchy into which the failure to plan or co-ordinate scientific advance has led us."[129]

On the other hand, *Nature* placed limits on planning. In discussing "the vexed question of pure as opposed to applied science," its joint editor, L. J. F. Brimble, attacked the extreme materialists who would have dropped branches of science that were not directly and immediately fruitful. But at the same time, Brimble also criticized the Society without mentioning it by name: "There are those who deliberately hold themselves aloof from any effect their science may have on human society; they are, to say the least of it, selfish, though how often has one heard them claim that they are the only champions of scientific freedom."[130] *Nature* continued to snipe at the Society's position. When the Biology War Committee was created in 1942, Brimble and G. E. Blackman, an ecologist at Imperial College and secretary of the Committee, commented that one of the Committee's effects might be to counteract "the attempts of some biologists to disturb the relations of modern science to society under the pretext of preserving the freedom of scientific research."[131]

Then during the second half of 1942, *Nature* published ideas that were anathema to the Society. Brimble advocated that the distinction between pure and applied science, which the planners wished to abolish and the Society wished to maintain, must go, "and a good riddance."[132] Brimble claimed that those who opposed planning believed that it would put scientists in chains. This he insisted "could never happen in the democracy for which we are fighting; but even if it did, then surely nothing but good could come of it provided such chains were guiding chains and not prison chains, and always provided that they were under the control of certain men of science themselves—trustworthy men who fully appreciated their duties to their fellowmen and who had the courage of their convictions."[133] To the Society, there was no place in pure science for shackles of any kind.

Brimble advocated that science must be properly planned and

organized so that it could play its part efficiently in the solution of postwar problems. He added that scientists should recognize their duties to society and be willing to be properly organized so that they could give the maximum service to community, state, and world.[134] To the Society, this was proof that their worst fears were not unwarranted.

The Society's greatest source of opposition was the Association of Scientific Workers, which requested its Oxford branch to keep an eye on the activities of the Society.[135] In March 1941, the *Scientific Worker* attacked "that reprehensible standpoint" that urged the complete withdrawal of scientists from social obligations on the plea that science has "a peculiar holiness, a purity which social contact will destroy."[136] The journal explained that: "This misconception or misrepresentation of the real significance of *scientific detachment* as it is exercised in analysis, observation and experimental verification in every sphere, including the social, is an escapist argument with which we are in fundamental disagreement. The writings of Bernal, Hogben, Crowther and others, the organisation of science as we know it in its dependence on the material circumstances of our society, have long exposed the meaninglessness of this vacuum conception of science."[137]

In reviewing a pamphlet, "Science and Socialism," in the *Scientific Worker* three months later, H. B. Nelson considered it especially welcome because "the counterattack of the idealists is developing afresh."[138] He quoted a recent statement of Polanyi's published in *Nature*—"Science exists only to the extent to which the search for truth is not socially controlled"—in explaining that the pamphlet "gives a crushing answer to nonsense of that kind and shows how impossible it is to maintain the isolation of science from the problems ... of poverty, dirt and disease, under-nourishment, slums, unemployment and war."[139]

In a review of Baker's *The Scientific Life*, J. G. Crowther intimated that the author would deny Britain victory in the war. He explained that during the war Britain had found it impossible to survive without the intelligent and widespread use of planned science. This finding had greatly fanned the planned science sentiment, and Crowther claimed that many of those who had been contemptuous of planned science ten years earlier were now demanding it as a necessity for survival. And thus, were Baker's "reactionary" attack on planned science to be successful it would deny Britain "the decisive weapon that can bring us victory, and him the opportunity to dream."[140] Baker's book was also reviewed in the *Scientific Worker*: "This book is the first public utterance of the Society [for Freedom in Science], of which little has been heard heretofore, and because it contains all the misconceptions and contradictions of the more thoroughgoing Anti-Planners (and for no other reason), it is desirable to devote some attention to it."[141] The anonymous reviewer went on to point out that the Association of Scientific Workers did not deny the importance of free inquiry in science, adding, however, that

> What we in the [Association of Scientific Workers] are working for is Freedom *for* Science, not only Freedom *in* Science. Freedom for science to get its giant forces to work on the needs and ills which surround us; freedom for scientific workers to direct their work where it is most needed instead of haphazard research to outstrip commercial competition or create an artificial demand; freedom for science to get its findings applied and not just left unheeded . . .; freedom to combat old prejudices, freedom from insularity and secrecy; and, lastly, freedom from want, both for science and the scientists, freedom to choose the best brains, and give them support without stint. If the freedom we work for comes to pass, there will be no need for the Anti-Planners to form a society to insist on choosing their own research. In their own field of endeavour, they will be part of a vast flood of discovery, and will realise how fully the end of scientific knowledge is the service of humanity.[142]

Tansley soon responded: "If 'nobody, least of all the AScW, would deny the importance in science of free inquiry and the chance discovery with unpredictable developments,' then the SFS has no quarrel with the AScW. Certainly the SFS has no quarrel with a campaign for Freedom *for* Science; it was founded because Freedom *in* Science, a prime necessity of healthy scientific life, was threatened, and not obscurely, by recent propaganda!"[143] The quarrel, however, had progressed too far for an easy reconciliation to be possible. About this time Bernal publicly alleged that Baker and the Society were "doing their best to frighten other scientists by bogies which they themselves have conjured up."[144] He protested that those who "are raising parrot cries for freedom in science and anti-planning are doing little service to science or civilization."[145]

In January 1943, the Association of Scientific Workers held a well-attended conference in London to consider not "whether science should be planned, but how it should be planned."[146] Throughout the event the Society for Freedom in Science was attacked. In his introduction to the published proceedings the Association's president, Sir Robert Watson-Watt, wrote:

> I am not one of those who apply to the declared anti-planners the unkind title of "The Chaotics." I believe in the perfectibility of the human mind; I believe—in moderation—in the inspirational value of hot baths. [Baker had written in *The Scientific Life* that scientists had sometimes had new ideas when taking hot baths.] And I believe that the anti-planners, when they have read this book, and thought a little, will realise that their campaign is directed, at best or at worst, towards Planned Sporadicism. If they can lead a crusade under that piebald banner they will at least advance a gaiety to which they have already made notable contributions.[147]

In requesting Blackett to prepare a paper for the conference, Reinet Fremlin, the Association's executive secretary, had asked him "to bring in a little 'anti-Bakerism.'"[148] Blackett and Polanyi were colleagues at the university in Manchester. On 28 October 1941, they had a heated meeting, and afterwards Polanyi wrote to Blackett:

> I am much worried by the hostile tone of your talk to me today. We have always disagreed, yet maintained an entirely genuine link of sympathy. I begin to doubt whether you still believe in this un-Marxist distinction. . . .
>
> I dare say if you and your friends had been as insistent on finding out the truth concerning the Soviet experiment, as I have been myself, we would now all face the future in a different world of mutual confidence. As it is, there seems to be little tradition left of sober and considerate argument, and only the desire left to deal a blow. Such is my profoundly sad impression today.[149]

In two drafts of Blackett's speech for the conference on planning, there are similar passages on the Society for Freedom in Science.[150] That in the later draft reads:

The Anti-Planners
A small but vociferous group of scientists who seem allergic to the word "planning." They appear to look forward to a world designed (not of course 'planned') to encourage them to make fundamental discoveries, while an admiring, but somewhat philistine world gaze on the new scientific Priestcraft practicing "The Scientific Life." The new revelation, now vouchsafed to them, that great discoveries often occur by chance, and are sometimes even stimulated by hot baths, seems to have blinded them to the real world or wars and revolutions in which we live. They apparently feel that the purity of their scientific virtue is likely to be ravished by the gangster planners of the AScW. We can almost imagine the future history of these unfortunates retold under some such title as "No hot baths for Dr. Baker."[151]

At the conference Hyman Levy also gave his assessment of the Society for Freedom in Science:

When I hear it argued that many of the most important scientific discoveries have been made by individuals quite unaware of the social importance and possible applications of their work, as if that were something to their credit, I cannot but think "Poor muts, that such clever people should be so ignorant." Those who would agitate from this, as some do, for scientific "freedom" in some abstract sense, agitate for anarchy and ignorance. They stress the uniqueness of scientists, as if uniqueness were not the most common of qualities.... Those who fear the destruction of the scientific spirit through regimentation have missed the meaning of science and the meaning of scientific history. We have always been regimented—albeit unconsciously—and we have always had a measure of freedom. To become conscious of these factors is to become more scientific, and to become more free in action. Those who would hide themselves from the new responsibilities thrown upon the scientific profession by seeking to remain individual workers within the four walls of their laboratory are mentally and emotionally still bound to an individualist stage of society. They do not stand aloof, as they may imagine: they obstruct. They are not free, but they are bound by an outdated tradition. In justifying their aloofness and their obstruction they oppose progressive development and are reactionary. It is a short step from that to organised opposition and fascism. Science has always been creative. In the past scientists have unconsciously fulfilled a social function. To-day they stand on the threshold of a period of consciously directed social effort. The greater their achievement in this respect, in its turn the more developed and the richer becomes their science.[152]

Throughout the war years, the debate on planning and freedom remained unresolved. On looking back at these times, Baker recalled that the Society for Freedom in Science suffered "misrepresentation, ridicule and active hostility."[153]

The Society for Freedom in Science was created in the winter of 1940-41 when a majority of British scientists favoured the left-wing view of planning of science. The Society's purpose was to combat the view, especially at the end of the war, when the Society feared that planning of science would be implemented by the government. During the war years, the Society remained a small, informal organization that officially accomplished little. However, through the individual contributions of its members, and especially of two of its leaders, John R. Baker and Michael Polanyi, it did much by way of criticizing the planning view and publicizing its own view of freedom in science. The Society's founders expected, and encountered, strong opposition. Its major and unrelenting opponents were found in the left-wing Association of Scientific Workers, but the Society was also opposed by *Nature*, and many prominent scientists including the leaders of the Royal Society refused to join it. Throughout the war years, the Society had little effect on the prevailing pro-planning view. Nevertheless, its efforts in those years constituted one of the more thought-provoking phases of the social relations of science movement.

---

1. Previous studies that discuss aspects of the Society for Freedom in Science include: Edward Shils, "A Critique of Planning: The Society for Freedom in Science," *Bulletin of the Atomic Scientists* 3 (1947): 80-82; Neal Wood, *Communism and British Intellectuals* (New York: Columbia University Press, 1959); Gary Werskey, "The Visible College: A Study of Left-Wing Scientists in Britain, 1919-1939," (Ph.D. diss., Harvard University, 1973); Robert Earl Filner, "Science and Politics in England, 1930-45: The Social Relations of Science Movement," (Ph.D. diss., Cornell University, 1973); Gary Werskey, *The Visible College: The Collective Biography of British Scientific Socialists of the 1930s* (New York: Holt, Rinehart & Winston, 1978).

2. J. D. Bernal, *The Social Function of Science* (London: Routledge and Kegan Paul, 1939). References here are to the first paperback edition (Cambridge, Mass.: MIT Press, 1967), p. xv.

3. Ibid.
4. Ibid.
5. Ibid.
6. Ibid., p. 242.
7. Ibid.
8. *The Objects of the Society for Freedom in Science*, 2d ed. (Oxford: Society for Freedom in Science, June 1946), p. 1.
9. A. G. Tansley, Speech given at informal meeting of Oxford members of the Society for Freedom in Science, New College, Oxford, 23 August 1941 (hereinafter referred to as: A. G. Tansley, 1941 speech), J. R. Baker Personal Papers (hereinafter referred to as: BPP).
10. Bernal, *The Social Function of Science*, p. 242.
11. J. R. Baker to Joseph Needham, 25 July 1942, Joseph Needham Papers, Cambridge University Library.
12. J. R. Baker, "Counterblast to Bernalism," *The New Statesman and Nation*, 29 July 1939, pp. 174–75.
13. Ibid., p. 174.
14. Ibid., 5 August 1939, p. 211.
15. M. Polanyi, "Rights and Duties of Science," *The Manchester School of Economic and Social Studies* 10 (1939): 175–93, 181–82. For a lengthy discussion of the conflicting views of Polanyi and Bernal, see M. D. King, "Science and the Professional Dilemma," in Julius Gould, ed., *Penguin Social Sciences Survey 1968* (Harmondsworth, England: Penguin Books, 1968), pp. 34–73.
16. Obituary of Michael Polanyi, *New York Times*, 24 February 1976, p. 38.
17. See list of documents relating to the Society for Freedom in Science made by J. R. Baker, 23 October 1974, Michael Polanyi Papers, Joseph Regenstein Library, University of Chicago (hereinafter referred to as: PP), Box 42.
18. J. R. Baker to forty-nine scientists, November 1940, BPP.
19. J. R. Baker to twenty-seven scientists, 6 December 1940, BPP.
20. Ibid.
21. J. R. Baker, note entitled "Society for Freedom in Science," 18 March 1941, BPP. Baker later wrote that Tansley "was always a good counsellor, using restraint when wild schemes would have damaged our cause, energetically supporting strong action whenever it could be successful. He was an eminently wise man, thoughtful, full of experience, moderate, optimistic" (J. R. Baker, "Sir Arthur Tansley, 1871–1955," *SFS Occasional Pamphlet* no. 16 [December 1955], p. 2).
22. J. R. Baker, note entitled "Society for Freedom in Science," 18 March 1941, BPP.
23. M. Polanyi and J. R. Baker, "Draft of statement to be handed to prospective members of the Society for Freedom in Science," (1 March 1941), PP, Box 42; A. G. Tansley, "Proposed Society for Freedom in Scienc," ([March] 1941), BPP; M. Polanyi, A. G. Tansley, and J. R. Baker, "Proposed Society for Freedom in Science," ([March] 1941), PP, Box 42.
24. An incomplete seventh draft is in BPP; an earlier one in PP, Box 21, SFS Folder.
25. *Proposed Society for Freedom in Science* (1941), PP, Box 21, SFS Folder.
26. Ibid., p. 2.
27. Ibid., p. 3.

28. A. G. Tansley, 1941 speech.

29. "SFS. Notice to Members," (October 1942), PP, Box 21, SFS Folder; "SFS. Notice to Members," (11 November 1943), BPP.

30. "SFS. Notice to Members," (January 1945), PP, Box 21, SFS Folder; "SFS. Notice to Members," (March 1946), BPP.

31. J. G. Crowther, *Science in Modern Society*, (New York: Schocken Books, 1968), p. 39.

32. *The Objects of the Society for Freedom in Science*, 2d ed. (June 1946), p. 5.

33. "SFS. Notice to Members," (March 1946), BPP.

34. J. R. Baker, "Note to SFS members," (April 1942); SFS Executive Committee minutes, December 1942; "SFS. Notice to Members," ([February] 1943), PP, Box 21, SFS Folder; "SFS. Notice to Members," (11 November 1943), BPP.

35. "SFS. Report on First General Meeting," (August 1942), PP, Box 21, SFS Folder.

36. "SFS. Statement of the Executive of the Provisional Committee," ([late 1941/early 1942]), p. 3, PP, Box 21, SFS Folder. Baker hoped that the branches, if formed, would consider such subjects as freedom in universities and research institutions and the financing of research in general (J. R. Baker to A. G. Tansley, 27 November 1941, BPP).

37. J. R. Baker, "The Objects of the Society for Freedom in Science," ([1943]) (being the first draft of the Society's 1944 circular of the same title), PP, Box 21, SFS Folder.

38. J. R. Baker to forty-nine scientists, November 1940, BPP.

39. J. R. Baker to twenty-seven scientists, 6 December 1940, BPP.

40. J. R. Baker to A. G. Tansley, 5 December 1940, BPP.

41. "SFS. Notice to Members" ([February] 1943), PP, Box 21, SFS Folder. The other chief function was to provide means for the discussion and publication of issues related to freedom in science. The two functions had been decided upon by the executive committee and agreed to by the general committee. See SFS Executive Committee minutes, December 1942, PP, Box 21, SFS Folder.

42. "SFS. Notice to Members," (11 November 1943), BPP; J. R. Baker, "The Objects of the Society for Freedom in Science," ([1943]) (being the first draft of the Society's 1944 circular of the same title), PP, Box 21, SFS Folder; *Nature* 154 (1944):48.

43. *Proposed Society for Freedom in Science* (1941), PP, Box 21, SFS Folder.

44. A. G. Tansley, 1941 speech.

45. Ibid.

46. J. R. Baker to A. G. Tansley, 22 November and 15 December 1941, BPP.

47. "SFS. Statement by the Executive of the Provisional Committee," ([late 1941/early 1942]), PP, Box 21, SFS Folder.

48. See also, "SFS. Report on First General Meeting," (August 1942), p. 2, PP, Box 21, SFS Folder.

49. A. G. Tansley to the editor, *Electronic Engineering* 15 (1942): 260. See also Tansley's remarks as recorded in "SFS. Report on First General Meeting," (August 1942), p. 1, PP, Box 21, SFS Folder.

50. *Proposed Society for Freedom in Science* (1941), p. 4, PP, Box 21, SFS Folder; M. Polanyi to Arthur Koestler, 23 May 1941, PP, Box 22.

51. M. Polanyi to Arthur Koestler, 15 May 1941, PP, Box 22.

52. A. Koestler to M. Polanyi, 13 August 1941, PP, Box 22.

53. Society for Freedom in Science, Executive Committee minutes, December 1942, PP, Box 21, SFS Folder.

54. J. R. Baker and A. G. Tansley, "The Course of the Controversy on Freedom in Science," *Nature* 158 (1946):574.

55. J. R. Baker, *Science and the Planned State* (New York: Macmillan Co., 1945), p. 9.

56. J. R. Baker, *The Scientific Life* (New York: Macmillan Co., 1943), p. 134; Michael Polanyi, *The Contempt of Freedom: The Russian Experiment and After* (London: Watts & Co., 1940). This was a collection of previously published essays including "The Rights and Duties of Science" and "Collectivist Planning" (1940), a criticism of central planning.

57. J. R. Baker, *The Scientific Life*, pp. 7-8, 7, 133.

58. Ibid., p. 140.

59. A. G. Tansley, 1941 speech.

60. "SFS. Statement by the Executive of the Provisional Committee," ([late 1941/early 1942]), PP, Box 21, SFS Folder.

61. Ibid.

62. J. R. Baker, "Memorandum to Committee Members," (April 1942); "SFS. Report on First General Meeting," (August 1942), pp. 1, 3, PP, Box 21, SFS Folder.

63. "SFS. Notice to Members," (October 1944), PP, Box 21, SFS Folder.

64. In 1942, Tansley gave the Herbert Spencer Lecture at Oxford University on "The Values of Science to Humanity." This was later published as a pamphlet and extracts from it were carried in *Nature* (published by George Allen & Unwin, 1942; see *Nature* 150 [1942]:104-10). Although neither the SFS nor any of its members were mentioned, the lecture nevertheless presented the SFS case in discussing and criticizing the planning movement.

65. *SFS Bulletin*, no. 6 (December 1949); no. 15 (December 1955); *SFS Occasional Pamphlet*, no. 17 (November 1957), p. 2; *Nature* 158 (1946):574. In connection with an issue yet to be mentioned below, Polanyi wrote: "It was difficult to get a hearing for opposing views. Those who knew about the persecution of biologists in Soviet Russia would not divulge their information. My writings and those of J. R. Baker which, from 1943 on, exposed this persecution were brushed aside as anti-Communist propaganda" (*Science, Faith, and Society* [Chicago: University of Chicago Press, 1964], p. 9).

66. "SFS. Notice to Members," (11 November 1943), BPP.

67. M. Polanyi, "The Automony of Science," *Memoirs and Proceedings of the Manchester Literary and Philosophical Society* 85 (1943):19-38.

68. In his presidential address to the council of the Association of Scientific Workers in May 1943, Watson-Watt said that there was no alternative to planning but the dream world of "planned sporadicism." *Nature* 152 (1943):141-42. See also below, p. 298.

69. The British Way Pamphlets, no. 7 (Glasgow: Craig & Wilson, 1943).

70. Material from this lecture was subsequently published as parts of chapter 4 ("Science under Totalitarianism") and chapter 5 ("The Duties of Scientists to Society") of his *Science and the Planned State* (1945).

71. An expanded version of this address became chapter 2 ("The Values of Science") of *Science and the Planned State*.

72. J. R. Baker, "The Scientist and the Community," *Electronic Engineering* 15 (1942-43): 147-48; J. D. Bernal, "The True Meaning of Planned Science," Ibid., 242-43; J. R. Baker to the editor, Ibid., p. 348.

73. M. Polanyi, "The Autonomy of Science," p. 31.

74. Ibid., p. 34.

75. Ibid., p. 31.

76. Ibid., p. 37.

77. *SFS Bulletin*, no. 6 (December 1949), p. 3.

78. See Herman Ould, ed., *Freedom of Expression: A Symposium* (London: Hutchinson International Authors, 1944), pp. 118–22.

79. "SFS. Message from the Executive Committee to all members of the Committee and to Professor P. W. Bridgman and Dr. A. Pijper," (29 September 1944), PP, Box 21, SFS Folder.

80. "SFS. Notice to Members," (January 1945); "SFS. Statement by the Executive of the Provisional Committee," ([late 1941/early 1942]), p. 2, PP, Box 21, SFS Folder; J. A Crowther, A. G. Tansley, and J. R. Baker to the editor, *Science* 101 (1945):273.

81. J. R. Baker, *Science and the Planned State*, p. 20.

82. *Nature* 148 (1941):739.

83. In 1946, when Bernal was still claiming that the Soviet Union had assisted and upheld freedom of investigation, *Nature* replied that when it thought of the Soviet genetics controversy it felt rather doubtful (*Nature* 158 [1946]:566).

84. J. R. Baker, personal interview with the author, Oxford, 12 August 1975.

85. M. Polanyi, "Science and the Decline of Freedom," *The Listener*, 1 June 1944, p. 599.

86. See also, M. Polanyi, "Science—Its Reality and Freedom," *Nineteenth Century and After* 135 (1944):83.

87. M. Polanyi, "Science and the Decline of Freedom," p. 599.

88. F. A. Hayek to M. Polanyi, 1 May 1941, PP, Box 22.

89. F. A. Hayek to J. D. Bernal [June 1939], J. D. Bernal Papers, Box 24, Folder B1.2.3.

90. F. A. Hayek to M. Polanyi, 1 July 1941, PP, Box 22.

91. J. G. Crowther, *The Social Relations of Science*, 2d rev. ed. (Chester Springs, Pa.: Dufour, 1967), p. 170.

92. Ibid., pp. 247–48.

93. Ibid., pp. 387–88.

94. Ibid., p. 382.

95. M. Polanyi, "The Growth of Thought in Society," *Economica* 8 (1941):437.

96. Ibid., p. 438.

97. Ibid., p. 448.

98. Ibid., pp. 441–42.

99. *SFS. Occasional Pamphlet*, no. 16 (December 1955), p. 2.

100. "SFS. Message from the Executive Committee to all members of the Committee and to Professor P. W. Bridgman and Dr. A. Pijper," (September 1944), PP, Box 21, SFS Folder.

101. "SFS. Notice to Members," (April 1945), PP, Box 21, SFS Folder.

102. "SFS. To all Committee Members, from the Secretary," (25 March 1945), PP, Box 21, SFS Folder.

103. "SFS. Notice to Members," (March 1946), BPP.

104. See *SFS Bulletin*, no. 1 (August 1946).

105. A. G. Tansley to Sir Henry Dale, 27 May 1941, Sir Henry Dale Papers, The Royal Society, Box 11, Folder 6.

106. A. V. Hill to A. G. Tansley, 6 June 1941, Dale Papers, Box 11, Folder 6.

107. Ibid.

108. Ibid.

109. Sir Henry Dale to A. G. Tansley, 1 August 1941, Dale Papers, Box 11, Folder 6.

110. Ibid.

111. A. G. Tansley to Sir Henry Dale, 4 August 1941, Dale Papers, Box 11, Folder 6.

112. Alan D. Beyerchen, *Scientists under Hitler: Politics and the Physics Community in the Third Reich* (New Haven, Conn.: Yale University Press, 1977), p. 19.

113. M. Polanyi to M. Born, 29 July 1941, PP, Box 20, "Memorabilia" Folder.

114. M. Born to M. Polanyi, 31 July 1941, PP, Box 20, "Memorabilia" Folder.

115. M. Born to M. Polanyi, 30 January 1942, PP, Box 20, "Memorabilia" Folder.

116. M. Born to M. Polanyi, 14 June 1944, PP, Box 21, SFS Folder.

117. Baker prepared and issued to the SFS membership a confidential report on the Conference—"SFS Newsletter, October 1941. Meeting of the British Association," PP, Box 21, SFS Folder.

118. Sir Richard Gregory to M. Polanyi, 29 January 1942, PP, Box 21.

119. Solly Zuckerman, *Scientists and War* (London: Hamish Hamilton, 1966), pp. 141–42. Although Zuckerman suggested that the Society had to some extent been tilting at windmills (p. 142), he nevertheless found the distinction between pure and applied science which the Society, and particularly Polanyi, had fought to preserve, useful and necessary (pp. 144, 147).

120. *Nature* 149 (1942):261.

121. J. R. Baker to Joseph Needham, 18 June 1942, Joseph Needham Papers, Cambridge University Library, Section 4.

122. J. R. Baker to Joseph Needham, 25 July 1942, Needham Papers, Section 4.

123. Ibid.

124. C. P. Snow, "John Desmond Bernal," in Charles Coulston Gillispie, ed., *Dictionary of Scientific Biography* 15, supplement 1 (New York: Charles Scribner's Sons, 1978), p. 18.

125. A. G. Tansley, 1941 speech.

126. *Nature* 148 (1941):679.

127. SFS Bulletin, no. 15 (December 1955), p. 1; *The Society for Freedom in Science: Its Origins, Objects and Constitution*, 3d ed., (1953), p. 9.

128. *Nature* 144 (1939):973.

129. *Nature* 147 (1941): 637–38; for additional criticism in an editorial by Brightman see ibid., pp. 551–53.

130. *Nature* 148 (1941): 497.

131. *Nature* 149 (1942):228.

132. *Nature* 150 (1942):302. Polanyi repeatedly defended the distinction. See, for example, his "Rights and Duties of Science"; "Science—Its Reality and Freedom"; and "The Planning of Science," *Political Quarterly* 16 (1945):316–28.

133. *Nature* 150 (1942):302.

134. Ibid., p. 305.

135. Association of Scientific Workers, Executive Committee minutes, 4 January 1942, ASW/1/2/23/1.

136. *Scientific Worker* 13 (1941):117.

137. Ibid.

138. Ibid., p. 146.

139. For correspondence in relation to Nelson's review see ibid., pp. 157, 163, 179.

140. *New Statesman and Nation*, 15 August 1942, p. 114.

141. *Scientific Worker*, July 1942, p. 26.

142. Ibid., p. 27.

143. *Scientific Worker*, October 1942, p. 56.

144. J. D. Bernal, "The Meaning of Planned Science," *Electronic Engineering* 15 (1942):242.

145. Ibid., p. 243.

146. Sir Robert Watson-Watt, "Introduction" in the Association of Scientific Workers, *The Planning of Science* (London: The Association of Scientific Workers, 1943), p. 4.

147. Ibid.

148. Reinet Fremlin to P. M. S. Blackett, 16 December 1942, P. M. S. Blackett Papers, Royal Society Library, E 7.

149. M. Polanyi to P. M. S. Blackett, 28 October 1941, Blackett Papers, J 65.

150. Both are in the Blackett Papers, E 7.

151. P. M. S. Blackett, notes for the Association of Scientific Workers' conference, 30 January 1943, Blackett Papers, E 7. In the published proceedings of the meeting Blackett's paper merely mentioned "a certain number of scientists who oppose any idea that people should plan anything at all" (Association of Scientific Workers, *The Planning of Science*, p. 81). However, a report on the conference reported him as saying: "There's a small but vociferous group of anti-planners who fear that the purity of their scientific virtue is likely to be ravished by the gangster planners of the Association of Scientific Workers" (*Picture Post* 18 [13 February 1943]:17).

152. Association of Scientific Workers, *The Planning of Science*, pp. 85–86.

153. *SFS Bulletin*, no. 6 (December 1949), p. 2; no. 15 (December 1955), p. 1.

10

The Central Organization of Science in Peacetime

The fifth and final phase of the social relations of science movement occurred during the period 1943–47 and concerned the central organization of science in peacetime. The establishment of some form of central organization was desired not only by most scientific parties, including the Royal Society, Association of Scientific Workers, and Parliamentary and Scientific Committee, but also, and most importantly, by the postwar Labour government. The government's readiness to strengthen the bonds between science and the state was a welcome change to almost all participants in the social relations of science movement. Given the general gratitude felt toward science at war's end, a Conservative government would probably have been just as willing. In 1943, Churchill had written to A. V. Hill: "It is the great tragedy of our times that the fruits of science should by a monstrous perversion have been turned on so vast a scale to evil ends. But that is no fault of science. Science has given to this generation the means of unlimited disaster or of unlimited progress. When this war is won we shall have averted disaster. There will remain the greater task of directing knowledge lastingly towards the purposes of peace and human good. In this task the scientists of the world, united by the bond of a single purpose which overrides all bounds of race and language, can play a leading and inspiring part."[1]

This chapter relates how the Royal Society, and especially its president, Sir Henry Dale, endeavored from 1943 to ensure the continuation in peacetime of an improved form of the wartime Scientific Advisory Committee. From 1943, the Association of Scientific Workers also gave increasing thought to the postwar organization of science, and it too wanted to have some central scientific body established within government in peacetime. The views of the Society and Association were known to the new Labour government when in November 1945 it established the Committee on Future Scientific Policy, charged with considering the policies that should govern the use of scientific manpower and resources during the next ten years. Because of the representations of the Society and the Association, the Committee was also instructed to consider the effectiveness of the government machinery for the formulation of scientific policy. Although the representatives of the Royal Society, Association of Scientific Workers, and government science who sat on the Committee believed that a policy-making body should be created, they held conflicting views about what its nature should be. In January 1947, on the Committee's recommendation, the Labour government formed the Advisory Council on Scientific Policy, as the peacetime successor to the Scientific Advisory Committee, to advise the Lord President on the formulation and execution of government scientific policy. The Council was the principal achievement of the social relations of science movement, and its creation was widely applauded within the scientific community.

The demise of the Engineering Advisory Committee had coincided with the appointment of the three scientific advisers to the Ministry of Production. At that time, A. C. G. Egerton, fearing that a similar fate might befall the Scientific Advisory Committee, discussed the possibility with Sir Henry Dale, who also wanted to see the Committee maintained. Dale raised the matter with the committee's chairman, R. A. Butler. He pointed out that although the three advisers would most certainly deal with some matters that otherwise would have come to the Committee, there were other issues concerning the function of science and its value to the government, both during the war and in the postwar period, on which the Committee could still give valuable advice, and which would be quite unsuitable for the three advisers to consider.[2] One such matter, already under study by the Committee, was the question of governmental and other research workers taking out patents, and Dale believed there were others of greater importance. Dale and Egerton's concern was unnecessary, however, for the government shared the Royal Society's belief in the usefulness of the Committee, which functioned throughout the war.

In December 1942, Butler resigned the chairmanship of the Committee. With his time increasingly occupied by the work of the Department of Education, he had found it necessary to give up a number of other interests. He told the Committee that he was sorry to relinquish the chairmanship, which was hardly true as he had not relished the position.[3] In his memoirs, Butler refers only once to the Committee. He explains how in March 1943 he spoke with Lord Cherwell about it. "The Prof said that it was a pity that this Committee had ever been appointed, that it had been called into being only to appease the *amour propre* of the scientific establishment, and that he himself did not think it worth a moment of his worry."[4] What Butler himself may have thought or replied at the time he does not reveal, but he has since stated that the Committee "did useful work in acting as a focus for scientists, and that Lindemann's polemical remarks were not only unjustified but ungenerous."[5]

With Churchill's approval, Anderson directed that the president of the Royal Society should henceforth be chairman of the Committee and the Lord President of the Council its president.[6] These changes made the Scientific Advisory Committee's position in government similar to that of the economic section of the War Cabinet that Dale had approvingly spoken of earlier.[7] Now scientists had direct access to the War Cabinet, and their potential influence appeared to be much enhanced.

When in 1943 the Official Committee on the Machinery of Government began to prepare for the postwar period, Dale was invited to submit a note on the Scientific Advisory Committee's future. Soon thereafter he decided that the Committee should review again, after four years of war, the working of government research organizations and recommend which parts of these organizations should be maintained in peacetime. To carry out the review, Dale considered it necessary to increase the membership of the Committee, so he circulated to the members a memorandum that was part of the note he had submitted to the Official Committee. It contained suggestions on the Committee's composition and future, including its possible continuation after the war.

The memorandum noted that the Committee had originally been appointed "after prolonged discussions and negotiations" and that its constitution might be said to represent "an acceptable residuum, rather than a theoretical ideal." Although its constitution of ex officio members had enabled the Committee to come into existence and to get to work under war conditions, Dale continued, it could not be considered suitable for permanent adoption in peacetime. The first and most obvious anomaly was that one-half of the membership consisted of the secretaries of the three research councils—the Department of Scientific and Industrial Research, the Medical Research Council, and the Agricultural Research Council. It was clearly not in theory desirable that a committee that was likely to be called upon in the future, as it had been in the past, to give independent advice to the government on the efficiency of the three research councils, and to make recommendations concerning the scope of their activities, should consist as to one-half its membership of their chief executive officers.[8] These reasonable views were not, however, accepted by all Committee members. One of the chief executive officers suggested that the Committee should not regard it as part of its function to inquire into the domestic affairs of the councils.[9] If the activities of one of these organizations were under consideration, he said, the secretaries of the other two could be counted upon to give unprejudiced opinions.

Dale's memorandum also suggested, in regard to the Royal Society's representation, that the scientific knowledge and experience represented on the Committee be widened and the proportional over-representation of official scientific administration reduced by the appointment of four or five additional nongovernment scientists, each serving for a fixed term and retiring in rotation. This received a mixed reception. A. V. Hill favored the appointment of three additional members, and

Egerton and Sir Edward Appleton two, but Sir Edward Mellanby and W. W. C. Topley were opposed to any additions.

In regard to the chairmanship of the Committee, the members agreed that for its permanent constitution it would be desirable to recommend that there should be a ministerial chairman, with the understanding that the president of the Royal Society would be vice-chairman. This would suggest that the Committee members found the post-Butler constitution less satisfactory than the earlier one, at least when Hankey was chairman. There was also general agreement that after the war the Committee should be made responsible either to the Prime Minister or to the Lord President of the Council.

Dale reported the various views of the Committee members when he appeared before the Official Committee on the Machinery of Government. He also informed it of the Committee's intention to make a review of government research organizations. The Official Committee requested that the report be made available to it as soon as possible.[10] From late 1943, much of the time of the Scientific Advisory Committee was given to this work, and a report was submitted in August 1944.[11] In this task the Committee secured, on Dale's suggestion, the help of the three scientific advisers to the Ministry of Production, who naturally had detailed knowledge of the actual working of the research organizations of the service and supply departments.[12]

The Committee continued with various other matters for the remainder of the war, during which time it did not concern itself with either its continuation or composition in the postwar period. But in June 1945 an incident occurred that led to these questions being considered once more. Immediately after the war ended in Europe, the Academy of Sciences of the USSR invited a party of British scientists to the celebration of its 220th anniversary. At the last moment, the government refused exit permits to eight of the party of twenty-nine, saying that they were engaged in work of the greatest importance in the production of war materials and in research. The eight included J. D. Bernal and P. M. S. Blackett.[13] This action angered many scientists, including Dale, Hill, and Egerton, who requested a meeting with the Lord President, now Lord Woolton.[14] Prior to this meeting, Woolton was briefed by an assistant secretary in the Lord President's office, Martin Teall Flett, who advised him that Dale would raise the question of the role of scientists in general, and of the Scientific Advisory Committee in particular, in the formulation of government policy.[15] Although Dale accepted the necessity for the decision to withhold exit permits from the scientists invited to the USSR, he nevertheless thought that the matter could have been better handled if the government had had a proper respect for the scientific world. In particular, Dale and his Royal Society colleagues felt that the Scientific Advisory Committee was being neglected, and consequently they con-cluded that the government was regarding it more as a conven-ient piece of window dressing than as a source of advice.

Flett assured Woolton that window dressing "was not absent from the mind of the Governmet in setting up the Committee but it has been (and continues to be) of more substantial value." In recent months the Committee had, for example, usefully considered the organization of medical research in the three services after the war, the preparation of an account of the nation's scientific work during the war,[16] the control of radium, and the conditions necessary to attract staff to research work in biological warfare. The Committee's value in dealing with comparatively long-term questions like the first was that it could take its time, consult all recognized experts in the field concerned, and produce truly authoritative recommendations. Moreover, it was currently fulfilling a useful function in bringing to the notice of the government proposals that originated in the Royal Society and other nongovernmental scientific organizations.

Flett thought, however, that Dale, Hill, and Egerton wanted the Committee to participate in major political decisions affecting science and in the selection of candidates for the more important scientific appointments (as provided for by the Committee's second term of reference). Flett suggested to Woolton that it must be made clear to the Royal Society that the Committee could not demand to be consulted on these questions as a matter of right. The minister responsible for a given matter must, as Sir John Anderson had said in 1942, be free to seek his advice from whichever body he might consider most suitable to provide it. The Committee, Flett added, would often not be the most suitable body to advise on either of these types of

The Central Organization of Science in Peacetime · 313

questions. For example, in regard to the first, the Royal Society's three representatives were not in sufficiently close touch with the working of the government machine to be able to advise very usefully on questions of day-to-day policy. Flett did not think, for instance, that the Committee could have been of any use in considering the case of the eight scientists who were not permitted to go to the USSR. In regard to the recommending of candidates for important appointments, there was the difficulty that in any given case at least some of the Committee's members would be ignorant of the subject involved. In addition, departments tended to prefer using ad hoc selection boards whose members had some substantial contribution to make to the ultimate decision.

Flett suggested how Woolton might reply if Dale were to raise these matters. No minister could be fettered in reaching a decision for which he was responsible by being bound to seek advice from one particular quarter. The Committee could not therefore claim to be consulted on any specific question or type of question as a matter of right. Woolton should say, however, that he appreciated the value of the services that the Committee was performing for the government and that he was eager that it should be used to the greatest possible extent. Finally, if Dale wished, Woolton should say he was prepared to remind his colleagues of the existence of the Committee and to invite them to make full use of it in considering any scientific problems with which they might be confronted.

These considerations show that although the Committee had been created by the government in 1940 as a matter of expediency, its performance over almost five years had demonstrated that it fulfilled a real need. They also contradict the view that the Committee was of little use and importance.[17] In spite of the war in Europe being over at this time, there was no suggestion that the Committee be disbanded. The nature of its future role was still, however, not settled. Scientists and government officials alike accepted that its role should be advisory, but the Royal Society wanted the Committee to be able to offer advice in areas wider than the government thought it necessary or desirable to grant. Dale considered the government much too prone to seek advice from government scientists such as Appleton and not to pay sufficient regard to the Royal Society and the universities.[18]

When the Royal Society contingent of Dale, Hill, and Egerton met with Woolton on 22 June, Dale referred to the constitutional position of the Scientific Advisory Committee.[19] He said that the Committee could be of considerably more use to the government than it was. One of the reasons for its creation had been for it to advise the government on policy questions affecting science, but very few questions concerned with the larger issues of policy had been referred to it. Its advice had been sought on occasion by the service departments and by such bodies as the Machinery of Government Committee, but for the most part, it had had to find its own work and had served mainly as a channel through which proposals, originating outside the government, could be brought to the attention of the Lord President and through him to his colleagues in government. Dale said that there were a number of problems upon which the Committee could now usefully be asked to advise, and cited the question of the attitude to be adopted towards German scientists and fundamental science in Germany. He also mentioned his concern that the German scientific instrument industry might be regarded as a dangerous war potential and so destroyed. Dale finally stated, and Hill concurred, that the composition of the Committee was unsuitable.

The matter of British policy toward German industry serves to illustrate one of the sources of the Royal Society's frustration. A policy did exist, but the Scientific Advisory Committee had not been officially informed of it, let alone asked to contribute to it. In July 1944, Cherwell had told Dale that a panel of scientists was to be appointed to advise the government on the policy to be adopted toward German industry with a view to restricting Germany's future war potential. Cherwell had suggested the names of certain eminent scientists for the panel and had asked Dale as chairman of the Scientific Advisory Committee for his opinion as to their suitability for this work. The Committee subsequently considered the proposed panelists and suggested some changes.[20] However, this method of operating could hardly have pleased the members of the Committee, who were clearly in an inferior position to Cherwell. The incident serves to illustrate Hill's assessment of the Committee: "The War Cabinet Scientific Advisory Committee undertook in fact a number of important tasks; though it could have been much more useful if a scientific courtier had not monopolized the grace and favour of the Prime Minister."[21]

In replying to Dale, Hill, and Egerton, Woolton said that he was most eager to advance by every means in his power the harnessing of scientific knowledge to the government machine. One question was how to ensure that the Committee received appropriate material with which to work. He welcomed Dale's suggestion that he should attend the Committee's next meeting in his capacity as its president and discuss with it ways and means of increasing the contributions that it might make to the formulation of policy. Woolton suggested that at that meeting the Committee should also consider the question raised by Dale concerning Britain's attitude to German scientists and fundamental science.[22]

Woolton subsequently asked for the record of evidence that Dale had previously given before the Machinery of Government Committee on the place of science in government. He also instructed Flett to find out what steps were being taken to formulate British policy in regard to fundamental science, scientists, and the scientific instrument industry in Germany. In writing to the DSIR on 22 June regarding these matters, Flett informed Appleton of Woolton's meeting with Dale, Hill, and Egerton. Of the meeting of the Scientific Advisory Committee that Woolton was to attend, he wrote: "As, however, it seems probable that this meeting cannot take place until early in August, we may have to advise the Lord President that, at least as far as our short term policy [regarding German science] is concerned, a decision cannot wait until [the Scientific Advisory Committee] has met."[23] Here was an urgent matter in which the Committee had indicated a strong interest and which could have been referred to it.

When August arrived, Woolton and the other members of Churchill's Caretaker Cabinet were no longer in office, the Conservatives having surprisingly been defeated by Labour in July's general election. The new government, with Clement Attlee as Prime Minister and Herbert Morrison as Lord President, was expected to be most friendly to science, which it was. Morrison has stated that as Home Secretary, and particularly as Minister of Home Security in the War Cabinet, it was his duty to attend many meetings of scientists chaired by Churchill and that it was a revelation to him to see the extent to which science had been mobilized by the government for the achievement of victory.[24] By the time he became Lord President, it was his belief that science and scientists had a great contribution to make to the well-being of the country in peace as well as in war.[25] Soon thereafter he drew cheers in the House of Commons when he said that the government was determined to do all it could, not in a sense of charity to science, but in the sense of helping, encouraging, and even inciting science to play its great part in the adventurous days ahead.[26]

During the closing years of the war, the Association of Scientific Workers, like the Royal Society, concerned itself with the postwar organization of science in Britain. In November 1943, the Association's executive committee endorsed the proposal of A. H. Bunting, an assistant chemist at the Rothamsted Experimental Station, that in view of the numerous postwar plans being advanced, the Association should issue a policy statement indicating that it stood for increased national production, not to provide profits for the few but to improve the welfare of the people.[27] The executive committee agreed that although the Association's objectives adequately showed "which side it was on in the social struggle," its position should be made very clear in any plans it issued. The social relations committee, enlarged by the addition of executive committee members including Bunting, W. A. Wooster, E. D. Swann, and N. W. Pirie, now a virus physiologist at the Rothamsted Experimental Station, was charged with drawing up a statement on the postwar planning of science.

The committee prepared a lengthy memorandum, "A Post-War Policy for Science," for the meeting of council in May 1944. The hope had been to have it discussed and accepted as Association policy, but wartime conditions delayed its printing and circulation to branches. Consequently, it was adopted only as a preliminary statement.[28] The council, however, did adopt a statement by the executive committee concerned largely with international postwar policy for science. Its central idea was that "science in the service of the people can create for them wealth and prosperity hitherto unknown."[29] P. M. S. Blackett was now the Association's president, and he told the council that for scientists to evolve an effective policy for postwar science it was essential that they have a thorough understanding of politics and economics.[30]

In February 1945, the social relations committee sponsored a two-day public conference on "Science in Peace." This, the third national conference held by the Association during the war, was designed to make the Association's postwar policy for science known to scientists, trade unionists, and the public.[31] Blackett chaired the first session on "Science and Production," Sir Robert Watson-Watt the second on "The Future Development of Science," and Hyman Levy the third on "Science in Everyday Life."[32] As was now customary, J. D. Bernal summed up the proceedings. A resolution, carried by acclamation, declared in part the conference's conviction that "in this country, the advance and efficient application of science requires democratic planning largely by scientists themselves. For this purpose we propose a Central Research and Development Council, under the authority of the Lord President."[33]

Later in 1945, the social relations committee's pamphlet *The Co-Ordination and Use of our Scientific Resources* elaborated upon the Association's idea of a central council.[34] By then, and much to the Association's pleasure, Labour was in power, and the pamphlet argued that the government's positive policy of national economic development required "an equally positive and parallel policy for science."[35] Elsewhere Blackett declared that in general the Association's task "is to give all possible help to the Labour Government in the carrying out of its progressive political programme and its expansionist economic policy. This needs the wider application of science in all branches of our society, and this again demands the better organisation at the government level of the scientific resources of the country."[36]

As responsibility for the various sectors of the national economy rested with ministers, the social relations committee's pamphlet argued, so each minister required the appropriate scientific organization within his own department. These needs could be met by the adoption of the existing system and the addition to it of a central scientific office at Cabinet level and further machinery to coordinate government, university, and industrial science. The Association rejected the alternative idea of a ministry of science, which had the support of some members of the Parliamentary and Scientific Committee[37]: "If all scientific work were done in a single Ministry, the rest of the Ministries would be divorced from science and the valuable responsibility of Ministers for an essential element of their own work would be lost."[38] The proposed central office's functions would be to inform the Cabinet of the progress of science and of the availability of scientific personnel, to suggest adjustments and new programs, and to arrange for the implementation of plans and the provision of adequate personnel. Three research councils of the physical, biological, and, significantly, social sciences would be created to do background research in areas of interest to more than one ministry. An interdepartmental committee would coordinate government science; its members would include senior scientists from each ministry, the secretaries of the three science councils, and a chairman who would also be head of the central office. Finally, the central office would have its own scientific staff that would be advised by a small committee of leading independent scientists conversant with government policy.

In the summer of 1945, a report on postwar science in the United States, *Science—the Endless Frontier*, drawn up by the head of the American wartime science organization, Vannevar Bush,[39] was studied with interest in Britian. The American proposals intensified the desire to have a sound postwar organization for British science. Meeting on the evening of August 20, the executive committee of the Association of Scientific Workers resolved that its president, Blackett, should write to the Lord President, Morrison. On the following day, the Association's general secretary, Reinet Fremlin, wrote to Blackett:

> Sir Robert Watson-Watt telephoned this morning to say that he had spoken to [Sir Stafford] Cripps last night after leaving our meeting. Cripps advised that your letter to Morrison should be sent in immediately, to-day if possible. It appears that Sir Robert had passed to Cripps a copy of the Bush report some time ago, and Cripps had read this with great interest and brought it to the attention of the Prime Minister, and that within the next couple of days there would be a Cabinet meeting on the subject. Cripps agreed that it was Morrison to whom you should write, but suggested that it should be very soon so that Morrison would be primed.[40]

Blackett wrote to Morrison the following day, August 22:

I have been asked by the Executive Committee of the Association of Scientific Workers to express to you their views as to the desirability of setting up within the Government machine some central organisation for the better planning and direction of the scientific resources of this country as a whole. Much discussion has taken place in recent years on the need for this, particularly by the Parliamentary and Scientific Committee, the British Association, and the Association of Scientific Workers. Recently Dr. Vannevar Bush's report "Science, the Endless Frontier" to the President of the United States has given a valuable lead to much that should be done in this country, particularly in relation to University education. All these investigations suggest the need of some type of Research and Development Council, but there are clearly many different ways in which such an organisation could function.

We wish to suggest that a useful first step would be for the Cabinet to set up a small investigative commission to study the best form of organisation, and to outline its main functions. This commission should be asked to report at an early date, but should consult as many of the relevant bodies concerned with science as possible.

We feel that the commission might be most effective if it was composed of the relevant department representatives, together with a few independent scientists; these should we believe be chosen for their wide understanding of the increasing part that science must play in the life of the country if the programme of the Labour Government for social advancement is to be quickly fulfilled.[41]

Soon thereafter another approach was made to Morrison by a personal friend, Solly Zuckerman, now Professor of Anatomy at the University of Birmingham, who arranged for him to meet with a few members of the Tots and Quots, including M. L. E. Oliphant, Poynting Professor of Physics at Birmingham, who had played a major role in the development of the magnetron.[42] The meeting discussed the need for a "science secretariat" to direct the country's peacetime scientific efforts, and Morrison requested Zuckerman to prepare a plan for such a secretariat. In complying, Zuckerman had the help of Bernal, Blackett and C. H. Waddington in addition to Oliphant. Their four-page document, completed on 16 September, stated that it was probable that the Royal Society would suggest that an advisory body such as the Scientific Advisory Committee or the council of the Royal Society itself should undertake the task.[43] Similarly, the Department of Scientific and Industrial Research might suggest that it could perform the necessary duties. In the Tots and Quots group's opinion, however, the only profitable solution would be for the government to agree to a central secretariat composed of individuals acceptable to scientists because of their scientific attainment and to the government because they could be relied upon to give whole-hearted support in the implementation of its policies. The group's judgment of the Scientific Advisory Committee, with whose workings none of them was familiar, was unfairly harsh, in contrast to Whitehall's positive assessment: "This body . . . was found in practice to lack both initiative and authority, and its achievements were negligible. Its failure was virtually recognized in the appointment of three scientific advisers to the Ministry of Production." Neither did the group have a high estimate of the work of the advisers, with which they were again not familiar: "They were provided with insufficient staff and such limited powers that they proved effective only in dealing with a few detailed problems."[44]

Meanwhile, by early September the government had already decided to act, and on the seventh Morrison agreed that Sir Edward Bridges should approach Sir Henry Dale concerning the creation of a small, high-level committee to make recommendations as to how the country's available scientific manpower should be used.[45] A few days later, Bridges, who knew of Blackett's letter to Morrison, considered various other items bearing upon the demand for some central machinery to plan the best use of the country's scientific manpower.[46] A question to the Prime Minister, due to be answered in Parliament on October 10, asked about the government's research program and about arrangements for the release of scientists from the Forces. The Parliamentary and Scientific Committee had solicited the Lord President's comments on its proposal to appoint a committee to make a survey of research and development in British industry and to consider how industry's postwar demand for scientific personnel and laboratories could best be met. Finally, the Minister of Labour had invited the Lord President to summon a meeting attended by themselves, the Chancellor of the Exchequer, and the Minister of Education, to discuss implications arising from the recommendation of the Committee on Further Education and Training that Britain must increase its output of graduates by approximately fifty percent. The Minister of Labour believed that the prospective deficiency in science and engineering graduates over the next few years required urgent consideration.

On 14 September, Sir Henry Dale met with Bridges and Sir Alan Barlow, at their invitation, to discuss the creation of a committee to consider the question of scientific manpower.[47] Dale agreed that "a more high-level and influential body than the Scientific Advisory Committee" was needed. He added that he would have liked to have used the occasion to reconstitute the Scientific Advisory Committee but thought he would have difficulties with some of his colleagues, presumably the secretaries of the three research councils, in putting this across. Bridges suggested that there might be (a) a small high-level body of "outside" scientists of the type required for the job now proposed, and (b) allied to it a panel of government representatives, including the secretaries of the research councils, and perhaps a secretary of the Royal Society. The immediate concern was to think of the right people for the high-level body.

After consulting with colleagues informally, and after a second meeting with Bridges and Barlow, Dale forwarded a list of names on 24 September.[48] A week later, and following discussions with Barlow, Bridges submitted the following list of names to Morrison: Appleton, Blackett, Dale, Egerton, Sir George Nelson, chairman and managing director of English Electric, and Tizard. None of them, with the "possible exception of Tizard," said Bridges, would make a good chairman. "The ideal would be to get a layman, conversant with Governmental affairs and occupying an independent position, whose name would carry weight if and when the report [on scientific manpower] is published." The only really suitable person that Bridges and Barlow could think of, provided it was feasible to ask him, was Sir John Anderson. But Attlee thought that "there might be criticism if we put too much on Anderson."[49] About a month later, Morrison asked Barlow to act as temporary chairman and to initiate as quickly as possible such preparatory work as the new body, called the Committee on Future Scientific Policy, might consider necessary.[50] Barlow soon became the permanent chairman. The other members were Appleton, Blackett, Egerton, Geoffrey Crowther, editor of the *Economist*, and Sir George Nelson. Tizard declined an invitation to serve;[51] Dale had earlier withdrawn his name from consideration.[52] No panel was appointed to work with the Committee, which in any case included representatives originally suggested for a panel, namely, a secretary of the Royal Society, Egerton, and a secretary of a research council, Appleton. They were the only members of the Committee who were also members of the Scientific Advisory Committee.

On 14 November, a few days after the invitations to serve on the Committee went out, Dale wrote to Flett.[53] He and Hill would automatically retire from the Scientific Advisory Committee on November 30, when their successors in the Royal Society would take office, and he desired to speak with Morrison about this. A meeting was arranged for 27 November, and Flett prepared a minute for Morrison which he forwarded together with a copy of the note on the Scientific Advisory Committee that Dale had prepared for the Machinery of Government Committee in 1943. Flett summarized the objections Dale had made at that time against the Committee's constitution. First, Dale did not think that in choosing its president the Royal Society should have to take into account the fitness of candidates to also fill the entirely different office of chairman of the Committee. Second, Dale felt that the official element in the Committee was too large. The secretaries of the research councils already had the ear of the Lord President in their official capacities and they ought not to compose fifty percent of the body that was expected to give "independent advice to the Goverment regarding the efficiency of its research organisations—and to make recommendations regarding the scope of [the] Research Councils' activities." Third, as to functions, Dale considered that the Committee should be used to a much greater extent to advise the government on scientific appointments. Flett explained that Dale took particular exception to the present system whereby the Lord President was not expected to consult any authoritative body of scientists about candidates for the DSIR Advisory Council and did not in fact consult anyone but the president of the Royal Society.

Flett considered Dale's views to have "hardened" since 1943. Dale now supported a solution proposed by Tizard of the problem of securing authoritative scientific advice for the government. In Tizard's scheme there would be a standing committee with an independent (i.e. nongovernment) scientist as chairman, three or four other outsiders (appointed after consultation with the Royal Society and the Institution of Civil Engineers), and the three secretaries of the research councils. Its terms of reference would be along the following lines: it would advise the government on the scientific implications of its general domestic and international policies insofar as the issues involved were not clearly the concern of a particular government department, and also on the scale and form of support to be given to scientific and technical education and to research work generally; it would keep under review the relation between the supply of and demand for scientists and engage in other related pursuits. Flett believed that this scheme would not find universal and automatic support among scientists. In particular, he thought that Appleton would consider his position very carefully before accepting it.

Dale was most eager that the as yet publicly unannounced Committee on Future Scientific Policy, or Barlow Committee, should consider the position of the Scientific Advisory Committee and make recommendations to the government about its future. Flett hoped that Barlow would agree that this was the proper course, and if it was then there was everything to be said for not revamping the Scientific Advisory Committee until Barlow's committee had fully considered the matter.

Flett and Dale agreed that the best course to recommend to Morrison was that the Scientific Advisory Committee should not be reconstituted for the moment and that Morrison should see Sir Robert Robinson, Dale's successor at the Royal Society, and explain to him that the future position of the Committee was being considered by the Barlow Committee and that recommendations might be expected from it during the next few months. In the meantime, Dale should continue as chairman and Hill as a member of the Committee in order that the few outstanding questions before it might be dealt with. Robinson and Hill's successor should attend the Committee's meetings during this interim period so that, if the Committee were to receive a new lease of life in its existing form, they would be better prepared to assume their roles. At a subsequent meeting, Morrison and Dale agreed on these measures.[54] There Dale expressed mild annoyance that after the Committee had submitted recommendations it had great difficulty finding out whether any action was being taken on them.[55]

A few days later, Morrison announced in the House of Commons that he had appointed a committee to consider the policies that should govern the use of scientific manpower and resources during the next ten years. As manpower was the most urgent problem, the committee was asked to submit an interim report on broad lines at an early date so as to facilitate progressive planning in those fields dependent upon scientific manpower. At a later date, the committee was to be asked to make recommendations about the establishment of permanent machinery for carrying out surveys regarding the best use of scientific resources in the national interest. In this connection, Morrison instructed the Committee to consider the "effectiveness of the Government's machinery for scientific direction," thereby satisfying the desires of the Royal Society and Association of Scientific Workers.[56]

During a debate on scientific manpower and resources in the House of Commons on the following day, 30 November, Captain A. R. Blackburn, a member of the Parliamentary and Scientific Committee, argued that it was essential for the government to set up a "central scientific planning body" to direct and coordinate scientific manpower and resources on a national basis. He suggested that since Churchill had appointed Lord Cherwell to a position of high importance in the War Cabinet as scientific adviser, a similar appointment might now be made by the government.[57] Morrison replied that he did not believe that all government science could be centralized in one department, nor that this should be attempted. Each department should have a scientific staff adequate to its needs. The real problem was to get these staffs linked up with the government so that somebody was watching how matters as a whole were going. At this point, Blackburn asked if that was not what Cherwell had done. Morrison responded that "the less we say about Cherwell the better":

> It is a highly controversial subject especially among scientists. Believe me there is more than one side to the picture of Lord Cherwell as Minister of Science, which he never was. He was first of

The Central Organization of Science in Peacetime · 325

all Personal Assistant to the Prime Minister, and then he became Paymaster General and did anything that the Prime Minister wanted him to do. It was a very convenient arrangement for the Prime Minister and it worked. I do not want to say too much about it, but do not let the House think it was all milk and honey. It was not. There were some difficulties. Lord Cherwell did great work for this country and he has great mental powers, but there was more than one view in the scientific world about him and his activities in the scientific field. I do not want to say more about him, and I do not want to say anything deprecatory about Lord Cherwell. I only say that that experiment does not justify the belief in the doctrine of a Minister of Science.[58]

In the same speech Morrison answered questions regarding the relationship of the Barlow Committee to the Scientific Advisory Committee. He said that there was no direct relationship and that the work of the latter would continue as it had been "exceedingly valuable" in studying problems and giving advice to the government. There was no conflict between the two Committees. The Barlow Committee was undertaking a special piece of work that Morrison wanted handled from rather a different angle. But it might be, he said, that the present constitution of the Scientific Advisory Committee would be looked into in order to see whether it could be improved, not so much in its personnel and individual composition, but "on the kind of layout which is appropriate and the elements of which it is composed."[59]

At its first meeting on December 5, the Barlow Committee decided to invite C. P. Snow to serve as its scientific assessor so that it might have the benefit of his knowledge of scientific manpower.[60] Also, as the Committee contained three representatives of the physical sciences, but no biologist, the Lord President's permission was sought to add a biologist. In this way Solly Zuckerman, with the support of Barlow, Blackett, and Appleton, though not of Egerton, became a member of the Committee.[61] Zuckerman's appointment ensured additional support for the views of the Association of Scientific Workers, already represented on the Committee by Blackett, against those of the Royal Society, represented by Egerton.

The first report of the Barlow Committee, *Scientific Manpower*, was issued in May 1946. Its introduction began: "We do not think that it is necessary to preface our report by stating at length the case for developing our scientific resources. Never before has the importance of science been more widely recognized or so many hopes of future progress and welfare founded upon the scientist."[62] In regard to science graduates, the report stressed in heavy type that "the immediate aim should be to double the present output, giving us 5,000 newly qualified scientists per annum at the earliest possible moment."[63] Morrison announced in the House of Commons that the government was in general agreement with this and the other conclusions reached by the Committee.[64] He hoped that they would receive immediate and serious consideration by the universities, so that detailed proposals for giving effect to them could be formulated in consultation with the University Grants Committee.

When the Committee took up its second charge, consideration of the "effectiveness of the Government machinery for scientific direction," it had several suggestions to consider. At this time the two senior committees responsible for scientific advice, apart from the Atomic Energy Committee, were the Scientific Advisory Committee and its counterpart for military science, the Deputy Chiefs of Staff Committee. The Barlow Committee was apprised of Dale's criticisms of the constitution and functioning of the Scientific Advisory Committee.[65] It also learned of the views of Tizard who emphasized the vital role of science in war and peace and argued that broad problems of war could no longer be separated from those of peace. Appreciating the need to uphold the principle of direct ministerial responsibility for each department of state, Tizard drew the conclusion that the defense and general aspects of science policy required two distinct but overlapping central authorities. One of these would be a Committee on Defence Research Policy, reconstituted from the Deputy Chiefs of Staff Committee on the pattern of the Chiefs of Staff organization, which was composed of men of great tactical experience who no longer had immediate tactical responsibilities. The other would be a standing committee on national and international science policy replacing the Scientific Advisory Committee. As a link between the two authorities, Tizard recommended that the chairman of the Defence Research Policy Committee should also act as scientific adviser to the

The Central Organization of Science in Peacetime · 327

government. His primary responsibility would be for defense matters but he must also study and advise on broader issues of science policy. He should be empowered and expected to enlist the help and advice of independent scientists when this was deemed desirable.

On 8 December 1945, prior to Zuckerman's becoming a member of the Barlow Committee and after Dale had persuaded Morrison to have the Committee consider the subject now before it, Zuckerman had submitted proposals for the central direction of the country's scientific effort to Morrison.[66] It is not unlikely that in drawing up these proposals, which the Committee also considered, Zuckerman again had the assistance of Bernal, Blackett, and others from the Association of Scientific Workers. As the body responsible for central direction, Zuckerman envisioned a staff responsible to the Lord President composed of "perhaps a dozen scientific and a few administrative officials (of varied status) seconded from Departments and from the private scientific world for limited periods." One of the staff's functions would be to "provide a liaison with the principal non-governmental organisations and institutes, such as the Royal Society." Thus, the Society would not have the important position that it enjoyed on the Scientific Advisory Committee and which it naturally wished to maintain, as was evident from a letter that Egerton wrote to Barlow. Egerton wished to "approach changes more as a delicate operation of grafting than as an engineering operation of designing a new machine and scrapping the old," and so suggested that if any change was to be made in the Scientific Advisory Committee, it should be done "by adding two other scientific members, one appointed by one of the three major Engineering Institutions in turn, and the other the Chairman of the Scientific Committee for Defence."[67] One place, he added, should be kept open for the co-option of an adviser on a particular matter as a temporary member.

On 12 February 1946, the Committee agreed broadly with Tizard's view that the government's scientific machine, at least on the civil side, required a new committee to coordinate its activity and growth.[68] The Committee thought, however, that it was not correct to speak of "central direction" being necessary. The coordinating body must proceed by argument and persuasion, which "should be adequate to settle nine-tenths of interdepartmental differences." Other disputes would have to be taken in the usual way to the Cabinet for decision.

The Committee also thought that the coordinating committee should consist principally of official scientists who were in daily touch with the scientific problems of their departments. However, it agreed that a "sprinkling" of academic scientists would be valuable.[69] The question was then raised whether, with such a committee under the Lord President, there should be a separate body consisting principally of non-official scientists whose relative independence would enable them to criticize more freely the working of, and gaps in, the government machine. That led to the further question of the future of the Scientific Advisory Committee. Some members of the Barlow Committee doubted that the Scientific Advisory Committee should be continued in its present form "since it neither acted as a standing tribunal for the assessment of priorities, nor maintained continuous surveys over the development of Departmental programmes, nor impinged on the day-to-day activities of the Research Councils." The general feeling of the Committee was, however, that a reconstituted Scientific Advisory Committee would be the most effective means by which the opinion of the scientific community on questions not directly concerned with problems of administration could be brought to the government's attention. Thus at this stage, the Committee thought in terms of two committees: a coordinating one composed mainly of government scientists and a second one of outside scientists watching over government science and bringing the scientific community's views to the government's attention.

Over a three month period beginning in mid-April, the Committee heard the views of ten experienced individuals, the first of whom was Sir John Anderson. Anderson wanted the Lord President to (a) be responsible for the general aspects of science affecting the responsibilities of government for which no particular departmental minister was responsible, (b) have competent assistance of a nonspecialist character, organized on a permanent footing, and (c) have available a standing advisory committee that would keep him in close touch with the world of science outside government service.[70] Anderson explained that the advisory body should be somewhat on the lines of the

The Central Organization of Science in Peacetime · 329

Scientific Advisory Committee, but with the official element less in evidence and a wider representation of the main branches of science. He would not include in it any member of an existing scientific staff; the secretaries of the scientific departments and the various directors of research would attend on invitation, as required. The president of the Royal Society would be included, but none of the secretaries, unless they qualifed on account of eminence in some particular major branch of science. In Anderson's view, the establishment of such an advisory body would "go a long way towards convincing the whole body of scientific opinion that the Government were determined to put science in its rightful place in the planning of our educational, social and economic system."

Following Anderson's appearance, the Committee began to think about the possibility of having one instead of the two committees it had previously contemplated. A questionnaire drawn up by the Committee to focus the thinking of future witnesses appearing before it asked: "Could the new committee fulfill the functions at present fulfilled by the Scientific Advisory Committee to the Cabinet, or should there be a separate body to act as a link between the Government and the field of academic science?"[71]

Six further witnesses had been heard from when, on 4 June, Blackett suggested that it would be useful if the Committee could agree on a statement of their present conclusions before hearing further evidence.[72] The meeting of June 18 was devoted to this matter. A week later Lord Hankey gave his views to the Committee.[73] He thought that it was necessary to have a central committee and suggested that it could be similar to the Scientific Advisory Committee with the addition of Service representatives as required. Hankey believed that its Royal Society members would be sufficient to represent outside science but thought that there should be a large panel of outside scientists and engineers who should be called upon to give advice as required. He felt that with such a committee no separate body would be required to act as a link with academic science.

A list of preliminary conclusions drawn up following the June 18 meeting was considered by the Committee on 16 July.[74] The principal conclusion was that "A Civil Science Committee should be set up responsible to the Lord President, and covering the whole field of Government civil science." Barlow said that it might be thought that one man assisted by a scientific staff would be sufficient. The arguments in favor of a committee, however, were the needs to provide for representation of the scientific departments and to afford adequate direct links to outside science. In addition, he suggested, the appointment of a committee might also make a better impression on Parliament and the public. Egerton, realizing that the proposed committee would be quite different from his idea of a modified Scientific Advisory Committee, in that it would have fewer representatives of the Royal Society on it, questioned whether a committee was needed in addition to a scientific secretariat. He suggested that an ably led secretariat could successfully carry out the four principal responsibilities, namely, to see that scientific advice was given, to coordinate the scientific activities associated with government, to see that attention was directed to new scientific developments that promised fruitful applications, and to assist in the best allocation of scientific resources. In Egerton's view, the necessary links with outside science were already provided by the outside representation on the advisory councils of the scientific departments. Consequently, it seemed to him and, he added, to some of his colleagues in the Royal Society, that a central committee was unnecessary. Moreover, there was the danger that it might become too authoritative and begin to overshadow the responsibility that the Royal Society now carried for promoting all branches of natural knowledge. In reply, Barlow countered that there was little risk of this danger becoming a real one. He had in mind that independent members of the committee would be chosen in consultation with, though not nominated by, the Royal Society, and that one of them should be able to speak on behalf of the council of the Royal Society. In Barlow's view, the committee would in fact be more likely to help the work of the Royal Society than to come into conflict with it. The Barlow Committee expressed itself in favor of having a committee and "Egerton did not dissent from the majority view." The viewpoint of the Association of Scientific Workers, represented by Blackett and Zuckerman, had prevailed over that of the Royal Society which would now have a lesser official role in government science. It was suggested that the title of the proposed committee should correspond to its

The Central Organization of Science in Peacetime · 331

importance as *the* central scientific advisory committee of the government.

Appleton supported the idea of a central committee, but in the Barlow Committee's second major battle, concerning the nature of the scientific secretariat that would assist the central committee, he stood in opposition to Blackett and Zuckerman. The conclusions drawn up after the June 14 meeting included a statement concerning the secretariat. One week later, Edward J. C. Dixon, a member of the Barlow Committee's secretariat, wrote to Blackett:

> The need for a Central Scientific Committee seems now to be accepted but I think there are widely different opinions on the scale of the scientific staff to be associated with it. We have on the one hand Flett's idea of an assistant secretary and a few very junior scientists and on the other [the] proposal for the equivalent of 2 under secretaries, 4 assistant secretaries and six principals plus one administrative official at each level. Snow has said that if the staff cannot be established on an adequate scale, he means mental brilliancy rather than numbers, it would be better not to establish it at all but it is not clear whether the committee will endorse proposals for the larger establishment. Barlow is shocked by the second proposal which he regards as a waste of scientific manpower. Consequently I have formed the opinion that it is desirable to make a plain demonstration of the sort of work the central scientific staff would in fact do. It is not that I have lost faith in my prophet and seek a sign but that I believe that, as a matter of tactics, it is necessary to provide a convincing answer to Appleton's question "What would they do?" It is not, of course, possible to foresee the outcome of any scientific experiment or enquiry but an imaginative project can be made by an analogy from other work of the same character. For this purpose I believe that the imagination should not be allowed to travel beyond immediate practical realities, e.g. some of the examples in Bernal's Chapter 14 "Science in the Service of Man" are a good many years ahead of us whereas others are more nearly in reach.[75]

About two weeks later, a group of six persons, including Bernal, C. H. Waddington, and Dixon, met to discuss the contents of a draft note outlining essential tasks for the secretariat to be placed before the Barlow Committee by Blackett.[76]

Blackett introduced the resulting note to the Committee on July 16. By way of emphasizing that a large secretariat of high quality was required, he provided two extensive lists of subjects that he suggested the secretariat might consider initially. The twenty-five subjects of the first list were all of immediate importance and required coordinated study by several bodies. In connection with the second list, bold suggestions were made in regard to the secretariat's powers. This list contained thirty-three subjects

> which are already being studied by or are properly the responsibility of a Ministry, one of the Research Councils, or some ad hoc body but not on a sufficient breadth of front. In these cases the secretariat would stimulate the direction and co-ordination of the work. In other cases the secretariat would initiate new studies. In every case the secretariat would attempt to get the closest possible collaboration with the Ministries, Research Councils and independent bodies such as the Royal Society etc., concerned with the problem under consideration. The approach would often be "how can the secretariat help a ministry or some other body to do the work it wants to do?"[77]

Recognizing the provocative nature of Blackett's suggestions, Barlow tactfully said that as his document would be of assistance to the Lord President in considering the scale of the secretariat, it might therefore be attached as an annex to the Committee's report.[78]

When the Committee next met on 23 July to consider the first draft of its report, Blackett was absent. Considerable revisions were made, especially in the section dealing with the secretariat, and a further draft was to be drawn up. On the following day Zuckerman wrote to Blackett:

> Tuesday's meeting was rather sticky, as Appleton, aided and abetted by Flett, tried to reverse our understanding about the strength necessary for the Secretariat. I blocked pretty thoroughly, and, to my surprise, was supported by Crowther. Barlow appeared to be neutral.
>
> I do hope you will be at next Tuesday's meeting. Whatever lip-service may have been paid to the idea that some central direction is necessary, there are one or two who would undoubtedly like to see an endentulous organisation emerging from our deliberations.[79]

In sending Blackett a copy of the second draft of the Committee's report to which his lists were appended, Dixon counseled:

> Appleton was greatly disturbed by the list of projects and attacked the second lot particularly as overlapping Departmental responsibilities. I suggest that you should select one project that you think is clearly "a winner" and present a reasoned programme of work that could be done on it and the results that might be expected to flow from it (by analogy with a selected project in actual experience the results of which are well-known or at least are not arguable). Only in this way will you overcome the mental resistance of members of the committee. You can do nothing about the emotional resistance anyway but you might make some intellectual progress—in any event you should be prepared for an attack on the bare list.[80]

The Committee met twice more—on 30 July and after the parliamentary recess on 8 October—before its report was issued on 16 October. The report did not include Blackett's lists and made no recommendation concerning the size of the secretariat, but regarding the latter's powers stressed that its duty was "to serve and not to command the departments."[81] Thus, in this battle Appleton was the victor. The secretariat eventually created was led by Alexander King, the former head of the British Commonwealth Scientific Office in Washington, D.C.[82] He was assisted by a physicist, a philosopher, and an administrative assistant secretary.

The report proposed that:

> a) General ministerial responsibility for civil scientific policy should rest with the Lord President; departmental Ministers should remain responsible for the scientific work of their own Departments.
> b) The Scientific Advisory Committee of the Cabinet—a war-time body—should be dissolved and replaced by an Advisory Council on Scientific Policy with terms of reference sufficiently wide to cover not only departmental activities but also academic and industrial research in so far as these infringe on Government policy.
> c) The Council should be composed in equal numbers of scientists in the Government service and of eminent outside scientists; its Chairman should be chosen for his imagination, independence of judgment, and experience of the Government machine—he should not necessarily be eminent in the world of science although that would be an advantage.

d) The Council should be assisted by a full-time scientific secretariat.
e) In order to secure adequate liaison on scientific policy in the civil and defence fields the Chairman of the Council should be a member of the Defence Research Policy Committee which was established during 1946; the Chairman of the Defence Research Policy Committee should likewise be a member of the Council and there should be a common element in the secretariats of the two bodies.[83]

Tizard became the first chairman of the Defence Research Policy Committee, taking up his duties on 1 January 1947. The Committee's other nine members were senior service officers and senior scientists in the service and supply departments. The Committee advised the Minister of Defence in his responsibility for forming general policy governing research and development in the services.[84] Tizard was later to describe the Committee's creation as a revolution in organization. Not until a short time before World War II had scientists begun to exercise an influence on the tactical use of weapons, and only later had they influenced their strategic use. Indeed, the very idea that scientists should interfere in such matters had been repugnant to senior officers in the mid-thirties. It was the experience of the war that had caused the revolution, and the appointment of the Defence Research Policy Committee, far from being resisted by the Chiefs of Staff, was in fact initiated by them.[85]

The establishment of the Advisory Council on Scientific Policy, as recommended by the Barlow Committee, was announced in a parliamentary written reply by the Lord Privy Seal, Arthur Greenwood, on 29 January 1947. Greenwood paid a warm tribute to the work of the Scientific Advisory Committee. Its advice on a great variety of problems, both military and civil, had been invaluable, he said, and both the present and the former governments had been greatly indebted to it for its help.[86]

The Advisory Council's function was expressed in general terms, namely, "To advise the Lord President of the Council in the exercise of his responsibility for the formulation and execution of Government scientific policy."[87] Morrison made clear from the beginning that he would make full use of the Council. In a minute that he addressed to its chairman before its first meeting, he said he expected the council to discuss problems on its own initiative whenever it felt that it might suggest ways in which the application of scientific knowledge and experience could assist in the solution of national problems. At the same time, he invited the council to make recommendations on a number of specific items of policy.

The Council's first chairman was Tizard.[88] Of the other members, Appleton, Barlow, Sir John Fryer (Secretary of the Agricultural Research Council), Mellanby, Sir Reginald Stradling (Chief Scientific Adviser, Ministry of Works), and Dr. A. E. Trueman (Deputy Chairman, University Grants Committee) constituted the official representation; Sir Howard Florey (Professor of Pathology at Oxford), Sir Claude Gibb (Managing Director, C. A. Parsons and Co.), Sir Edward Salisbury, Sir Ewart Smith (Director of Imperial Chemical Industries), Professor A. R. Todd (Professor of Organic Chemistry at Cambridge), and Zuckerman formed the external scientific-industrial representation. Two of the Council's three joint secretaries were from the Lord President's office, the third from the Ministry of Defence. Thus, in accordance with the recommendations of the Barlow Committee, close liaison was arranged between defence and civil science by the appointment of Tizard as chairman of both the Council and the Defence Research Policy Committee and by the inclusion of one of the secretaries of the latter in the secretariat of the Council. Blackett later wrote that "widespread acclaim from both scientists and the military greeted the formation of these bodies and the choice of Tizard to head them."[89]

Compared with the Scientific Advisory Committee, the Council was a much larger body, having thirteen as opposed to seven members. The Royal Society's official role was greatly reduced, but the secretaries of the three research councils were still included, and with three other government officials composed one half of the Council. Of the balancing group of six nonofficial members, four were university scientists and two were representatives of industry. As for Tizard, he could be regarded as representing both official and outside interests. For, as he explained, in creating the Council the old mistake of separating scientific advisers from administrators was not repeated and he had access to any confidential information that he needed for the work of the Council.[90]

Dale spoke of the Council with satisfaction in his presidential address to the British Association in 1947: "This new opportunity for scientific and technological knowledge to make its views known, at the highest levels at which matters of state are discussed and decided, seems to be an innovation of the greatest importance in our peace-time structure of government."[91] Dale's predecessor as president, Sir Richard Gregory, was jubilant; he regarded the Council's creation as the last biggest step urged by *Nature* ever since he had initiated the journal's leading articles in October 1915.[92] Speaking to the Association of Scientific Workers as its president, Blackett congratulated the government on "this important, if rather belated step, which finally establishes a central machinery for the guidance of the scientific activities of this country."[93] He recalled with satisfaction that the Association had been "both the first and the most energetic of exponents of such a move." A great deal, he said, was expected of the Advisory Council. Much that concerned the future development of science in Britain and its applications to current needs would depend on the scientific judgment, practical understanding, and social responsibility of its members. "If they are tender with vested interests, whether scientific, educational, industrial or departmental, they will betray our hopes."[94]

As the *Times* observed with approval, the setting up of the Council was recognition, underlined by the experience of war, that science represents a body of knowledge and a method that have to be taken into account in the direction of national affairs. The Council and the Defence Research Policy Committee, whose creations formalized the highest advisory functions of scientists, had a special character in so far as they carried the implications that the advice of scientists, like that of economists, was now considered to be of immediate importance in the formulation of general national policies, in peace no less than in war.[95] A great change had occurred.

After the Royal Society had successfully lobbied for the creation of the Scientific Advisory Committee to the War Cabinet in 1940 and after the Society had participated in the Committee's useful work during the war, it was to be expected that the Society would want the Committee, or some improved form of it, to continue to serve the country in peacetime. The Association of Scientific Workers also wanted science to be centrally organized in peacetime. The independent approaches of the Society and Association to the Labour government, which was already persuaded of the importance of the cultivation of science in the national interest, resulted in the creation in January 1947 of the Advisory Council on Scientific Policy with the general task of advising the Lord President in the formulation and execution of government scientific policy. Before the war, no body had existed that officially brought together the heads of the three research councils, the Department of Scientific and Industrial Research, the Medical Research Council, and the Agricultural Research Council. The Scientific Advisory Committee had brought them together with the leaders of civil science as represented by the officers of the Royal Society. The Association of Scientific Workers had thought, however, that the representation of civil science, being only from the Royal Society, was too narrow. Now the Advisory Council on Scientific Policy brought the heads of the research councils together with the heads of other related government departments, a more widely represenative group of civil scientific leaders, and also industrial leaders, to consider and make recommendations to the government on scientific policy. In addition, in order to ensure adequate liaison on scientific policy in the civil and defence fields the Council's chairman was to be a member also of the Defence Research Policy Committee, established in 1946. Thus government science, both civil and defence, nongovernment science, and industry were officially linked as never before in peacetime.

1. W. S. Churchill to A. V. Hill, 30 October 1943, Sir Henry Dale Papers, Royal Society, Box 4, Folder 7.

2. Sir Henry Dale to A. C. G. Egerton, 28 August 1942, Dale Papers, Box 4, Folder 2.

3. Scientific Advisory Committee, minutes, 10 December 1942, Cab. 90/3. As this chapter shows, it is not correct to say, as Stephen Roskill does, that Butler's resignation from the Scientific Advisory Committee "effectively marked the end of its story" (Stephen Roskill, *Hankey: Man of Secrets*, 3 vols. [London: Collins, 1970-74], 3:548).

4. *The Art of the Possible: The Memoirs of Lord Butler K.G., C.H.* (Boston: Gambit, 1972), p. 110.

5. Philip Gummett, *Scientists in Whitehall* (Manchester: Manchester University Press, 1980), p. 31.

6. Note by P. H. F. Rickett, secretary to the Scientific Advisory Committee, 19 December 1942: Paper SAC (42) 63, Cab. 90/3.

7. See chapter 8, p. 240.

8. In its comprehensive review of governmental research organizations in 1941, the Committee had found the Agricultural Research Council to be much less efficient than the other two councils. Scientific Advisory Committee's First Report, Paper SAC (41) 11, paragraphs 84–95, Cab. 92/2. On 4 March 1941, Sir John Anderson informed Lord Hankey, then the Committee's chairman, that he was taking the matter up with the Agricultural Research Council (Cab. 21/1165).

9. Scientific Advisory Committee, minutes, 6 October 1943, Cab. 90/4.

10. Ibid., 5 January 1944, Cab. 90/5.

11. "Second Report on Government Research Organisations," Paper SAC (44) 14, Cab. 90/5. The structure of this report was as follows: Part I, Introduction; Part II, The Present Organisation (with sections on the Ministry of Production, Admiralty, Ministry of Supply and War Office, Ministry of Aircraft Production and Air Ministry, Inter-Service Organisations, Ministry of Home Security, and Petroleum Warfare Department); Part III, General Principles and Recommendations; Part IV, Recommendations with regard to individual Departments (with sections on the Admiralty, Ministry of Supply and War Office, Ministry of Aircraft Production, Ministry of Home Security, and Operational Research). A later report was devoted to medical research.

12. Scientific Advisory Committee, minutes, 1 December 1943, Cab. 90/4.

13. *Times*, 13 June 1945, p. 5; 14 June 1945, p. 8; 15 June 1945, pp. 2, 8.

14. Hill also wrote a letter to *Nature* protesting the government's action (*Nature* 155 [1945]: 753).

15. Minute to Lord President, 20 June 1945, Cab. 21/1166.

16. The Scientific Advisory Committee became concerned that the British scientific contribution to the war would be played down in comparison to the American contribution. It therefore engaged the science journalist J. G. Crowther and the physicist R. Whiddington to describe the British contribution. Their subsequent book, *Science at War* (New York: Philosophical Library, 1948), had chapters on radar, operational research, the atomic bomb, and science and the sea.

17. See, for example, Roskill, *Hankey: Man of Secrets*, 3:489; and Ronald W. Clark, *Sir Edward Appleton* (Oxford: Pergamon Press, 1971), pp. 129–30. Clark quotes Sir Henry Tizard's assessment that the Committee was "really ineffective" (p. 129). Elsewhere, Clark quotes this as "really very ineffective," but also says that on another occasion Tizard stated that the Committee "did some useful but not outstanding work" (Clark, *Tizard* [Cambridge, Mass.: M.I.T. Press, 1965], p. 275).

18. M. T. Flett to Lord President, 27 June 1945, Cab. 21/830.

19. Note by M. T. Flett, 22 June 1945, Cab. 21/830.

20. Scientific Advisory Committee, minutes, 25 July 1944, Cab. 90/5.

21. A. V. Hill, "Memories and Reflections" (manuscript, 2 vols.), 1:100.

22. Note by M. T. Flett, 22 June 1945, Cab. 21/830.

23. M. T. Flett to C. Jolliffe, DSIR, 23 June 1945. Cab. 21/830.

24. Association of Scientific Workers, *Science and Human Welfare* (London: Temple Forum Press, 1946), p. 6.

25. Herbert Morrison, *Government and Parliament* (London: Oxford University Press, 1954), p. 330.

26. *Times*, 1 December 1945, p. 2.

27. Association of Scientific Workers, Executive Committee minutes, 14 November 1943, ASW/1/2/24/12.

28. *Scientific Worker* (August 1944), p. 3.

29. Association of Scientific Workers, *Agenda for Twenty-Seventh Annual Council, 28-29 May 1944*, p. 47.

30. *Scientific Worker* (August 1944), pp. 1-2.

31. Association of Scientific Workers, *Agenda for Twenty-Eighth Annual Council, 19-21 May 1945*, p. 5.

32. Association of Scientific Workers, *Science in Peace* (London: Association of Scientific Workers, 1945), pp. 3, 6, 11.

33. Ibid., p. 11.

34. The pamphlet was one of two published together in Association of Scientific Workers, *Science and Government* (London: Association of Scientific Workers, 1945).

35. Ibid., p. 28.

36. *Scientific Worker* (December 1945), p. 3.

37. Association of Scientific Workers, Executive Committee minutes, 13 February 1944, ASW/1/2/25/3.

38. Association of Scientific Workers, *Science and Government*, p. 29.

39. On the origins of the Bush report see Daniel J. Kevles, "The National Science Foundation and the debate over postwar research policy, 1942-45: A political interpretation of *Science—the Endless Frontier*," Isis 68 (1977): 5-26.

40. Reinet Fremlin to P. M. S. Blackett, 21 August 1945, P. M. S. Blackett Papers, The Royal Society Library, F 11.

41. P. M. S. Blackett to H. S. Morrison, Lord President, 22 August 1945, Blackett Papers, F 11.

42. Solly Zuckerman, *From Apes to Warlords* (New York: Harper and Row, 1978), p. 365.

43. "Need for a Science Secretariat," (16 September 1945), Blackett Papers, F 11.

44. Ibid., p. 1.

45. Sir Edward Bridges to H. S. Morrison, Lord President, 1 October 1945, Cab. 124/532.

46. M. T. Flett to J. A. C. Robertson, 11 September 1945, Cab. 124/532.

47. Sir Edward Bridges, note for record, 14 September 1945, Cab. 124/532.

48. Sir Edward Bridges to H. S. Morrison, Lord President, 1 October 1945, Cab. 124/532.

49. C. R. Attlee to H. S. Morrison, 9 October 1945, Cab. 124/532.

50. H. S. Morrison to P. M. S. Blackett, 9 November 1945, Blackett Papers, F 11.

51. Sir Henry Tizard to Sir Alan Barlow, 14 November 1945, Cab. 124/532.

52. H. S. Morrison to C. R. Attlee, 8 October 1945, Cab. 124/532.

53. M. T. Flett to Lord President, 21 November 1945, Cab. 21/830.

54. M. T. Flett to Sir Alan Barlow, 27 November 1945, Cab. 21/830.

55. M. T. Flett to Miss Church (Treasury), 27 November 1945, Cab. 21/830.

56. Sir Alan Barlow to Lord President, 26 April 1946, Cab. 124/535.

57. *Parliamentary Debates* (Commons) 416 (1945): cols. 1837-38

58. Ibid., col. 1862. After the Conservative Party's defeat in 1945, Cherwell returned to his professorial chair at Oxford.

59. Ibid., col. 1856.

60. Committee on Future Scientific Policy, minutes, 5 December 1945, Cab. 132/51. During the war Snow "became successively a member of the Physics Panel of the Advisory Committee to the Central Register; Head of the Physics Section of the Central Register; and finally Technical Director (Junior Personnel) of the Appointments Department of the Ministry of Labour" (Ronald W. Clark, *The Rise of the Boffins* [London: Phoenix House, 1962], p. 68).

61. Sir Alan Barlow to Lord President, 14 January 1946, Cab. 124/532; Committee on Future Scientific Policy, minutes, 22 January 1946, Cab. 132/51.

62. *Scientific Manpower* (Cmd. 6824) (London: H.M.S.O., 1946), p. 3.

63. Ibid., p. 8.

64. *Scientific Worker*, n.s. 1 (1946): 21.

65. Committee on Future Scientific Policy, document F. S. P. (46) 11, "The Central Government Machine for Science. Note by Chairman" (11 pp., 31 January 1946), Cab. 132/52.

66. Committee on Future Scientific Policy, document F. S. P. (46) 2, Solly Zuckerman, "Machinery for Central Direction of Scientific Effort," (2 pp., 8 December 1945), Cab. 132/52.

67. A. C. G. Egerton to Sir Alan Barlow, February 1946, Cab. 124/534.

68. Committee on Future Scientific Policy, minutes, 12 February 1946, Cab. 132/51.

69. It was noted that a similar arrangement was being proposed in the United States. The Committee had before it a memorandum prepared by Dr. Alexander King, Director of the British Commonwealth Scientific Office, Washington, D.C., "Organisation of Scientific Research in the United States Government," (7 pp., 7 January 1946), Cab. 132/52.

70. Committee on Future Scientific Policy, minutes, 16 April 1946, Cab. 132/51.

71. Committee on Future Scientific Policy, document F. S. P. (46) 43, 27 April 1946, "Questionnaire on Government Scientific Organisation", Cab. 132/52.

72. Committee on Future Scientific Policy, minutes, 4 June 1946, Cab. 132/51. The six were, in order of appearance, Sir Henry Dale; Sir Oliver Franks, Permanent Secretary, Ministry of Supply; Sir Ben Lockspeiser, Director-General of Scientific Research (Aircraft), Ministry of Supply; Sir Reginald Stradling, chief scientist in the Ministry of Works; Sir Stanley Angwin, Assistant Director-General, Post-Office.

73. Committee on Future Scientific Policy, minutes, 25 June 1946, Cab. 132/51. On later occasions, Sir Thomas Merton (Scientific Adviser, Board of Trade), Dr. H. L. Guy (Institution of Mechanical Engineers), and Mr. A. P. Rowe (Deputy-Controller of Research and Development, Admiralty) would also give their views to the Committee.

74. Ibid., 16 July 1946, Cab. 132/51.

75. E. J. C. Dixon to P. M. S. Blackett, 21 June 1946, Blackett Papers, F 12. Dixon was referring to chapter 14 of Bernal's *The Social Function of Science* (1939).

76. E. J. C. Dixon to P. M. S. Blackett, 8 July 1946, Blackett Papers, F 12. C. H. Waddington was a friend of Bernal's, a biologist, a member of the Cambridge Scientists' Anti-war Group and Tots and Quots, and author of *The Scientific Attitude* (Harmondsworth, Middlesex: Penguin Books, 1941).

77. "Draft Programme for Scientific Secretariat," 15 July 1946, p. 1, Blackett Papers F 12.

78. Committee on Future Scientific Policy, minutes, 16 July 1946, Cab. 132/51.
79. Solly Zuckerman to P. M. S. Blackett, 24 July 1946, Blackett Papers, F 12.
80. E. J. C. Dixon to P. M. S. Blackett, 25 July 1946, Blackett Papers, F 12.
81. Committee on Future Scientific Policy, Document F. S. P. (46) 56, 16 October 1946, "The Central Government Machinery for Civil Science" (6 pp.), p. 5, Cab. 132/52.
82. Philip J. Gummett and Geoffrey L. Price, "An Approach to the Central Planning of British Science: The Formation of the Advisory Council on Scientific Policy," *Minerva* 15 (1977): 141. Gummett's and Price's subject is that dealt with in this chapter, but my treatment of it differs from theirs in several respects. For example, they write that whereas in September and October 1945, Bridges and Barlow were working only for an inquiry into scientific manpower, by November they were committed to a more widely-ranging inquiry. "It is difficult," say Gummett and Price, "to interpret this change." (p. 130) But in my account, the difficulty does not arise.
83. H. Morrison, *Government and Parliament* (London: Oxford University Press, 1954), pp. 330-31.
84. "Expenditure on Research and Development," *Third Report of the Select Committee on Estimates, Session 1946-47* (London: H.M.S.O., 1947), appendix 1, p. 37.
85. Sir Henry Tizard, *A Scientist in and out of the Civil Service* (22d Haldane Memorial Lecture delivered at Birkbeck College, London, March 1955. 21 pp.), p. 14.
86. "Expenditure on Research and Development," Appendix 1, p. 37.
87. *First Annual Report of the Advisory Council on Scientific Policy (1947-48)* (London: H.M.S.O., 1948), p. 2.
88. Ibid., p. 2.
89. P. M. S. Blackett, *Studies of War* (New York: Hill & Wang, 1962), p. 112.
90. Tizard, *A Scientist in and out of the Civil Service*, p. 17.
91. *Advancement of Science* 4 (1946-48): 280.
92. W. H. G. Armytage, *Sir Richard Gregory: His Life and Work* (London: Macmillan & Co., 1957), p. 214. *Nature*'s first leading article was on "Science in National Affairs": Sir Harold Hartley, "The Life and Times of Sir Richard Gregory, BT., F.R.S., 1864-1952," *Advancement of Science* 10 (1953): 280.
93. Typescript summary of Blackett's presidential address to the Association of Scientific Workers dated 24.5.1947, Blackett Papers, E 22.
94. Blackett was to be disappointed with the Advisory Council's performance. J. G. Crowther, *Science in Modern Society* (New York: Schocken Books, 1968), p. 119.
95. *Times*, 14 March 1947, p. 5.

Epilogue

The central organization of science was the principal manifestation of the Labour government's determination to further the development and application of science for the benefit of the nation. Such a government policy had been desired by most participants in the social relations of science movement. Consequently, the movement began to fade, as is seen in the rapid postwar decline in the activities of the British Association's Division for the Social and International Relations of Science. At the same time, a resolution of the freedom versus planning dispute among scientists was achieved. The epilogue deals with the Labour government's immediate postwar measures regarding science, the decline of the British Association's Division, and the resolution of the freedom versus planning dispute. It ends with concluding sections on the entire social relations of science movement and earlier scholarship on it.

The British Association's Division had from 1941 pursued the goal of persuading statesmen that society would benefit from the increased cultivation of science. Adopting a more direct approach, the Association of Scientific Workers, Parliamentary Science Committee, and Parliamentary and Scientific Committee had appealed to the government for specific innovations. The Royal Society, however, had been the most influential in strengthening the relations between science and government. When Herbert Morrison, Lord President in the Labour government, spoke of these relations in January 1948, the *Scientific Worker* noted with obvious pleasure that his "speech was unique in that never before had any Government spokesman so clearly recognised the essential role of science [in] the community."[1] Two years previously, Morrison had told the Parliamentary and Scientific Committee:

> You have noticed, I trust, practical proof that the present Government is fully alive to the importance and possibilities of science in every field of national endeavour. The Cinderella days of science are gone. Today there is a place for scientists side by side with civil servants, workers and managers in industry, farmers, professional men, and technologists of all kinds in the great work of national recovery and social planning upon which we are embarked.
>
> It is good administration which brings to its problems advisers from every section of the community. In the bad old foolish days only too often the help of scientists was ignored. There were occasional days when the scientist was almost the only one consulted, and that was just as foolish, if not more so. Today we realize that scientists must be included in the team. That is what the scientist wants and what you and I want. In that way we strengthen the administrative structure, and the scientist makes his contribution to social, economic and industrial advance.[2]

In 1948, Ritchie Calder remarked that the Labour government had given abundant proof that it wanted "all the science it can get."[3] One of its first actions had been to reorganize the Scientific Civil Service. The government was resolved, declared an official publication, "that the conditions of service for scientists working for the Government shall be such as to attract into the Civil Service scientifically trained men and women of high calibre, and to enable them after entry to make the best use of their abilities, in order that scientists in the Government Service may play their full part in the development of the nation's resources and the promotion of the nation's wellbeing."[4] Salaries for government scientists were increased to make them competitive with those in industry, and except where defense secrets were involved, government scientists were free to communicate the results of their researches.[5]

In the area of industrial research, greatly increased support from industry together with a corresponding increase in the government grant had extended the scale of the research associations' work to a notable degree. In 1937, the total income of the associations had been just over £400,000; in 1947 it was nearly £2,500,000. During the same period the number of associations had increased from twenty-two to thirty-five, and there were few industries that were of sufficient size and sufficient homogeneity to support a research association that did not do so. The annual income of the DSIR had also been substantially increased.[6] Standing at £682,180 in 1932, the figure had dropped to a low of £599,653 in 1943, but thereafter increased annually to reach £1,239,570 in 1946 and to go even higher in the next two years. The DSIR increased in number and value its grants to established researchers. It also contributed to the increased output of science graduates by enlarging the number of its grants to students.

To increase scientific manpower was a top priority of the Labour government. The Barlow Committee's recommendation that the output of science graduates should be doubled at the earliest opportunity was achieved by 1950, well before it had been thought possible. During the eight years following the publication of the Committee's report on manpower, government grants to the universities increased from £6.9 million to £20 million for recurrent expenses, and from zero to £5.7 million for capital expenses.[7] In his farewell presidential address to the Association of Scientific Workers in 1947, Blackett recalled with satisfaction that "the first serious study of the need for University expansion, of the future demands for scientists and of the adequacy of the supply of intelligence, is to be found in the A.Sc.W. pamphlet 'Science and the Universities'[1944] in which many of the recommendations of the Barlow Committee are presaged."[8]

In late 1947, the Committee on Industrial Productivity was created to supplement the work of the Advisory Council on Scientific Policy.[9] As chairman of both of these bodies and the Defence Research Policy Committee, Tizard had the whole of government science under constant review. The new Committee's function was to advise the Lord President and the Chancellor of the Exchequer on "the form and scale of research effort in the natural and social sciences which will best assist an early increase in industrial productivity and further to advise on the manner in which the results of such research can best be applied."[10] The *Scientific Worker* found it noteworthy that the Committee included, in addition to scientists, representatives from the trade union movement, industry, the Economic Planning Board, and the economic section of the Cabinet offices.[11] The Association of Scientific Workers was pleased that economic and scientific experts were working together, a practice that it had been advocating from early 1947.

By 1948, the British government was spending upwards of £70,000,000 a year on scientific research and development while increasing its scientific manpower in the expectation of spending even more. In the view of an old campaigner, Ritchie Calder, science was getting its "New Deal."[12]

The period of gradual resumption of the British Association's normal operations after the war formed a watershed in the life of its Division for the Social and International Relations of Science. In 1946, the Association found it impossible to have a full annual meeting, but a one-day gathering was held. Regular annual meetings were resumed the following year. By that time, the Division's activities were, for several reasons, already in decline.

Of these one of the principal was, paradoxically, that in spite of the use of the atomic bomb and other science-based weapons, there was no adverse social pressure on scientists as there had been during the 1930s. A victorious, though exhausted, nation was instead thankful for the science-based offensive and defensive systems that had helped ensure its victory in the war, and now looked to science as an indispensable contributor in the task of reconstruction.

A second principal reason was that the Division had seen its primary goal realized. It had endeavored to persuade politicians that government should make increased use of science and scientists. This the Labour government was doing; but more because of the experiences of the war than of the Division's persuasion. The creation of the Advisory Council on Scientific Policy had fulfilled Sir Richard Gregory's greatest wish.[13]

Furthermore, the Council was established during Sir Henry Dale's tenure as British Association president—he succeeded Gregory in that office. His own successor within the Association was Tizard, who in 1938 had been a member of the original Division Committee. Thus, the Association's leaders were intimately involved with the new developments in the relations between science and government that they heartily supported.

Now that its goal of the past several years had been realized, the Division required a new one. In 1946, however, its guiding light was eighty-two years of age, not usually an age for innovative efforts. In fact, Gregory quickly dropped from prominence in the affairs of the Association. In the postwar decade, he devoted himself, in the words of his biographer, to works of peace.[14] No younger men came forward to define new tasks for the Division.

To the contrary, there was criticism of it. This was primarily on the alleged grounds that the Division had tended to lead the Association in the direction of political controversy and to usurp the functions of sections.[15] The former it had hardly done, and the latter only inadvertently, as the sections had been inactive during the war. In the immediate postwar period, the Division cooperated with sections on two occasions in arranging conferences, but upon the resumption of annual meetings such cooperation ceased as the sections once more pursued their traditional roles.[16]

A further reason for the Division's decline in the postwar decade was the lack of finances.[17] As early as 1943, the Association's General Treasurer had noted that the active policy pursued by the Division was already proving costly and would probably increase in the future.[18] Unlike the expenses of annual meetings, those of Divisional meetings could not be met from a local fund. The hope had been that substantial grants-in-aid would be obtained from governmental, industrial, and private sources. However, after the war no efforts were made, perhaps because of the reasons just cited, to obtain such funds before the mid to late 1950s.

Down to that time, the Division organized occasional activities. In addition to the jointly sponsored conferences already mentioned, one further conference was arranged in 1946 before Gregory ceased to be president.[19] After that only two conferences were held—one in 1948 and the next seven years later, at which time the Division Committee's chair was occupied by Ritchie Calder.[20] Apart from these conferences, held at times other than those of annual meetings, the Division sponsored an occasional session or discussion at an Association meeting.[21] Thus, the Division became a shadow of its former self. Although it was eventually abolished in 1960, its spirit was to live on in the Association. The considerations that had led to the formation of the Division were regarded by the Association as being as valid in 1960 as they had been in 1938, and so its purposes were restated in the objects of the Association. To these were added the following: "to maintain, develop and extend the social and international relations of science."[22] This was the only addition made since the Association had been founded in 1831. The social relations of science movement had left its mark on the British Association.

The dispute among scientists concerning freedom and planning in science was largely resolved soon after war's end. By late 1942, John R. Baker of the Society for Freedom in Science had discerned encouraging signs of a change in attitude towards freedom in science. In a notice to members of the Society, he confidently explained that:

> When the first circular was sent out in November 1940 to suggest to scientists that such a Society as this should be formed, there was no organised opposition to the powerful propaganda directed against freedom in science. Since then, the situation has changed. It is not possible to say exactly to what extent this Society has been responsible for the change, but its influence has probably been considerable. The President [Sir Henry Dale] and the Biological Secretary [A. V. Hill] of the Royal Society made no pronouncement in favour of free science until after the Chairman of our Executive [A. G. Tansley] had corresponded with them. The whole trend of the editorial policy of *Nature* was formerly along the lines of the current propaganda. Now *Nature* holds the balance more fairly between the two opinions. . . .
> It is now unlikely that our case will go by default. An understanding

of the threat to freedom in science has spread beyond our membership. It is now certain that a very large number of scientists—probably the great majority of established research-workers—regard freedom as an indispensable condition of scientific life, though many do not join us because they are still not convinced of the danger of the threat. The Society is in fact fulfilling its objective of becoming a focus for the discussion and clarification of the issues involved.[23]

The change on the part of *Nature* became more apparent during 1943. Rainald Brightman observed in an editorial that all serious advocates of the more effective organization of scientific resources made it clear "that while the community may, and should, claim the right to determine within broad limits the extent of the resources to be devoted to scientific and technical research, and the broad allocation of those resources between different fields and on major products, that allocation must be subject to the advice of men of science themselves, and the technique and the manner in which the particular problems are attacked must be the affair of the scientific worker alone."[24] Polanyi found it entirely admirable that this view was shared by Sir Stafford Cripps, a member of the government.[25] Under Polanyi's influence Rainald Brightman came to doubt whether the Soviet Union had always understood the nature and limitations of scientific method. In a *Nature* editorial, Brightman admonished scientists that in addressing themselves to their wider tasks and in accepting wholeheartedly their social responsibilities, they must remember that it was their prime responsibility to guard and cherish that unfaltering quest for truth that is at the heart of science; he urged them to ensure that in all their organization and planning they consented to nothing that would impair their "freedom of inquiry or utterance and the ultimate but fundamental loyalty to truth."[26] A lecture by Baker to the London and Home Counties Branch of the Institute of Physics was reported favorably and at length in *Nature* under the heading "Freedom in Science." The writer concluded that it was clear that Baker had "opened up an important subject for consideration."[27] During 1944, when the Society for Freedom in Science launched its drive for new members, *Nature* this time described the function of the Society and published its five principles.²⁸ Two years later, *Nature*'s joint editor, L. J. F. Brimble, wrote of the "main (and only true) objective" of science, namely, the search for and exposition of the truth.²⁹

Encouraging developments of this nature were quickly communicated within the Society. On 12 May 1945, Polanyi wrote to Baker that on the previous day he had addressed the University of Manchester branch of the Association of Scientific Workers on the subject of science and welfare.³⁰ The leading members of the group had attended and had expressed, surprisingly, unreserved support for the views advanced by Polanyi concerning the independent standing of pure science and its proper pursuit on academic grounds. They spoke of Bernal and other contributors to an Association of Scientific Workers' rally on the planning of science as vociferous persons who should not be taken seriously. They also claimed that the most recent publications of the Association were quite in agreement with the Society's viewpoint.

This was so. A report on science in the universities prepared by the Association and submitted to the University Grants Committee of the Treasury in March 1944 argued that advances in applied science depended upon the progress of fundamental research. Furthermore, the continued development of the country in the postwar period would require the application of science to problems of industry, agriculture, education, and health, at the high pitch of intensity reached during the war; such a requirement could only be met continuously if applied science was backed by a corresponding expansion of fundamental research "since the disinterested research of to-day becomes the practical application of to-morrow. Therefore, the fullest provision must be made for the natural development of fundamental science."³¹ The Association's president, P. M. S. Blackett, privately criticized the memorandum's conception of fundamental science as being too narrow, being viewed only in terms of science's potential applications. He pointed out to a fellow member of the Association that more stress might have been placed on science's cultural and intellectual aspects. It is true, he wrote,

> that 'pure' science cannot flourish in a capitalist society with unemployment and wars, and that the major task of our generation

is the application of science, and the attainment of a society in which it can be fully applied. But, particularly in a Socialist society the pursuit of science as a worthwhile activity by itself will become more and more possible. A Socialist society might well devote an appreciable fraction of its resources to pure science (as it might also to music and art, etc.) even though it had perhaps temporarily decided that further rapid technological development was undesirable. This time may be a long way off, but I think that some modern progressive thinking goes a bit too far, in the reaction from the science for science sake attitude, in the direction of tying up science with technological improvement. They are tied up now—they may be less so in the future e.g. astronomy and cosmology etc. tend to get forgotten especially, though I know that this is not intended. I do not lay much stress on this point, but it may be worth bearing in mind.[32]

During the Association of Scientific Workers' conference on science and peace in February 1945, Blackett expressed his concern that fundamental research should be adequately funded.[33] He differentiated between fundamental and applied research in terms of time, defining the former as a long-term social investment that could be expected to pay high but uncertain dividends at some indefinite future time. Only over short periods of emergency, such as the war, could the country afford to neglect this long-term investment in favor of the quicker returns of applied research. The task of world-wide reconstruction after the war, he continued, would be a long-term matter that would require the resources of world science for a generation; it would not be accomplished by attacking only day-to-day requirements. For the foreseeable future, Blackett argued, fundamental research had a vitally useful part to play; that was the justification of his demand for its support. Looking further ahead to better times when the problem of providing the material necessities of life might be solved, he thought that fundamental research might be considered valuable for its own sake in common with the arts. It was mistaken, he said, to hold that fundamental research must always be geared to the eventual usefulness of its applications.

The freedom in science viewpoint received a further boost in Association of Scientific Workers' circles early in 1946 when N. F. Mott, professor of theoretical physics at the University of

Bristol, published an article in the *Scientific Worker* on the subject of secrecy in scientific research. At the time, restrictions were being imposed particularly on scientists doing nuclear research. Mott attacked Blackett's notions of fundamental and applied science which he thought ignored an important difference between the two, namely, that while applied science was a proper subject for detailed planning, "it is a characteristic of pure research, on the other hand, that the results of a research programme cannot be guessed when the work is begun. Planning of pure research is therefore scarcely possible, except in so far as funds are allocated to this or that subject or to a certain group of investigators on the grounds that their research shows promise and that something interesting is sure to come out of it. The research worker himself is guided by his own judgment on the value and interest of his problems, and it is on the judgment of the man concerned that the direction of advance depends."[34]

Another exponent of the wrong view, in the eyes of the Society for Freedom in Science, was the British Association's Division. Its first postwar conference, held in December 1945, focused on the subjects of the planning of research and the relationships between pure science and the community. Prior to the conference, the Society sent copies of a pamphlet expressing its views to all fellows of the Royal Society.[35] On this occasion the Society's three leading figures—Baker, Polanyi, and Tansley—were invited to speak, and they readily accepted. To their astonishment and pleasure, their views were generally enthusiastically received. In Polanyi's own words: "My address renewed my criticism of planning and upheld the traditional independence of scientific inquiry. I had expected a hostile reaction, but, to my surprise, speakers and audience showed themselves in favour of science pursued freely for its own sake."[36] In an address entitled "The Social Message of Pure Science," Polanyi noted that it was no longer generally accepted, as it had been until the 1930s, that science should pursue knowledge for its own sake, regardless of any advantage to the welfare of society. He argued, however, that the most vital service that scientists could perform was to restore their discredited scientific ideals. They had to reassert that the essence of science is the love of knowledge and that the utility of knowledge does not concern them primarily. Scientists, he said, "are pledged to a higher obligation, to values more precious than material welfare; to a service far more urgent than that of material welfare."[37]

Baker had anticipated the conference with curiosity and impatience. Half expecting a repetition of what he had found at the Division's 1941 conference on science and world order, he found instead that "nothing less than a revolution in thought" had occurred.[38] Baker must have found particular pleasure in the words of another participant, Herbert Morrison, who referred with pride to Britain's achievement in pure science and promised that the government would encourage scientists "to pursue knowledge for its own sake."[39] Scientists, he said, must have an independent life of their own.

In writing of the conference in *Nature*, Rainald Brightman noted that one of its most striking features was the *rapprochement* between the more vigorous protagonists of planning and the defenders of the freedom of science. Polanyi's "fine" address at the opening session had, said Brightman, "struck a note which was generally welcomed, and the fundamental importance of freedom of investigation and of communication was emphasized from all quarters; so much so that the staccato enumeration by Prof. J. R. Baker, at the end of the second session, of discoveries which could not have been planned struck a jarring note."[40] *Nature* had found no dissent from the view that any necessary planning regarding priorities in fundamental research should be the responsibility of scientists themselves. Also, the conference had stressed that in regard to fundamental research the proper policy was to see that it was adequately endowed and allowed to develop along the lines that the search for new knowledge alone directs, leaving the investigator the widest possible freedom in selecting his subject.[41] *Nature* itself favored these views, and so the winter of 1945-46 was a triumphal one for the Society. It had seen its fundamental fears vanish, its membership swell, and its views widely embraced. Upon reading in *Nature* an extract from Bernal's speech at the Divisional conference, Polanyi wrote to him: "I quite agree that the divergent views which you and I have represented in the past few years on the subject of freedom in science have now been sufficiently clarified to allow for active co-operation between the two parties holding them. A union of efforts is also urgently needed in view of the great problems confronting us throughout the world."[42]

During 1946, Baker and Tansley published in *Nature*, at its invitation, an article on the history and purpose of the Society for Freedom in Science.[43] In explaining the invitation, the journal's joint editor, L. J. F. Brimble, stated that "while not necessarily agreeing entirely with all the points of view put forward on behalf of the Society, it seems desirable that its aims should be set forth before the world of science, for there is undoubtedly sound *raison d'être* for the Society at the present time, and its main objectives are worth striving for."[44] In even stronger language, he added that the principles of the Society as stated in its five propositions are of "cardinal importance and worthy of full support."[45] Brimble admitted that *Nature* had not always seen eye to eye with the Society in the past, but he declared that now the aims and objectives of both were "so similar."[46]

In their article, Baker and Tansley described the Association of Scientific Workers as one of the chief opponents of freedom in science. The Association's executive committee resolved that a reply should be published in *Nature*.[47] The committee also agreed that eminent members of the Association should be told about the nature of the reply and encouraged to write to *Nature* so that a "broad controversy" might be developed. It rejected the suggestion of one of its members, A. H. Bunting, that the Association should cooperate with the Society for Freedom in Science to study the question of secrecy in nuclear research, on the grounds that as the Society was "mainly anti-Soviet and anti-A.Sc.W." the Association could not collaborate with it.

By mid-January 1947, a reply to Baker and Tansley written by Bernal with the collaboration of Blackett, Bunting, and W. A. Wooster had been sent to *Nature*.[48] When the editors refused to publish it, they were asked to accept a shorter version.[49] However, after *Nature* published a full report of the Association's meeting on "Science and the Real Freedoms," no further attempt was made to have a reply to Baker and Tansley published.[50]

A draft of the shorter version contained the surprising statement that "the five principles enunciated by the Society for Freedom in Science can . . . all be accepted by the Association, though some change of emphasis would make for improvement, particularly in the third,[51] which could be considered socially irresponsible as it stands, although the suggestion it contains of a conscious planned democracy of science is undoubtedly of value."[52] The Association nevertheless recognized a major difference between itself and the Society regarding the means to be employed in achieving their common aims of the greatest development of science under the most ideal conditions for its pursuit. The Association regarded the Society as demanding a minimum of interference from, and even of contact with, outside interests; whereas it desired national planning of research and its integration with social needs. As the Association explained elsewhere: "Within a general framework which would emphasize the fields of science needing special attention, individual scientists should be given the maximum freedom in choice of subjects, opportunity for frequent discussion and revision of programme."[53] The difference between the Association and the society had become small indeed.

A remarkable change in attitude had occurred within the British scientific community. The strong emphasis of the preceding years on the central planning of science in the interests of the community was now modified to a desire to use science, and a desire no less strong, to protect the freedom of science. The Society's numerous efforts must have aided the process; but other causes of the change were to be found in Britain's experience in the war.[54] There was a general realization of the indebtedness of the country's victory to its science and scientists. A science, freely developed before the war, had been successfully exploited by scientists, and not by bureaucrats, during the war. This had led to victory over a scientific power of the first rank which had, however, forced many of its best scientists into exile before the war and had regimented those remaining during the war. Within the British government, Herbert Morrison noted that this regimentation had proved suicidal in Germany.[55] This circumstance had been realized before the war in Europe was over and before the atomic bomb was exploded. To flourish, science required freedom; and scientists rather than bureaucrats knew best how to make the fullest use of science. The catastrophic developments in the Soviet Union in the fields of genetics and plant breeding—publicized, it is true, by the Society for Freedom in Science—taught the same lessons. Also, as *Endeavour* observed, the word *planning* had acquired a suspiciously dictatorial ring in the ears of free men.[56]

Although the Society did not alone create the new attitude, it had adumbrated it most effectively.

By 1947, the social relations of science movement which had begun in 1931 had seen its concerns largely satisfied. Its animating spirit lived on within the Association of Scientific Workers which now watched with considerable satisfaction over a much changed and still fluid national science organization. When Bernal succeeded Blackett as the Association's president in 1947, he admonished the executive committee that

> we have a large share in impressing on the Government the necessity of the machinery which now exists. Our steady drive for a central scientific staff [has] been largely successful with the setting up of the Advisory Council on Scientific Policy. We must now follow up and intensify our work in the Association. ... We must not try to duplicate or do what is proper to Government Departments, but we should be in the position of a pressure group or pilot group and should indicate fields in which we know good results can be got and we must draw them to the attention of the appropriate people and get them implemented. We should be able to mobilize not only our own members but scientists who are willing to work although they are not members. If it can be shown that the Association knows its job and means business, some form of semi-official relationship with Government science machinery ... may be achieved.[57]

At the international level, the creation of UNESCO with one of the Association's past-presidents, Julian Huxley, as its first director, and the establishment of the World Federation of Scientific Workers also gratified Association members.

The social relations of science movement was but one phase, albeit an important one, of the continuing concern in Britain with the social relations of science. As the movement faded, the distressing prospect of the proliferation of nuclear weapons and the curbs being imposed on scientific freedom, particularly in the field of nuclear research, were engaging the attention of concerned scientists.[58] A movement of British atomic scientists was begun within the Association of Scientific Workers. A broader movement initiated at a later date and concerned about world population growth, environmental pollution, air travel, and computerization in addition to nuclear weapons, continues to express itself today through the Society for Social Responsibility in Science.[59]

The principal characteristic of the social relations of science movement that began in Britain in 1931 and lasted some sixteen years was that in addition to their traditional investigation of the natural world, scientists in the leading scientific organizations became concerned with a series of issues involving the relationships between science and society. The issues were shaped by international developments including the economic depression of the 1930s, the cultivation and use of science in the USSR, the rise of fascism in Europe, and the Second World War.

As the issues involved different questions about the social relations of science, I have regarded the movement as being composed of five related phases. All of them occurred within the period from 1931 to about 1947; they built upon one another; many scientists, through their memberships in different scientific organizations, were involved in several of the phases; the phases shared a common concern that science should be used for the benefit of society; and, finally, most of the organizations involved came to seek satisfaction of their concerns in greater integration of science and government.

The first phase of the movement began in 1931, initiated by the economic depression. The depression's socially devastating effects, including technological unemployment and the unemployment of scientists, allegedly brought about by "the machine," led scientists to reexamine their relationships with society. By 1934, the British Association, which for half a century had largely ignored the social relations of science, was encouraging its members to study the social impact of science. Because of the catholic nature of the Association, this new social consciousness involved most British scientists. Although the Association came to study the social relations of science, it maintained its practice of not becoming involved in national politics concerning science and its social uses. That practice had led to the British Science Guild being formed in 1905. During its early years, the Guild had actively involved itself in national affairs concerning science. In the late twenties, the Association of Scientific Workers recognized that issues involving science were increasingly coming before Parliament and attempted to create a parliamentary science committee as a means whereby scientists might contribute to the discussion and raising of issues related to science in Parliament. In 1933, the Guild and Association succeeded in establishing an independent Parliamentary Science Committee. The Committee concerned itself with parliamentary and governmental issues involving science, and by 1939, when it became the Parliamentary and Scientific Committee, it had made some modest contributions. More importantly, its work focused attention on matters that would become of increasing interest to the entire social relations of science movement, namely, the relationships between science and government. Such relationships had been greatly developed earlier in the century through the establishment of the national research councils. With their formation, the official upper level of the scientist within government had been determined, namely, as adviser to cabinet ministers. During the social relations of science movement, the advisory machinery at this level was developed and integrated to cover all government science and establish formal links with academic science and industry. All phases of the social relations of science movement became concerned in one way or another with the relationships between science and government, and when in the immediate postwar world the relations were regarded as being satisfactorily established, the movement declined.

Concern with the relationships of science to government was seen in a third aspect of the first phase of the social relations of science movement. Many younger scientists were impressed in the early thirties with the enlightened cultivation and use of science within the USSR, contrasting it with the "frustration" of science in the industrialized capitalist countries, including Britain. They also came to view science in Marxist terms. From 1932, they joined the Association of Scientific Workers and soon dominated it. Their acknowledged leader, J. D. Bernal, seized the opportunity of preparing a memorandum for the Parliamentary Science Committee on the financing of British scientific research which had suffered during the depression. The memorandum called for the government to create an endowment fund for research and an independent central authority to administer it, but the government rejected these proposals in 1937.

The persecution of scientists in Nazi Germany, Germany's rearmament, the use of mustard gas by Italy in Ethiopia in the winter of 1935-36, and the fear of the further misuse of science in warfare, all underlay the second phase of the social relations of science movement. This phase is best seen in the activities of the British Association from 1936. The question of the scientist's responsibility for the use of science in making weapons became an insistent and troubling one. By 1938, British scientists were agreed that all members of society, and not only scientists, bear responsibility for the uses of science, including its use in warfare. By the same year, interest in the social relations of science had become so intense that the British Association, in an unprecedented action, created a Division for their study in the belief that that study was as important as the scientist's traditional investigation of the natural world. However, the Division soon abandoned the study of the social relations of science and instead attempted to educate the public, and more particularly its elected representatives, about the powers of science. Thus, this phase of the movement also came to focus attention on the relationships between science and government. The Division maintained the Association's practice of not becoming directly involved in national affairs. It attempted to influence the government indirectly by inviting government leaders to participate in its conferences which addressed subjects of national importance concerning science. It is impossible to say what the effect of this practice was. The war was much more effective in teaching the public and politicians to appreciate the powers of science, but at least the Division's efforts reinforced that teaching.

The third phase of the movement saw, in contrast, direct involvement in national affairs by the Royal Society and Association of Scientific Workers. The international events that underlay this phase were the Munich crisis and the Second World War. In the mid-thirties, the Association of Scientific Workers, under the influence of its left-wing members, had repeatedly condemned the misuse of science in warfare. However, in recognizing the growing threat of Nazi Germany, the

Association came to regard fascism as a greater evil than war, and following the Munich crisis, it advocated that the nation make the fullest use of its scientific resources in the war against fascism that seemed inevitable. It was this sentiment that drew a hitherto aloof Royal Society into the social relations of science movement. Having successfully initiated the register of scientific personnel available for war work, the Society's leaders appealed to the government to create a central organization that would coordinate the country's scientific effort in the war. However, only after public criticism had been heard that the government was not making the best use of the country's scientific resources in the war did Churchill authorize, as a matter of political expediency, the creation of the Scientific Advisory Committee to the War Cabinet in September 1940. The less successful Engineering Advisory Committee was formed in April 1941. It was to exist for just over a year, but the Scientific Advisory Committee operated throughout the war years and had a peacetime successor. In bringing together the heads of the three national research councils and the president and two secretaries of the Royal Society, the Scientific Advisory Committee effected an integration of government science as well as of government and civil science that had not existed before. The Committee reported to the Lord President and oversaw almost the entire scientific effort of the war. It was an important innovation in the evolution of the relationships between science and government and for that reason was universally welcomed by scientists.

However, scientists were still only advisers to the political and military leaders who determined the strategy of the war effort. The supreme strategy was decided by the Defence (Operations) Committee of the War Cabinet chaired by Churchill, who was his own Minister of Defence. The planning of operations was centralized in the Chiefs of Staff Committee and the planning of supply in the Ministry of Production. In a wave of criticism of the government's use of science in the war, which began in mid-1941 and intensified after the Allied reversals in North Africa in the early summer of 1942, A. V. Hill called for the creation of a scientific general staff that would determine the strategic uses of science in the war. In mid-1942, Hill raised the matter in Parliament and the press during the most political period of the social relations of science movement. The Association of Scientific Workers, the Parliamentary and Scientific Committee, and the members of the Scientific Advisory Committee and the Engineering Advisory Committee all supported Hill. The government responded in September 1942 by appointing three scientific advisers to the Ministry of Production, which resulted in the Engineering Advisory Committee becoming defunct. Although the Association of Scientific Workers continued into 1943 to call for the creation of a scientific general staff, the Scientific Advisory Committee was to remain the greatest gain made by scientists within government during the war. There were no further waves of criticism by scientists of the government's use of science in the war. The Royal Society and the Association of Scientific Workers began instead to think about the postwar organization of science.

Fears about the possible nature of that organization had already initiated a fourth phase of the social relations of science movement which involved an important polemic among scientists themselves concerning planning and freedom in science. Influenced by the practice of the USSR, left-wing scientists and the Association of Scientific Workers had from the mid-1930s advocated the planning of science in Britain. By 1940, their view enjoyed wide support in British scientific circles. Fearing that at war's end the government might introduce the planning of science, John R. Baker and others formed the Society for Freedom in Science during the winter of 1940–41 to prepare to resist such an action. Once again, attention was focused on the relationships between science and government. Throughout the war years the Society, an informal organization, officially accomplished little, but its individual members, especially Baker and Michael Polanyi, criticized the planning view and upheld their own view of the necessity of freedom in science in books, articles, speeches, and letters to editors. Nevertheless, the majority of scientists continued to favor planning. However, in the immediate postwar years a complete reversal occurred in the general view, and even the Association of Scientific Workers came to agree with the principles upheld by the Society for Freedom in Science.

After the government advisory machinery for science had been developed and the nation's science integrated in wartime, scientists naturally wanted science to be centrally organized in peacetime. The Labour government elected immediately after the war was most sympathetic to the suggestions of the Royal Society and the Association of Scientific Workers that permanent machinery be established, and in January 1947 it created the Advisory Council on Scientific Policy as the successor to the Scientific Advisory Committee. The Royal Society had favored the establishment of a modified Scientific Advisory committee on which it would continue to enjoy a prominent position, but the Association of Scientific Workers wanted a body that would be more broadly representative of scientists. The Council's composition was a victory for the Association. The creation of the Council was the greatest achievement of the social relations of science movement. It was an entirely new peacetime organization within British government, and like its forerunner, the Scientific Advisory Committee, was widely welcomed. Sir Richard Gregory, the leader of the British Association's Division and a former editor of *Nature*, remarked with pleasure that the Council was the type of body whose creation he had been advocating ever since he had initiated the leading articles in *Nature* in 1915.

From 1942, Gregory and the Division had endeavored to educate politicians concerning the powers of science. In creating the Advisory Council on Scientific Policy, the Defence Research Policy Committee, and the Committee on Industrial Productivity, in increasing the funds for industrial scientific research, the output of science graduates, and the grants to universities, and in reorganizing the scientific civil service, the Labour government showed that it fully appreciated the powers of science. Consequently, the activities of the Division quickly declined in the postwar period. Although it was finally abolished in 1960, its spirit was kept alive within the Association through the incorporation in the Association's objects of the intention "to maintain, develop and extend the social and international relations of science." The social relations of science movement had left its imprint on the British Association.

The period of the social relations of science movement saw the growth among British scientists, politicians, and laymen of a deeper appreciation of the importance of the proper cultivation and use of science in national life and the related development and centralization of government machinery for science.

The social relations of science movement has frequently been equated with the activities during the 1930s of a handful of prominent left-wing scientists and the organization that they joined and then dominated, the Association of Scientific Workers. But, as has been shown, this is a much too restricted view of the movement that involved a far broader spectrum of scientists and scientific organizations and which continued into, and indeed saw its greatest political activities and achievements during, the 1940s.

In *The Social Relations of Science*, J. G. Crowther gives a brief and incomplete sketch of the new interest in the social relations of science during the 1930s.[60] In the sequel to it, *Science in Modern Society*, Crowther does not discuss the social relations of science movement as such, although he briefly mentions the Society for Freedom in Science, the British Association's Division for the Social and International Relations of Science, and the immediate postwar developments in government science in Britain.[61]

Hilary and Steven Rose in their sweeping account of science in British society, *Science and Society*, also do not discuss the social relations of science movement as such, although in their third and fourth chapters they mention many of the events that comprised it.[62] Nevertheless, from these chapters one can draw only a mere sketch of the movement, which in many respects is incomplete, in some incorrect and in others misleading.

Neal Wood devotes one chapter, "Utopians of Science," in *Communism and British Intellectuals*, to the social relations of science movement. He correctly sees that the movement began in the early thirties and ended shortly after the war's end, but his initial sixteen-page account of it based on printed works is, as any such account must be, inadequate in numerous respects. Consequently, his subsequent analysis of the movement in the remainder of the chapter is flawed. Wood focuses on the radical writings of scientists including Bernal, Haldane, Levy, and Needham, and pays scant attention to the actions of scientists as expressed through their organizations. Thus, for example, the economic crisis of the early thirties and its effects on the British Association, British Science Guild, and Association of Scientific Workers are overlooked, as is the increasing concern among scientists down through the thirties about the misuses of science in warfare. Wood incorrectly characterizes the movement as "a loosely integrated *intellectual* movement, never formally organized or institutionalized,"[63] (Italics added) forgetting the formation of the British Association's Division for the Social and International Relations of Science and the domination by the radical scientists of the Association of Scientific Workers, organizations whose activities he does not examine. Viewing the movement as an intellectual one, Wood is led to state, again incorrectly, that its decline "was largely due to the growth of a small but extremely articulate and hard-hitting opposition among the scientists [the Society for Freedom in Science]."[64] He misses completely the sustained drives by, in their various ways, the Association of Scientific Workers, British Association, and Royal Society to foster relations between science and government, drives that scientists considered to have met with success in the early postwar years, and whose success led in large part to the natural termination—and not, as Wood would have it, the "collapse"[65]—of the movement. Had Wood examined Bernal's participation in these drives throughout the thirties and forties instead of looking at what Bernal had written in 1929 in *The World, the Flesh and the Devil: An Inquiry into the Future of the Three Enemies of the Rational Soul* he would hardly have offered as one explanation of the movement "the insatiable impulse of some to control nature and society."[66] Wood's fundamental error is to construe the movement too narrowly as an intellectual one principally involving radical scientists. That is certainly an important part of the story, but the movement embraced much more.

In an article published in 1971, Gary Werskey takes a broader view than Wood of the social relations of science movement, seeing it as more than an intellectual movement and as involving a "Reformist" faction as well as Wood's "Radical" faction.[67] However, it is not true that, as he states, the movement "involved no more than a few members of the scientific community."[68] Werskey sees his two groups operating through different kinds of scientific organizations prior to 1938—the Reformists, led by Sir Richard Gregory, working through the British Association and the Radicals through the Association of Scientific Workers and the Cambridge Scientists' Anti-War Group. But he is wrong in maintaining that their principal shared concern during the early thirties was that of "raising the social status of scientists."[69] He has not examined the activities of the British Association and Association of Scientific Workers and has overlooked the formation and work of the Parliamentary Science Committee. Werskey sees the Reformist and Radical factions, despite their differences, "paradoxically" coming together in 1938 to found the British Association's Division for the Social and International Relations of Science. This view is true insofar as radical scientists, including Bernal and Levy, were members of the British Association and participated in a minor way in the Division's founding. However, the Division must be seen to be a Reformists' (to use Werskey's term) creation; Gregory was the driving force in its formation. Werskey makes no study of the work of the Division, neither does he examine what the Association of Scientific Workers or Royal Society attempted by way of getting scientists into government during and after the war years, although he does briefly mention the formation and outlook of the Society for Freedom in Science. He sees a split occurring between the Radicals and Reformists following the war and the movement fading away, not as Wood would have it because of the opposition of the Society for Freedom in Science, but principally "because there was a gradual improvement in the Government's treatment of science and scientists."[70] Like Wood's account of the movement, Werskey's briefer one relies mainly on printed sources and is limited and incomplete, but his recognition of Reformist and Radical factions is a constructive step beyond Wood's treatment in the direction of an understanding of the movement in its full complexity. However, when one researches the unpublished sources relevant to the movement and examines the actions of scientists in their historical contexts, Werskey's categories, particularly that of Reformists, are seen to be of little use, and I chose not to borrow them.

Werskey himself later eschewed the terms "Radical" and "Reformist" in *The Visible College*, his fine collective biography of five of the leading scientific socialists: Bernal, Haldane, Hogben, Levy, and Needham. "I have written," Werskey explains in the foreword, "a collective biography—as opposed to a general and more abstract account of the scientists' movement in the thirties—because I believe that no significant social phenomenon can be understood apart from the motives and aspirations of the persons who shape it."[71] Nevertheless, it comes as something of a surprise that the social relations of science movement is not one of the three major themes that run throughout the book, and, further, that no reference is made to it in the index. It is also curious that in the above quotation Werskey regards the movement as occurring only in the thirties, whereas it extended into the postwar years as he himself described in his earlier article.

Werskey inevitably deals from time to time with aspects of the movement. He describes the British Association's Division for the Social and International Relations of Science as the institutional locus of an alliance of radical and liberal scientists in which radicalism was "still very much a minority affair."[72] However, he does not proceed beyond the Division's founding to examine its activities. Later, he briefly mentions the Division's most noteworthy conference—that on science and world order in 1941—describing it as "one small step for the scientific Left; a giant step for the B.A., at least compared to its rather sleepy existence of a decade before."[73] But he has overlooked the social concern and related activities that were a major feature of British Association life from 1931, and so has failed to understand that the novel aspect of the 1941 conference was the Association's active pursuit of its new goal of impressing upon statesmen the importance of cultivating science for the benefit of the nation.

Werskey briefly sketches the formation and views of the Society for Freedom in Science and concludes that the clash between the Society and its opponents, the radical scientists, could "be judged a draw."[74] But as I have shown, the general support for planning in science that had prevailed in scientific circles when the Society for Freedom in Science had been founded in 1940-41 to combat it, had been replaced by war's end by equally widespread support for freedom in science. Thus, I disagree with Werskey's contention that "the scientific Left was still, as of 1945, in its ascendancy."[75] Significantly, the Association of Scientific Workers, which from the mid-thirties had been dominated by the radical scientists, had by 1947 come to support the principles upheld by the society for Freedom in Science.

Werskey has nothing to say about the Parliamentary Science

Committee, although at one point he incorrectly ascribes its most ambitious project[76]—the drawing up and submission to the government in 1937 of the so-called Bernal Memorandum on the financing of scientific research—to the Parliamentary and Scientific Committee that succeeded the Parliamentary Science Committee in 1939.

Beyond mentioning the Tots and Quots' *Science in War* (1940),[77] Werskey does not discuss what the scientific Left attempted, through the Association of Scientific Workers, to accomplish during wartime. Nor does he examine the Royal Society's more forceful and fruitful efforts in this area. Relying on the Roses' *Science and Society* he merely remarks misleadingly that "not even the officers of the Royal Society could resist the urge to meddle, albeit most delicately, in various affairs of state."[78] But Werskey's five subjects and other leaders of the scientific Left both knew and greatly appreciated the efforts made by officers of the Royal Society, notably Hill, to have scientists brought into the executive levels of government. Finally, Werskey remarks that the appointment in 1945 of the Barlow Committee on Future Scientific Policy "represented a minimal and highly orthodox response to a set of ideas whose time had already come."[79] But he does not examine the Committee's work, and yet it was this committee, with Blackett and Zuckerman on it, that recommended establishing the Advisory Council on Scientific Policy, whose subsequent creation delighted the entire scientific community and especially the scientific Left.

Finally, to adopt a broader perspective, Keith Middlemas has recently written that by the later 1930s the power of outside groups to affect British government policy had evaporated.[80] While Middlemas has the Anglican Church particularly in mind, it is clear that his remark is intended to apply to any group in British society, for he adds that "only in the relatively unorganized scientific profession can there be found evidence of a politically hostile current of opinion, among scientists appalled either at the destructive power of modern weapons or the technical inefficiences of British industry and management."[81] He obviously has the radical scientists in mind, and is apparently unaware of the existence of the Parliamentary Science Committee which, as has been described, achieved some modest successes before it was succeeded by the more influential Parliamentary and Scientific Committee. The latter's influence was in turn exceeded by that of first, the Royal Society's leaders Bragg, Hill, and Egerton, and the Tots and Quots group led by Zuckerman, in the establishment of the Scientific Advisory Committee to the War Cabinet; second, the various national engineering societies in the subsequent formation of the Engineering Advisory Committee to the War Cabinet; third, Hill and the Association of Scientific Workers in having the three scientific advisers appointed to the Ministry of Production; and, fourth, Dale of the Royal Society and Blackett and Zuckerman representing the views of the Association of Scientific Workers in shaping the creation of the Advisory Council on Scientific Policy. These instances show that in regard to determining the relationships between science and government, the power of groups and individuals had not evaporated. They also serve as a reminder, in regard to Middlemas' remark, that in Britain the authority of the church was weakening while that of science was increasing during the first half of the twentieth century. Outside groups representing interests perceived as vital could still affect government policy.

1. *Scientific Worker*, n.s. 3 (February 1948): 3.

2. H. Morrison, Lord President, speech to Parliamentary and Scientific Committee, 30 January [1946], Cab. 124/525.

3. Ritchie Calder, *Science and Socialism*, Labour Party's Towards Tomorrow Discussion Pamphlet #3, 1948, p. 7.

4. *The Scientific Civil Service: Reorganisation and Recruitment during the Reconstruction Period*, Cmd. 6679 (London: H.M.S.O., September 1945), p. 2.

5. Calder, *Science and Socialism*, p. 7.

6. DSIR, *Report for the Year 1947-48: With a Review of the Years 1938-48*, Cmd. 7761 (London: H.M.S.O., 1949), pp. 2,3.

7. *Seventh Annual Report of the Advisory Council on Scientific Policy (1953-54)*, Cmd. 9260 (London: H.M.S.O., 1954), pp. 2,1.

8. Typescript summary of Blackett's presidential address to the Association of Scientific Workers, dated 24.5.1947, p. 4, Blackett Papers, Royal Society Library, E 22.

9. *Scientific Worker*, n.s. 3 (February 1948): 24.

10. Ibid.

11. Ibid.

12. Calder, *Science and Socialism*, p. 11.

13. W. H. G. Armytage, *Sir Richard Gregory: His Life and Work* (London: Macmillan & Co., 1957), p. 214.

14. Ibid., p. 215.

15. *Advancement of Science* 4 (1946–48): 72.

16. A conference with the education section was held on 25 July 1946 on "UNESCO and Universities" (*Advancement of Science* 4 [1946–48]: 36–68); and one with the engineering section was held on 12 and 13 September 1946 on "London Traffic and the London Plan" (Ibid., pp. 96–139).

17. "Report of the BA Council to the General Committee, 1946–47," ibid., p. 370; *Nature* 178 (1956): 658, 659, 665.

18. *Advancement of Science* 2 (1942–43): 361.

19. This was a well-received, one-day conference on "The Place of Universities in the Community" held at the University of Manchester on 10 May 1946 (Ibid. 4 [1946–48]:89; 5 [1948–49]: 4–20).

20. The 1948 one-day conference on "Human Factors in Industry" was held at Leamington Spa (Ibid. 5 [1948–49]: 96-118). The 1955 two-day conference on "Land Use" was held at Exeter (Ibid. 12 [1955–56]: 28).

21. Sessions were held on: "Science across the Frontiers" (1948); "Food and People" (1949); and "The Dearth of Science Teachers" (1950) (Ibid., 5 [1948–49]: 285; 6 [1949–50]: 394; 7 [1950–51]: 361). A discussion was held on "Science and the Unpredictable" (1953) (Ibid., 11 [1954–55]: 343).

22. Ibid., 17 (1960): 339.

23. "SFS. Notice to Members," (October 1942), Polanyi Papers, Box 21, SFS Folder.

24. *Nature* 152 (1943): 143.

25. Ibid., p. 217.

26. Ibid., p. 423.

27. *Nature* 151 (1943): 295–97, 297.

28. *Nature* 154 (1944): 48.

29. *Nature* 158 (1946): 566.

30. M. Polanyi to J. R. Baker, 12 May 1945, J. R. Baker's personal papers.

31. Association of Scientific Workers, *Science in the Universities* (March 1944), p. 16.

32. P. M. S. Blackett to unidentified member of the Association of Scientific Workers [May 1944], Blackett papers, H 9.

33. Association of Scientific Workers, *Science in Peace* (London: Association of Scientific Workers, 1945), p. 7.

34. *Scientific Worker*, n.s. 1 (April 1946): 6.

35. Society for Freedom in Science to all fellows of the Royal Society, 24 November 1945, Dale Papers, Box 11, Folder 6.

36. M. Polanyi, *Science, Faith and Society* (Chicago: Phoenix Books, 1964), p. 7.

37. *Advancement of Science* 3 (1944–46): 289.

38. *Time and Tide* 26 (1945): 1052.

39. *Advancement of Science* 3 (1944–46): 296–99.

40. *Nature* 157 (1946): 2.

41. Ibid.

42. M. Polanyi to J. D. Bernal, 8 January 1946, Bernal Papers, Cambridge University Library, Box 86, Folder J 182.

43. J. R. Baker and A. G. Tansley, "The Course of the Controversy on Freedom in Science," *Nature* 158 (1946): 574–76.

44. *Nature* 158 (1946): 565.

45. Ibid., p. 566.

46. Ibid.

47. Association of Scientific Workers, Executive Committee minutes, 10 November 1946, ASW/1/2/27/23.

48. Association of Scientific Workers, Science Policy Committee minutes, 11 January 1947, ASW/1/2/28/1/iv; Executive Committee minutes, 12 January 1947, ASW/1/2/28/1/ii.

49. Association of Scientific Workers, Science Policy Committee minutes, 8 March 1947, ASW/1/2/28/3/iii.

50. *Nature* 159 (1947) 511–12; Association of Scientific Workers, Executive Committee minutes, 12 April 1947, ASW/1/2/28/4/iv.

51. See above, p. 272.

52. "Freedom and Organisation in Science," (8 April 1947), p. 1, Bernal Papers, Box 77, I 7.1.

53. Association of Scientific Workers, *Science and the Nation* (Harmondsworth, England: Penguin Books, 1947), p. 163.

54. Neal Wood, however, regards the Society for Freedom in Science as the principal cause. See *Communism and British Intellectuals* (New York: Columbia University Press, 1959), p. 134.

55. *Advancement of Science* 3 (1946): 298. Morrison also showed his appreciation of freedom in fundamental research in a speech in the House of Commons on 29 November 1945. *Parliamentary Debates* (House of Commons) 416 (1945): col. 1859.

56. *Endeavour* 4 (1945): 82.

57. Association of Scientific Workers, Executive Committee minutes, 14 September 1947, ASW/1/2/28/9/ii.

58. In 1947, the executive Committee of the Association of Scientific Workers agreed to place a motion on the following broad lines before the TUC: "That the *major problem* facing science and the use of science is the question of secrecy both nationally and industrially with particular reference to the impact of atomic energy development and the motion to state what we think should be the aims of the Trade Union movement in the reduction of secrecy in science" (Executive Committee minutes, 8 June 1947, ASW/1/2/218/6/ii).

59. Sir Solly Zuckerman, *Beyond the Ivory Tower: The Frontiers of Public and Private Science* (London: Weidenfeld & Nicolson, 1970), p. 6.

60. J. G. Crowther, *The Social Relations of Science*, rev.ed. (Chester Springs, Pennsylvania: Dufour, 1967), pp. 427–39.

61. J. G. Crowther, *Science in Modern Society* (New York: Schocken Books, 1968), parts one and two.

62. Hilary Rose and Steven Rose, *Science and Society* (Harmondsworth, England: Penguin Books, 1969).

63. Neal Wood, *Communism and British Intellectuals* (New York: Columbia University Press, 1959), p. 137.

64. Ibid., p. 134.

65. Ibid., p. 137.

66. Ibid., p. 138.
67. Paul Gary Werskey, "British Scientists and 'Outsider' Politics, 1931-45," *Science Studies* 1 (1971): 67-83, 71.
68. Ibid.
69. Ibid.
70. Ibid., p. 81.
71. Gary Werskey, *The Visible College: The Collective Biography of British Scientific Socialists of the 1930s* (New York: Holt, Rinehart & Winston, 1978), p. 14.
72. Ibid., p. 246.
73. Ibid., p. 271.
74. Ibid., p. 285.
75. Ibid., p. 262.
76. Ibid., p. 236.
77. Ibid., p. 263.
78. Ibid., p. 273.
79. Ibid., p. 275.
80. Keith Middlemas, *Politics in Industrial Society: The Experience of the British System since 1911* (London: Andre Deutsch, 1979), p. 365.
81. Ibid., p. 366.

Index

Academic Assistance Council, 102
Academic Freedom Committee, 81
Admiralty, 15, 168, 169, 172, 177, 182, 219, 235, 238, 241, 243
Advancement of Science, 146
Advisory Committee for Aeronautics, 230
Advisory Council on Scientific Policy, 4, 7, 308, 333–37, 345, 347, 356, 362, 367, 368
Aeronautical Research Committee, 230
Agriculture Bill, 60
Agricultural research, 16
Agricultural Research Council, 17, 85, 165, 167, 180, 184, 294, 310, 337
Air Council, 235
Air Ministry, 168, 169, 182, 223, 226, 238
Air raids, 127
Allied Post-War Requirements Bureau, 143
Allied Technical Advisory Committee on Agriculture, 143
American Association for the Advancement of Science, 6, 95, 105, 106, 108, 109, 110, 111–14, 115, 119, 123, 124, 125
Amsterdam Royal Academy of Sciences, 106, 107

Anderson, Sir John, 197, 198, 202–6 passim, 218, 234, 237, 238, 239, 241, 242, 243, 246, 248, 250, 251, 252, 309, 312, 321, 328–29
Appleton, Sir Edward, 168, 172, 173, 177, 178, 180, 181, 183, 193, 195, 196, 198, 205, 227, 250, 252, 311, 313, 315, 321, 322, 323, 325, 331, 332, 333, 335
Areopagitica, 282
Armstrong, Henry E., 30, 31, 53
Association of Scientific Workers, 4–7 passim, 9, 18–19, 20, 22, 23, 28, 35, 42, 51, 52–59, 64, 65, 66, 71, 72, 76, 80, 81, 82–84, 85, 90, 91, 99, 106, 119, 120, 122, 124, 141, 142, 155–67, 196, 199–201, 206, 216, 220–22, 224, 230, 236, 253, 254, 257, 265, 280, 286, 296–300, 307, 308, 316–19, 325, 327, 330, 336, 337, 343, 345, 346, 350, 351, 354, 355, 356, 358, 359, 360–68 passim
Atlantic Charter, 144
Atomic bomb, 205, 346, 355
Atomic Energy Committee, 326
Atomic scientists movement, 356
Attlee, Clement, 184, 190, 191, 315, 321

Bacon, Francis, 3, 38
Baker, John R., 265, 267–77, 279, 280,

282–83, 291, 293, 296–99 passim, 300, 348, 349, 352, 353, 354, 361
Baldwin, Stanley, 16, 62, 100, 168
Balfour, Lord, 16
Barker, Ernest, 136
Barlow, Sir Alan, 181, 183, 187, 191, 193, 194, 195, 321, 323, 325, 327, 330, 331, 332, 335
Barlow Committee. *See* Committee on Future Scientific Policy
Battle of Britain, 188
Beard, James R., 201, 202, 204, 232, 239, 241, 244
Beaverbrook, Lord, 217
Beneš, Eduard, 140
Bernacchi, Commander L. C., 58, 99
Bernal, J. D., 5, 71, 72, 73, 74, 75–78, 80, 81, 83, 84, 90, 128, 129, 131, 132, 136, 142, 157, 161, 162, 166, 167, 222, 224, 225, 265–67, 268, 274, 277, 278, 280, 285, 288, 289, 291, 292, 293, 296, 297, 311, 317, 319, 327, 331, 353, 354, 356, 358, 363, 364, 365, 367
"Bernalism," 267
Bernal Memorandum, 85–90
Besa machine gun, 240
Bevan, Aneurin, 239
Bevan, Ernest, 145
Biological warfare, 312
Biology War Committee, 295
Birkhoff, George D., 124
Blackburn, Captain A. R., 324
Blackett, P. M. S., 79, 81, 83, 90, 101, 128, 132, 162, 183, 252, 288, 289, 311, 316–21 passim, 325, 327, 329–33 passim, 335, 336, 345, 350, 351, 354, 356, 367, 368
Blackman, G. E., 295
Blackman, V. H., 58
Board of Science, 11
Board of Trade, 14
Born, Max, 289–91, 292
Boswell, P. G. H., 32, 42, 99, 122, 123, 124
Bragg, Sir William, 32, 121, 122, 158, 160, 166–75 passim, 178, 180, 183, 184, 187, 190–99 passim, 218, 232, 368
Bragg, W. L., 159, 160, 183
Brave New World, 33, 77
Bridges, Sir Edward, 170, 172–77 passim, 179–82 passim, 203, 320, 321

Brightman, Rainald, 36, 37, 44, 80, 96, 97, 98, 101, 104, 105, 106, 129, 137, 295, 349, 353
Brimble, L. J. F., 255, 295, 350, 351, 354
British Association for the Advancement of Science, 4, 5, 9–13 passim, 17, 18, 20, 22, 27, 28–46, 51–57 passim, 66, 71, 72, 76, 81, 83, 91, 95–100, 104–6, 110–14 passim, 119–48, 161, 164, 190, 191, 197, 207, 319, 336, 346, 347, 357, 359, 363–66 passim; Division, 6, 7, 119–48, 155, 165, 221, 222, 290, 291, 294, 343, 346–48, 352, 353, 359, 362–66 passim
British Broadcasting Corporation, 87, 145
British Institute of Social Service, 52
British Medical Association, 103–4
British Museum, 15
British Science Guild, 4, 5, 9, 13–20 passim, 22, 23, 28, 35, 43, 51–56 passim, 58, 59, 64, 65, 66, 72, 76, 84, 85, 91, 95, 99–100, 115, 120, 199, 357, 363
British scientific research, 79–80
Brookings Institution, 111
Brooks, F. T., 123, 125
Building Research Board, 16
Bukharın, Nikolai, 75
Bunting, A. H., 316, 354
Burgers, J. M., 106, 107, 108
Burgin, Leslie, 171
Bush, Vannevar, 225, 318, 319
Butler, Sir Edwin, 180, 193, 195, 196, 198
Butler, R. A., 219, 220, 228, 232, 233, 234, 236, 237, 239, 243, 244, 246, 247, 249, 308, 309, 311

Calder, Ritchie, 36, 37, 38, 40, 41, 45, 96, 109, 110, 112, 126, 129, 131, 132, 138, 344, 346, 348
Cambridge Scientists' Anti-War Group, 156–57, 162, 364
Carnegie Endowment for International Peace, 106
Carnegie Institution, 106
Carr-Saunders, A. M., 131
Central Bureau of Scientific, Professional, Technical, and Administrative Staffs, 160, 161, 162
Central organization of science, 156, 163, 164, 215, 232, 307–37

Central Register, 161, 162, 164, 166, 167, 178, 187, 222
Central scientific and technical board, 248 253–54, 255, 257
Chamberlain, Neville, 168, 177, 181, 182, 183, 188, 190, 191, 193, 194, 195, 197
Channel tunnel, 237, 261 (n. 60)
Chapman, Sydney, 107, 122, 123
Chatfield, Lord, 170–78 passim
Chemical Defence Establishment, 241
Cherwell, Lord, 185, 186, 203, 206, 222, 223, 226, 227, 228, 234, 249, 250, 309, 314, 324, 325
Chief of technical staff, 224
Chiefs of Staff, 206, 228, 229, 236, 326
Chiefs of Staff Committee, 226, 237, 243, 244, 258, 326
Chorlton, Alan E. L., 62, 201
Church, Major Archibald G., 19, 20, 22, 35, 36, 53
Churchill, Winston, 140, 144, 156, 176, 177, 181, 182, 183, 187, 189, 191, 195, 196, 198, 202–6 passim, 218, 219, 227, 235, 250, 251, 258, 307, 309, 315, 324, 360
Cockcroft, John, 252
Committee against Malnutrition, 82
Committee for Scientific Survey of Air Defence, 168, 230
Committee for Scientific Survey of Air Warfare, 185
Committee of Civil Research, 16
Committee of Imperial Defence, 169, 170, 171, 173, 174, 175, 181; Man Power (Technical) Committee of, 172, 173, 176
Committee of Privy Council for Scientific and Industrial Research, 14, 85; Advisory Council of, 85, 88–89, 322
Committee on Further Education and Training, 320
Committee on Future Scientific Policy, 308, 321, 323, 325–34, 335, 367
Committee on Industrial Productivity, 345–46, 362
Communism and British Intellectuals, 363
Conant, James Bryant, 98, 109
Conklin, Edward G., 105, 108, 109
Contempt of Freedom, The, 277
Court of wisdom, 109, 112

Cripps, Sir Stafford, 144, 249, 253–57 passim, 318, 349
Crowther, J. A., 273
Crowther, J. G., 29, 80, 138, 147, 166, 277, 284, 285, 290, 291, 292, 296, 321, 332, 363
Cullen, William, 230

Daily Herald, 36, 41, 129
Dale, Sir Henry, 42, 232, 234, 239–42 passim, 248, 251, 252, 287, 288–89, 293, 294, 308–15 passim, 320, 321, 322, 326, 335, 347, 348, 368
Darkness at Noon, 276
Darling, Frank Frazer, 270
Darwin, C. G., 183
Davies, Clem, 239
Defence Research Policy Committee, 334–37 passim, 345, 362
Department of Government Chemist, 15
Department of Scientific and Industrial Research, 14, 15, 16, 23, 63, 64, 65, 74, 83, 84, 85, 88, 89, 162, 165, 167, 178, 180, 181, 184, 258, 294, 310, 315, 320, 337, 345
Deputy Chiefs of Staff Committee, 326
Desch, Cecil Henry, 30, 132, 138
Development Commission, 16
Devonshire Commission, 12, 13
Devonshire, Duke of, 12
Dialectical materialism, 76, 280, 281
Discovery, 45, 130
Dixon, E. J. C., 331, 333
Donnan, F. G., 124, 162, 168
Dudley, Earl of, 58, 59

Economic Advisory Committee, 17
Economic planning, 289
Economic Planning Board, 346
Eden, Anthony, 140
Egerton, A. C. G., 161, 162, 166, 167, 169, 170, 172, 173, 175, 181, 184, 185, 188, 195, 196, 197, 205, 287, 293, 308–12 passim, 314, 315, 321, 322, 325, 327, 330, 368
Electronic Engineering, 280
Elliot, Wing Commander William, 173–76 passim, 179, 181, 182, 188, 191, 193, 194, 196
Endeavour, 356

376 · Index

Engineering Advisory Committee, 156, 203, 204, 215, 216–18, 219, 220, 225, 226, 227, 231, 232, 233, 237–46 passim, 248, 249, 250, 255–59 passim, 308, 360, 361, 368

Engineers' Study Group on Economics, 82, 120

Essential Work Order, 220

Ewing, Sir Alfred, 31, 32, 33, 35, 37, 38, 104

Falmouth, Lord, 182, 183, 192, 196–99 passim, 202, 204

Fascism, 4, 5, 75, 76, 95, 100, 104, 157, 159, 221, 299, 357, 360

Federation of British Industries, 53

Ferguson, Allan, 42, 43, 121, 122, 123, 125, 126

Finance of British scientific research, 63–66 passim, 71, 82, 85–90, 358

Fleming, A. P. M., 204, 232, 233, 239, 242, 244

Fletcher, Sir Walter, 32

Flett, Martin Teall, 312, 313, 315, 322, 323, 331, 332

Florey, Sir Howard, 335

Fowler, Sir Henry, 42

Fowler, Sir Ralph, 232, 239, 241

Freedom in science, 265, 266, 269, 271, 275, 276, 279, 282, 285, 287, 289, 290, 291, 295, 299, 351, 353, 354, 355, 361, 362

Fremlin, Reinet, 162, 163, 165, 200, 254, 255, 298, 318

French organization of science, 161

Frustration of science, 128, 130, 142

Frustration of Science, The, 77, 78

Fryer, Sir John, 335

Gale, A. J. V., 236

Geneva Gas Protocol, 101

Geological Survey, 15

George VI, King, 133

Germany, 14, 276, 288, 289, 314, 315, 355, 359

Gibb, Sir Claude, 335

Gilson, Étienne, 109

Glazebrook, Sir Richard, 32, 230

Government grant, 11, 12

Granville, Edgar, 239

Greenly, Sir John, 146

Greenwood, Arthur, 334

Greenwood, Major, 169

Gregory, Sir Richard, 18, 30, 34, 35, 36, 38, 39, 43, 52, 55, 65, 99, 100, 105, 120–25 passim, 131, 132, 133, 136, 138, 139, 140, 145–48 passim, 190, 197, 239, 291, 292, 336, 347, 348, 362, 364, 365

Guy, Dr., 222

Halcrow, W. T., 204

Haldane, J. B. S., 73, 80, 127, 282, 284, 288, 363, 365

Haldane, Lord, 13, 14, 230

Halifax, Lord, 88

Hall, Sir Daniel, 105, 136

Hankey, Lord, 181, 182, 183, 191–97 passim, 202–6 passim, 216–19 passim, 225, 233, 241, 246, 247, 311, 329

Harcourt, Rev. William Vernon, 10

Harries, E. P., 255

Harvard University, 109

Hayek, F. A., 284

Heilbron, Ian M., 252, 253, 257

Henderson, Sir James B., 58

Herring Industry Bill, 60

Hessen, Boris, 72, 76

Hill, A. V., 101, 123, 158, 161, 162, 166, 167, 169, 170–75 passim, 181, 184, 185, 188–91 passim, 193–98 passim, 205, 206, 216, 218, 224, 225, 227, 228–30, 231–36 passim, 239, 241, 243, 244, 250, 254, 255, 287–88, 290, 291, 293, 294, 307, 310, 311, 312, 314, 315, 322, 323, 348, 360, 361, 367, 368

Hitler, Adolph, 101, 188, 288

Hoare, Sir Samuel, 219

Hogben, Lancelot, 73, 80, 113, 284, 285, 288, 291, 296, 365

Holland, Sir Thomas, 123

Holman, B. W., 19, 22, 52, 58, 59, 64, 65, 80, 81, 82, 99, 162, 200, 204

Hopkins, Sir Frederick Gowland, 37–40 passim, 42, 57, 81, 96

Horabin, T. L., 239

Hore-Belisha, Leslie, 171

House of Commons, 224, 229, 234, 324, 326

House of Lords, 225, 246

Howard, Albert, 58, 65, 99

Howarth, O. J. R., 32, 125, 136, 147
Hume, Captain, 55
Humphrey, John, 254
Huxley, Julian, 74-75, 78-80, 81, 114, 131, 132, 138, 165, 167, 356

Ideals, 283-84, 285, 287, 290
Imperial Chemical Industries, 79, 146
Imperial Institute, 15
Indian Science Congress, 110
Inskip, Sir Thomas, 168
Institute of Civil Engineers, 198, 323
Institute of Electrical Engineers, 197, 198, 201, 202
Institute of Mechanical Engineers, 198
International Council of Scientific Unions, 6, 106; Committee on Science and its Social Relations, 95, 107-8, 115
Ives, Herbert E., 125

Jeans, Sir James, 44
Joint Chiefs of Staff (U.S.), 225
Joint Committee on New Weapons and Equipment (U.S.), 225, 227, 229, 233, 234
Joint-Engineering Council, 202
Joint General Staff, 225
Joint technical board, 238, 239, 243

Kaempffert, Waldemar, 109, 110, 112
King, Alexander, 333
Klatzow, Leonard, 162, 165, 166, 167
Koestler, Arthur, 276
Koo, Wellington, 140

Labour government, 7, 19, 29, 307, 308, 315, 316, 317, 319, 337, 343, 344, 346, 362
Lane, Allen, 188
League of Nations, 101
Lee, Rear Admiral W. A., 225
Left Book Club, Scientists' Group, 128
Left-wing scientists, 5, 42, 71-91 passim, 265, 361, 364, 366, 367
Levy, Hyman, 78, 80, 81, 128, 129, 131, 132, 137, 162, 299, 317, 363, 365
Libya, 226, 228
Lindemann, F. A., *See* Lord Cherwell
Listener, 78

Locarno Pact, 101
Lockyer, Lady, 99
Lockyer, Sir Norman, 13, 99, 100
Lysenko, T. D., 280, 281
Lysenko affair, 283
Lyttelton, Oliver, 219, 220, 232, 234, 238, 239, 240, 242, 243, 246, 248, 249, 250-54 passim, 256

MacDonald, Ramsay, 29, 58, 61, 85, 127, 128, 129
McGowan, Lord, 146
MacLeod, Roy M., 12
Maisky, Ivan Mikhailovich, 140
Manchester Guardian, 39, 98, 104, 231, 279, 280, 284
Manchester Literary and Philosophical Society, 136
Markham, S. F., 199, 200
Matthews, Bryan, 252
M.A.U.D. Committee, 205-6
Medical Research Committee, 16, 23, 83, 85, 165, 167, 179, 184, 294, 310, 337
Mellanby, Sir Edward, 179, 181, 193, 195, 196, 198, 205, 257, 311, 335
Merton, Thomas R., 183, 239, 241, 252, 253, 258
Michurin, I. V., 281
Middlemas, Keith, 367
Milk, 133-35
Million Fund, 63, 64
Minister of Air, 223
Minister of Defence, 219, 334
Minister of Labour, 222, 320-21
Minister of Production, 222, 225, 229, 230, 236, 237, 239, 243, 244, 248, 249, 256, 258, 259
Minister of science, 36, 324, 325
Ministerial Military Co-ordination Committee, 181
Ministry of Agriculture, 62
Ministry of Aircraft Production, 205, 219, 240, 241, 242
Ministry of Labour, 222
Ministry of Production, 7, 216, 219, 241, 253, 255, 258, 360; Scientific advisers to, 216, 249, 250, 252, 253, 255, 256, 257, 259, 308, 311, 320, 361, 368
Ministry of science, 156, 163, 165, 166, 168, 196, 317, 318

378 · Index

Ministry of science and education, 12
Ministry of Supply, 177, 182, 189, 194, 219, 220, 221, 226, 233, 235, 241, 242
Moore-Brabazon, Colonel, 205
Morell, J. B., 12
Morrison, Herbert, 140, 315, 318–23 passim, 325, 326, 327, 334, 344, 353, 355
Morrison, W. S., 172
Moses, Brigadier General R. G., 225
Mott, N. F., 351
Moulton, F. R., 112, 114, 125, 126
Moulton, Harold G., 111, 112, 125
Munich agreement, 158
Munich crisis, 155, 156, 157, 359, 360
Museum of Practical Geology, 15
Mustard gas, 100, 115, 359
Myres, J. L., 42

National Association of Scientific Writers, 109
National Physical Laboratory, 16, 176
National science council, 53–56, 156
National Union of Scientific Workers, 18, 19
Natural History Museum, 15
Nature, 13, 18, 58, 80, 102, 114, 120, 121, 123, 126, 134, 140, 141, 145, 187, 188, 236, 255, 265, 279, 293–96 passim, 300, 336, 348, 349, 350, 353, 354
Needham, Joseph, 81, 83, 167, 267, 284, 292, 293, 363, 365
Nelson, Sir George, 321, 322
Nelson, H. B., 296
New Fabian Research Bureau, 82
New York Times, 104, 110–14 passim, 125, 126
Nutrition, 134

Official Committee on Machinery of Government, 309, 310, 311, 314, 315, 322
Oliphant, M. L. E., 319
Organization for Scientific Research and Development (U.S.), 225
Orr, Sir John, 114, 131, 132, 134

Palmerston, Lord, 11
Parker, Harold, 176
Parliamentary Science Committee (first), 19–21, 22, 28, 53, 54

Parliamentary Science Committee (second), 4, 5, 6, 52, 57–63, 65, 66, 71, 80, 84, 85, 87–91 passim, 136, 163, 199, 200, 201, 343, 358, 365, 366, 367
Parliamentary and Scientific Committee, 4, 6, 199–202 passim, 216, 222–25 passim, 228, 229, 236, 246, 248, 250, 253, 254, 255, 257, 307, 318, 319, 320, 324, 343, 344, 358, 361, 367, 368
Patent Office, 62
Patent Office Library, 60–61
Patents, 237, 309
Paterson, C. C., 204
Pedler Lecture, 100
PEN Club, 282
Pendred, Loughnan, 230
Phillip, J. C., 105
Picture Post, 238, 239
Pirie, N. W., 81, 316
Planning, 38, 278
Planning of science, 74, 80, 257, 265, 266, 271, 275, 276, 280, 285, 287, 293, 295, 296, 297, 300, 316, 350, 352, 353, 355, 361, 366
Plugge, Captain Leonard, 200, 225, 228, 229, 239, 248
Poison gas (*see also* Mustard gas), 103–4
Polanyi, Michael, 265, 268, 270, 271, 273, 275, 276, 277, 279, 280, 281–86 passim, 289, 290, 291, 294, 295, 296, 298, 300, 349, 350, 352, 353, 361
Political and Economic Planning, 82, 120, 136, 138
Post Office 15, 16; Advisory Committee of, 61, 62
Powell, Christopher, 201
Production Advisory Committee, 255
Progress, 52
Progress and the Scientific Worker, 52, 64
Provisional Committee on Academic Freedom, 81

Radar, 224
Radford Mather Lecture 127, 129
Rayleigh, Lord, 20, 110, 125, 129, 138, 158, 230
Redmayne, Sir Richard, 65
Register of scientists (*see also* Central Register), 158, 161, 360

Research associations, 63, 64, 65, 86, 345
Responsibilities of scientists, 30, 36, 37, 40, 119, 128, 129, 130, 132, 147, 158, 359
Responsibility for science, 6, 139
Ricardo, H. R., 204, 232, 233, 239, 242, 244
Riley, D. P., 142, 147
Ripon, Bishop of, 22
Ritchie, A. D., 279
Robbins, L. C., 240
Robertson, A., 204
Robinson, J. J., 15
Robinson, Sir Robert, 239, 323
Rommel, General, 226
Roosevelt, Franklin D., 144, 225
Rose, Hilary and Steven, 363, 367
Royal Air Force, 223, 230
Royal Observatory, 15
Royal Ordnance Factories, 221
Royal Society, 3, 4, 6, 7, 9, 10, 11, 16, 22, 31, 35, 38, 39, 53, 71, 81, 105, 121, 139, 155, 156, 158, 161, 162, 164, 165, 166, 167–70, 172, 175–80, 183, 184, 185, 188–92, 194, 196, 202, 206, 207, 219, 241, 258, 273, 281, 288, 294, 300, 307–14 passim, 316, 319, 321, 322, 323, 327, 329, 330, 332, 335, 336, 337, 344, 352, 359, 360, 361, 362, 364, 365, 367, 368
Russell, Bertrand, 39
Russell, Sir John, 42, 143
Rutherford, Lord, 85, 88, 102, 110

Salisbury, Sir Edward, 335
Samuel, Lord, 236
Science, 111, 125, 279, 292
Science
    applied, 285, 295, 351, 352
    as a dynamic order, 285–86
    freedom in. See freedom in science
    frustration of, 142, 358
    fundamental, 350, 351, 352, 353
    Liberal view of, 268
    Marxist view of, 5, 71, 72, 78, 79, 80, 84, 90, 284, 358
    misuse of, 128, 156, 158
    planning of. See planning of science
    pure, 79, 285, 289, 290, 295, 350, 351, 352, 353
    Socialist view of, 268
    totalitarian view of, 282
    in universities, 350
    values of, 104
    and war effort, 155–99, 201–7, 215–59
Science council, 12, 13
Science—the Endless Frontier, 318, 319
Science in Modern Society, 363
Science Museum, 15
"Science in Parliament," 59, 201
Science and the Planned State, 282
Science and Society, 363, 367
Science in War, 187–88, 194, 367
Science and World Order, 147
Scientific Advisory Committee, 4, 6, 7, 139, 140, 156, 196–98, 201, 202, 203, 204–7, 215, 216, 218, 220, 225, 226, 231–34 passim, 237, 238, 239, 241–44 passim, 246, 248, 249, 250, 255–59 passim, 308–15 passim, 319–23 passim, 325–29 passim, 333–37 passim, 360, 361, 362, 368
Scientific Advisory Council, 53
Scientific Advisory Council on Fishery Research, 16
Scientific Advisory Council to Parliamentary Science Committee, 20
Scientific advisory councils, 189
Scientific Civil Service, 344
Scientific general staff, 216, 225, 226, 229, 246, 247, 250, 360, 361
Scientific Life, The, 277, 282, 296, 297, 298, 299
Scientific Manpower, 325
Scientific manpower, 321, 324, 325–26, 345, 346
Scientific missions, 235, 237
Scientific Worker, 52, 82, 83, 90, 124, 142, 158, 161, 166, 196, 265, 296, 297, 344, 346
Scientists and war, 102–3, 128, 129, 130
Sea-Fish Commission, 61, 62
Second International Congress of the History of Science, 72, 73, 91
Secrecy in scientific research, 352, 354
Shinwell, Emanuel, 239
Sinclair, Sir Archibald, 185, 186
Slater, William, 167
Smuts, General Jan Christiaan, 28, 32, 218

Smith, Sir Frank, 178
Snell, Lord, 247
Snow, C. P., 130–31, 224, 293, 325, 331
Social Function of Science, The, 136, 265–66, 267, 268, 274, 277, 284, 331
Social Relations of Science, The, 277, 284, 285, 290, 291
Society for Freedom in Science, 4, 7, 265, 269–79, 282, 286, 287, 291, 293, 295–300 passim, 348–56, 361, 363–66 passim
Society for Protection of Science and Learning, 102
Society for Social Responsibility in Science, 357
Society of Chemical Industries, 106, 230
Soddy, Frederick, 78
Spanish Civil War, 127
Stamp, Sir Josiah, 39–42 passim, 98, 106, 109, 110
Stanhope, Earl, 171, 180, 181
Stanier, W. A., 251, 253, 257
Stone, Henry W. J., 21, 55, 58, 59, 60, 84, 200
Strabolgi, Lord, 246
Stradling, Sir Reginald, 335
Strange, Colonel Alexander, 12
Strategic bombing controversy, 226
Stratton, F. J. M., 42, 107
Street, Sir Arthur, 226
Swann, E. D., 142, 254, 255, 257, 316
Swinton, Lord, 223

Tank Board, 257
Tanks, 217, 226, 228, 230, 240, 247
Tansley, Arthur George, 270, 274, 275, 277, 278, 287, 288, 289, 291, 293, 297, 348, 352, 354
Taylor, F. Sherwood, 286
Technocracy, 36, 39
Television Development Commission, 61, 62
Thomson, G. P., 205
Times, 10, 32, 33, 36, 39, 44, 120, 228, 229, 231, 236, 253, 336
Tizard, Sir Henry, 42, 168, 169, 175, 181, 182, 183, 185, 186, 204, 219, 222, 223, 224, 226, 227, 230, 232, 234, 235, 247, 250, 321, 322, 323, 326, 327, 334, 335, 345, 347

Tobruk, 226
Todd, A. R., 335
Topley, W. W. C., 184, 196, 311
Tots and Quots, 167, 187, 319, 320, 367, 368
Trades Union Congress, 221, 255
Treasury, 12, 62, 85, 176, 181
Treaty of Versailles, 101
Trueman, Arthur Elijah, 270, 279, 335

UNESCO, 356
U.S., finance of research, 86
U.S.S.R., 276, 280, 281, 288, 289, 293, 311, 312, 313, 349, 361; genetics in, 280–81, 282, 355; science in, 72–73, 74–75, 86, 282, 357, 358
University Grants Committee, 350

Vavilov, N. I., 72, 281
Visible College, The, 365

Waddington, C. H., 81, 319, 331
Walker, Miles, 33, 35, 36, 58
War Cabinet, 175–80 passim, 182, 184, 185, 186, 189, 223, 224, 226, 227, 229, 232, 236, 237, 240, 241, 244, 245, 247, 248, 252–57 passim, 309, 315, 324; Defence (Operations) Committee of, 219, 227, 228, 247, 258; Economic section of, 240
War Office, 168, 169, 177
Watson-Watt, Sir Robert, 221, 279, 298, 317, 318
Watts, W. W., 99
Weaver, Warren, 286
Webb, Beatrice, 73
Webb, Sidney, 73
Wells, H. G., 78, 98, 122, 140
Werskey, Gary, 364–67
Western, R. W., 58, 83, 84, 165
Williamson, J. W., 87
Wilson, Sir Arnold, 32, 58, 59, 61, 63, 85
Wilson, Sir Horace, 181, 182, 183, 194
Wimperis, H. E., 230–31, 239
Winant, John Gilbert, 140
Wingfield, Dr., 165
Wood, Sir Kingsley, 171
Wood, Neal, 91, 148, 363–64, 365
Woolton, Lord, 146, 311, 312, 313, 314, 315

Wooster, W. A., 81, 82, 122, 156, 157, 160, 316, 354
World association of science, 126
World Power Conference, 106
World, the Flesh and the Devil, The, 364
World Federation of Scientific Workers, 356

Wrottesley, Earl of, 11, 12, 13

Year Book of Agricultural Co-operation, 143

Zuckerman, Solly, 157, 167, 187, 188, 292, 319, 325, 327, 330, 331, 332, 335, 367, 368